Atlas of Advanced Endoscopy

Springer Nature More Media App

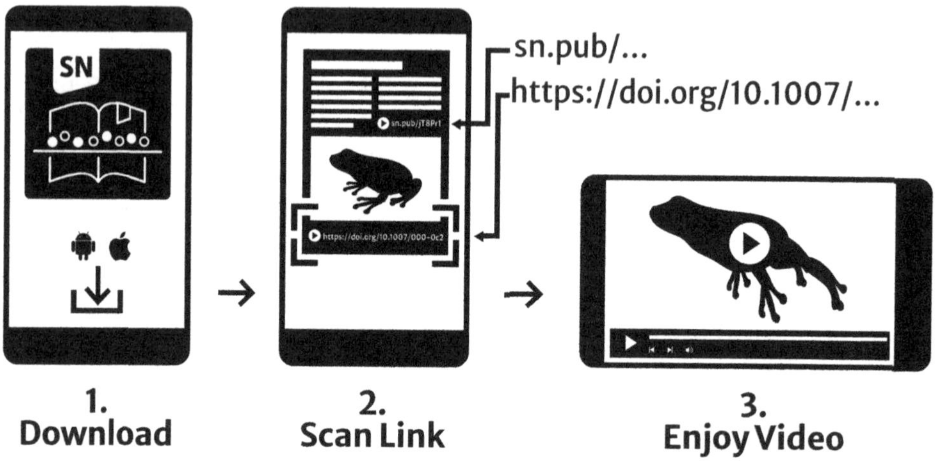

Support: customerservice@springernature.com

Yasushi Sano • Philip Chiu • Rajvinder Singh
Noriya Uedo • Kenichi Goda • Chikatoshi Katada
Editors

Atlas of Advanced Endoscopy

Springer

Editors
Yasushi Sano
Gastrointestinal Center
Sano Hospital
Hyogo, Japan

Rajvinder Singh
Department of Gastroenterology
Lyell McEwin Hospital
University of Adelaide
Adelaide, SA, Australia

Kenichi Goda
Gastrointestinal Endoscopy Center
Dokkyo Medical University Hospital
Tochigi, Japan

Philip Chiu
Department of Surgery
Faculty of Medicine, Chinese University of Hong Kong
Sha Tin, China

Noriya Uedo
Department of Gastrointestinal Oncology
Osaka International Cancer Institute
Osaka, Japan

Chikatoshi Katada
Department of Therapeutic Oncology
Kyoto University Graduate School of Medicine
Kyoto, Japan

This work contains media enhancements, which are displayed with a "play" icon. Material in the print book can be viewed on a mobile device by downloading the Springer Nature "More Media" app available in the major app stores. The media enhancements in the online version of the work can be accessed directly by authorized users.

ISBN 978-981-97-2731-5 ISBN 978-981-97-2732-2 (eBook)
https://doi.org/10.1007/978-981-97-2732-2

This Springer imprint is published by the registered company Springer Nature Singapore Pte Ltd.
The registered company address is: 152 Beach Road, #21-01/04 Gateway East, Singapore 189721, Singapore

Foreword 1

The only lesion in the world

The globally accepted diagnostic classification of gastrointestinal tumors includes the Japan Esophageal Society (JES) Classification and the BING Classification for the esophagus, the VS Classification/MESDA-G for the stomach, the JNET Classification for the colon, and the Paris Classification for the entire gastrointestinal tract. When finding a lesion, we may think we have made a correct endoscopic diagnosis by classifying the lesion according to these widely accepted classifications. However, there are two problems inherent to this.

The first problem is that classifying lesions is not the goal in itself. We need to classify lesions to understand them better. Relating this to appropriate pathology and treatment after classification is the true diagnosis. Only when utilizing such classification to comprehend pathophysiology of the lesion or relating such classification to treatment, we can say that we have made a correct diagnosis.

Secondly, there are always exceptions to classification. We inevitably try to force a lesion into one of the existing classification categories, which puts us at risk of making a wrong diagnosis. An unclassifiable lesion should be diagnosed as unclassifiable, and the individuality of the unclassifiable lesion can be understood by carefully reading the characteristic findings of the lesion. An appropriate diagnosis should be possible without the classification process in such unclassifiable cases.

The components of the diagnostic classification of gastrointestinal tumors are mainly color tone or staining and changes in surface structure and microvascular architecture. Observing the lesion under conditions that allow optimal determination of these diagnostic factors is essential for appropriate classification. Editor-in-Chief Yasushi Sano and Secretary General Mineo Iwatate have meticulously compiled this world's highest-quality endoscopy atlas. It contains a selection of the best shots from among many shots observed by professional endoscopists under optimal conditions, which were considered most suitable to express the individuality of the lesion in the eyes of a professional. It is the very individuality in the only lesion in the world and a beautiful work of art that reflects highly refined truth.

In recent years, the number of journals publishing case reports as scientific articles has dramatically decreased due to the negative effects of meritocracy based on impact factors and other indicators. All clinical research components are generated based on individual patients and lesions. If each case is inappropriately diagnosed, the clinical data collected from inappropriate individuals will be unreliable. No matter how much state-of-the-art technology like next-generation sequencing or cutting-edge statistical methods like omics analysis is used, the research results will be unreliable, of low quality, and in danger of leading the world in the wrong direction. In endoscopy, which begins with observing, it is crucial to diagnose each case thoroughly, decide on a treatment strategy based on the correct diagnosis, and conduct various studies.

With these reasons in mind, *Atlas of Advanced Endoscopy*, written by members of the ANBIIG, the world's leading photographers and endoscopists, is a must-read for current and future endoscopists worldwide. Herein, I confidently recommend this endoscopy atlas.

Mitsuhiro Fujishiro
Department of Gastroenterology, Graduate School of Medicine
The University of Tokyo
Tokyo, Japan

Foreword 2

The world's best atlas of endoscopic diagnosis of digestive tract.

Currently, endoscopic diagnosis of the digestive tract is performed using high-resolution videoendoscopy. In addition to white light observation, various technological innovations such as image-enhanced endoscopy, magnifying endoscopy, and endocytoscopy have been developed, and recently, AI/CAD has also become available. Extensive learning and training are essential for the accurate clinical application of international diagnostic classifications, such as the Japan Esophageal Society Classification, the BING Classification for the esophagus, the VS Classification (MESDA-G) for the stomach, the NICE Classification, and the JNET Classification for the colon, and the Paris Classification for the entire digestive tract. While various learning methods exist, listening to excellent lectures multiple times does not necessarily lead to true diagnostic expertise. It is crucial to gain experience through numerous case discussion, particularly under the guidance of competent clinical and pathological instructors, using invaluable cases and their high-quality endoscopic images. The comparison between endoscopic and pathological images is the most critical process during an effective learning method. Understanding how endoscopic findings are formed based on pathological tissue structures leads to diverse applications to various clinical cases. Learning with low-quality endoscopic images does not improve diagnostic ability. The *Atlas of Advanced Endoscopy*, the world's best atlas of endoscopic diagnosis, is now available. This issue, under the leadership of Editor-in-Chief Yasushi Sano and Secretary General Mineo Iwatate, with experts mainly from The Asian Novel Bio-Imaging and Intervention Group (ANBIIG), provides sufficient comparison with the highest-quality endoscopic and pathological images (histological structures) and commentary including tips on diagnostic procedures. This atlas is highly beneficial to supplement individual case experience and apply to clinical case discussions.

Recently, in the era of AI/CAD, standardized images such as X-ray, CT, MRI, and pathological images can be simply evaluated by AI; however, endoscopic images are entirely different. Endoscopic image quality varies greatly depending on the endoscopist's skill level. Endoscopists can only take high-quality images if they can control various challenging situations such as folds, flexures, and blind spots and conditions such as endoscopic operability, colonic preparation (residues, mucus) quality, observation angles for lesions, distance from lesions, focus, air volume during observation, and light intensity (halation). To achieve correct endoscopic diagnosis with AI/CAD, we must be aware that sufficient skill level of individual endoscopists is required. This atlas can also help learn what kind of endoscopic images should be captured for efficient AI diagnosis.

Finally, various endoscopic techniques, classifications, and the emergence of AI/CAD have brought many learning opportunities in the endoscopic diagnosis of the digestive tract. I express my respect and gratitude to Editor-in-Chief Yasushi Sano and Secretary-General Mineo Iwatate for publishing this timely *Atlas of Advanced Endoscopy* and to the many experts for providing variable clinical cases with detailed explanation and reviews. I sincerely hope this atlas becomes the bible for endoscopists worldwide to elevate the global level of endoscopic diagnosis of the digestive tract and contributes to as many endoscopists as possible.

Shinji Tanaka
Japan Gastroenterological Endoscopy Society (JGES)
Tokyo, Japan

Omichi General Hospital, Onomichi
Hiroshima, Japan

Hiroshima University
Hiroshima, Japan

Foreword 3

Recently, with regard to endoscopic diagnosis of the gastrointestinal tract, a perceptible discordance of diagnostic quality has emerged between Japan and the Western world, particularly the United States. It is as if these two worlds now occupy disparate spheres. On occasion, Western researchers express bemusement regarding the detailed Japanese approach to endoscopy, deeming it excessively thorough and akin to a pastime. What led to this significant divergence of endoscopic diagnosis between Japan and the West? This disparity may be attributable to the distinct developmental trajectories of endoscopic diagnosis in Japan and the West.

Gastrointestinal endoscopic diagnosis in Japan traces its lineage back to the deployment of gastric cameras, which captured images of the stomach without direct observation. Consequently, the fundamental diagnostic procedure involved painstaking interpretation of the captured images, including areas beyond the primary target, much akin to the interpretation of X-ray images. In contrast, the Western approach, commencing with fiber endoscopes, prioritized observation, relegating image capture to specific findings. This initial divergence likely played a pivotal role in shaping subsequent diagnostic methodologies. With its recorded images, the Japanese approach facilitated retrospective diagnosis and fostered extensive feedback, including discussions on missed early cancerous findings and their associated features. This made it possible to investigate the significance of more detailed findings as a viable research agenda, along with advances in biopsy diagnosis. Undoubtedly, the increased incidence of gastric and colorectal cancer in Japan contributed to the evolution of this meticulous approach to diagnosis. Nevertheless, the consensus among Japanese endoscopists regarding the importance of recording—viewing it not as a burden but as an essential element—elevated endoscopic diagnosis to the status of photographic art.

Another hallmark of Japanese-style endoscopic diagnosis is its synergy with pathologists. Interpretation of fine endoscopic details based on a one-to-one comparison with corresponding pathological results from biopsies or resected specimens became the gold standard. The enthusiasm of endoscopists to reveal the true nature of diseases drove them to demand more detailed diagnostic modalities, leading to the development of dye endoscopy, magnifying endoscopy, image-enhanced endoscopy (IEE), and endocytoscopy (endoscopic microscopy).

This atlas embodies the goal of detailed endoscopic diagnosis, striving to clarify the intrinsic nature of gastrointestinal lesions and provide the highest-quality cases within the novel diagnostic field of "endoscopic pathology." While specific technical terminology and interpretations may pose challenges for Western endoscopists, the cases presented herein unquestionably represent the various ways in which endoscopy can be a valuable tool for diagnosis of gastrointestinal lesions.

As mentioned at the outset, it is acknowledged that various criticisms have been leveled at this Japanese-style painstaking approach to endoscopy, akin to questioning the rationale for high-speed bullet trains. The history of high-speed rail travel can be traced back to Japan's Shinkansen in 1964, followed by the TGV in France (1981) and the ICE in Germany (1991). While these high-speed railroad systems have now become commonplace in numerous European countries, their value has yet to be fully embraced in the United States. In other words, the development of a high-speed railroad necessitates substantial technological, labor, and capital investment (Japan National Railways received a significant loan from the World Bank). However, once established, their utility becomes an everyday fixture, and high-speed travel is no longer perceived as extraordinary.

With these reflections in mind, I conclude this foreword, expressing the hope that detailed endoscopy, in pursuit of more accurate endoscopic pathology, will become commonplace in Japan and elsewhere worldwide, and that readers will better appreciate the dedication of endoscopists to reveal in greater detail the nature of gastrointestinal lesions.

Shigeaki Yoshida
Aomori Prefectural Hospital Administration
Aomori, Japan

Preface

To the readers,

I would like to express my sincere gratitude to you for picking up and reading this book.

This book is a collaborative work by the top experts in the field of gastrointestinal endoscopy in the Asia-Pacific region, and I am proud that it was published in a timely manner. We created this endoscopy atlas to provide the latest knowledge on the classification, diagnosis, and treatment in gastrointestinal endoscopy fields, and to provide contents that can be immediately useful in clinical practice.

1.1 Project Background

Last year, in 2022, we agreed to publish an atlas of advanced endoscopy using new endoscopic technologies, mainly with members of The Asian Novel Bio-Imaging and Intervention Group (ANBIIG, URL: ▶ https://anbig.org/), which was originally formed under the leadership of Koji Fujio (Olympus). After several rounds of communication with Springer Publishing, we officially signed a publishing agreement. In producing this book, over 90 experts from the Asia-Pacific region in the fields of pharynx, esophagus, stomach, duodenum, colon, and anorectum agreed to write this book. Despite the very short period of manuscript writing, the generous cooperation of the authors made it possible to create the rich content of this book. I would like to thank them again for their enthusiasm and contribution.

1.2 Japanese-Style Endoscopic Photo-Documentation

I have been involved in the production of international atlases in the past, but unfortunately, there was no book that I was satisfied with all the endoscopic photos. One reason for this is that when it comes to international co-authorship, there are problems with deadlines, languages, and differences in thinking, and often the uniqueness and beauty of Japanese-style endoscopic photography are not fully shared and expressed. However, in this project, we overcame the international barriers by communicating closely with the authors and reviewing all the endoscopic photos ourselves, and sometimes asking for resubmission. We were able to complete a book that shows the true beauty of Japanese-style endoscopic photos to the world. I am confident that this book symbolizes the fusion of Japanese-style endoscopic aesthetics and endoscopic medicine.

1.3 Endoscopy and Artificial Intelligence (AI)

In an era where AI technology is rapidly spreading, endoscopy atlases may be considered outdated. However, even when AI becomes widely used, if you cannot accurately capture lesions and take beautiful photos of areas of interest, you will not be able to obtain AI-assisted diagnosis. I would like you to take a look at the section on AI overview and how to take beautiful photos in this book and understand its importance. Perhaps this excellent atlas of endoscopic photos and videos produced by members from the Asia-Pacific region will retain its value for the next 10 years. I leave this book as a testament to the excellent photography techniques of our predecessors for future generations. I hope you enjoy the beautiful photos in this book.

Finally, without the dedicated cooperation of my respected doctors and friends who are editors, and my colleagues at Sano Hospital, it would have been impossible to publish this book in such a short period. I am truly grateful. I hope that this atlas will become an essential book in endoscopy rooms around the world and contribute to the development of the endoscopy field. Let us continue to move forward for the future endoscopy and contribute to the health and happiness of the patients.

Thank you very much.

Yasushi Sano, MD, PhD

Chairman, The Asian Novel Bio-Imaging and Intervention Group

Clinical Professor, Kansai Medical University, Osaka, Japan

Director, Gastrointestinal Center, Sano Hospital, Hyogo, Japan

Contents

Editors and Contributors

1.1 About the Editors

Yasushi Sano
is a Clinical Professor at Kansai Medical University as well as the Director of Gastrointestinal Center at Sano Hospital, Kobe, Japan. He is one of the world leaders in the field of diagnostic and therapeutic colonoscopy. He has presented many lectures regarding image-enhanced endoscopy (IEE) technique and demonstrated endoscopy live with EMR/ESD technique at several countries. He skillfully uses magnifying colonoscopy with advanced imaging technique. He has published more than 180 peer-reviewed manuscripts and 9 book chapters. He is a member of several advisory boards including Gastroenterological Endoscopy (JGES, JGA, Japan), Gastrointestinal Endoscopy (GIE, USA), and Chairman of Asian Novel Bio-Imaging and Intervention Group (ANBIIG). He has participated in more than 230 workshops/symposiums across various countries.

Philip Chiu
is Professor of Division of Upper GI and Metabolic Surgery, Department of Surgery, Director of Multi-Scale Medical Robotics Center, Director of Endoscopy Center, Institute of Digestive Disease and Director of CUHK Chow Yuk Ho Technology Center for Innovative Medicine and Dean (External Affairs), Faculty of Medicine, Chinese University of Hong Kong. He is the first to perform ESD for treatment of early GI cancers in Hong Kong in 2004. In 2010, he performed the first peroral endoscopic myotomy (POEM) in Hong Kong as well as pioneered the world first robotic gastric ESD in 2011, followed by the world first robotic colorectal ESD in 2020. His research interests include esophageal cancer management, minimally invasive and robotic esophagectomy, novel endoscopic technologies for diagnosis of early GI cancers, endoscopic surgery as well as robotics for endoluminal surgery. He has published more than 370 peer-reviewed manuscripts and 6 book chapters. He is currently co-editor of endoscopy and subject editor for surgical endoscopy.

Rajvinder Singh
is a Professor of Medicine with the University of Adelaide and the Director of Gastroenterology at the Lyell McEwin and Modbury Hospitals, South Australia. He is an Editorial Board member of Clinical Endoscopy and past Editorial Board member of Endoscopy, Digestive Endoscopy, and Endoscopy International Open. He has an h-index of 41. He has authored more than 150 publications and book chapters on advanced endoscopic mucosal imaging and resection techniques. He has been involved in various committees drafting national and international guidelines on this subject. He was the past chair of the Australian Gastrointestinal Endoscopic Association (AGEA). He is frequently invited to conduct Basic and Advanced Endoscopy Workshops and has participated in more than 200 workshops/symposiums across various countries.

Noriya Uedo
is a Vice-Director, Department of Gastrointestinal Oncology, Osaka International Cancer Institute. He graduated from the School of Medicine, Kagoshima University, in 1992 and started his training in gastrointestinal oncology and endoscopy in the current institution from 1994. He is a Councilor of several Japanese medical societies. He is serving as a Visiting Professor in Fukuoka University Chikushi Hospital and used to serve as a Visiting Professor in Lund University, Malmö University Hospital, and Chinese PLA General Hospital.

Kenichi Goda
graduated from National Defense Medical College in 1993. He is a gastroenterologist specialized in gastrointestinal endoscopy and the Professor and Director of Gastrointestinal Endoscopy Center of Dokkyo Medical University Hospital. He is the chairperson-elect of Barrett's esophagus society of Japan. He has more than 90 peer-reviewed international publications and has delivered numerous oral and poster presentations in numerous international meetings. He has received awards of Distinguished Paper Award of Japan Gastroenterological Endoscopy Society 2004 and Distinguished Presentation of JDDW 2012.

Chikatoshi Katada
graduated from Kitasato University School of Medicine in 1998. From 2001 to 2005, he worked at National Cancer Center Hospital East. From 2005 to 2022, he worked at Kitasato University Hospital. From 2018 to 2019, he participated in the visiting clinician program at Mayo Clinic. From 2022 to the present, he worked at Kyoto University Hospital. He has more than 120 peer-reviewed international publications and has delivered numerous presentations in international meetings. His areas of expertise are clinical research on carcinogenesis and prevention of alcohol-related cancers of the esophagus and head and neck. Therefore, he is well versed in the endoscopic diagnosis and treatment of cancers arising in this area.

1.2 About the Contributors

Mitsuhiro Fujishiro
was born in Toyohashi city, Aichi prefecture, Japan in 1970. He graduated from the School of Medicine, The University of Tokyo, in 1995. After finishing 2 years of training as a general internist, he became a resident of Endoscopy Division and Department of Gastroenterology, National Cancer Center Hospital, Tokyo, to learn fundamental skills of diagnostic and therapeutic GI endoscopy, especially endoscopic submucosal dissection (ESD). He returned to The University of Tokyo in 2000 as a graduate student at the Department of Gastroenterology and intensively investigated innovation of ESD, especially in injection solution and endoscopic electrocautery device and got a PhD in 2004. He became an Assistant Professor at the Department of Gastroenterology in 2005, an Associate Professor at the Department of Endoscopy and Endoscopic Surgery in 2009. In the meantime, he established the clinical application of ESD, while educating young endoscopists at The University of Tokyo. Recognized for his achievements, he became a Professor at the Department of Gastroenterology and Hepatology, Nagoya University Graduate School of Medicine in January 2019. In July 2021, he became a visiting professor at Nagoya University and a professor at the Department of Gastroenterology, The University of Tokyo, to further advance research and education in the wide field of gastroenterology with his colleagues. He is a director of the Japanese Society of Internal Medicine and the Japan Gastroenterological Endoscopy Society and a councilor of the Japanese Society of Gastroenterology. He is a board-certified specialist in these scientific societies. He also serves as a director of the Asian Pacific Digestive Week Federation and a councilor of the Asian Pacific Society of Digestive Endoscopy. He is primarily interested in endoscopic diagnosis and treatment for early-stage GI neoplasms, having more than 550 English publications and more than one hundred times of foreign invitations.

Shinji Tanaka, MD, PhD, FJGES, FASGE
is President of JA Onomichi General Hospital; he is a Professor Emeritus, Hiroshima University. He serves as a visiting professor of Gastroenterology at the Graduate School of Biomedical and Health Sciences, Hiroshima University. He graduated from Hiroshima University School of Medicine in 1984. Dr. Tanaka served as Associate Professor at the Department of Endoscopy, Hiroshima University Hospital in 1998. He also serves as Professor at the Department of Endoscopy, Hiroshima University Hospital and Endoscopy and Medicine, Graduate School of Biomedical and Health Sciences, Hiroshima University in 2007. He became President at JA Onomichi General Hospital in 2023. Dr. Tanaka is President of the Japan Gastroenterological Endoscopy Society (JGES), Board Certified Trainer; President of the Japanese Association for Capsule Endoscopy (JACE), Board Certified Trainer; Executive councilor of the Japanese Society of Gastroenterology (JSGE), Board Certified Trainer; Executive of the Japanese Society for Cancer of the Colon and Rectum (JSCCR), etc.

Shigeaki Yoshida
graduated from the Faculty of Medicine at Hokkaido University in 1971. After working at NCCH as Head of the GI Endoscopy Division (1985–1995), he moved to NCCHE as Head of the Division of Endoscopy and GI Oncology (1992–1995), Deputy Director (1995–2004), and Director (2004–2007). He has inspired and trained many world-renowned endoscopists with leading global endoscopy innovations such as NBI. In 2007, he became the Chief Executive Officer of the Aomori Prefectural Hospital Administration and has worked to balance administration and community healthcare, focusing on cancer control.

1.3 Contributors

Seiichiro Abe Endoscopy Division, National Cancer Center Hospital, Tokyo, Japan

Yoichi Akazawa Department of Gastroenterology, Juntendo University School of Medicine, Tokyo, Japan

Tiing Leong Ang Department of Gastroenterology and Hepatology, Changi General Hospital, Singapore, Singapore

Amol Bapaye Shivanand Desai Center for Digestive Disorders, Deenanath Mangeshkar Hospital & Research Center, Pune, India

Naresh Bhat Department of Gastroenterology, Hepatology and Clinical Nutrition, Aster CMI Hospital, Bengaluru, India

Patricia Anne I. Cabral-Prodigalidad Institute of Digestive and Liver Diseases, St Luke's Medical Center Global City, Taguig, Philippines

Philip Chiu Department of Surgery, Faculty of Medicine, The Chinese University of Hong Kong, Shatin, NT, Hong Kong SAR, China

Jonard T. Co Institute of Digestive and Liver Diseases, St Luke's Medical Center Global City, Taguig, Philippines

Hisashi Doyama Department of Gastroenterology, Ishikawa Prefectural Central Hospital, Kanazawa, Ishikawa, Japan

Takahiro Fujii Takahiro Fujii Clinic, Tokyo, Japan

Takahiro Fujimori Department of Pathology, Sano Hospital, Kobe, Hyogo, Japan

Mikio Fujita Gastrointestinal Center, Sano Hospital, Kobe, Hyogo, Japan

Yasuaki Furue Department of Gastroenterology, Kitasato University School of Medicine, Sagamihara, Kanagawa, Japan

Kenichi Goda Gastrointestinal Endoscopy Center, Dokkyo Medical University Hospital, Tochigi, Japan

Santa Hattori Gastrointestinal Center, Sano Hospital, Kobe, Hyogo, Japan

Hirokazu Higuchi Department of Medical Equipment, Kyoto University Hospital, Kyoto, Japan

Toshiaki Hirasawa Department of Gastroenterology, Cancer Institute Hospital, Tokyo, Japan

Daizen Hirata Gastrointestinal Center, Sano Hospital, Kobe, Hyogo, Japan

Shiaw-Hooi Ho Department of Medicine, Faculty of Medicine, Universiti Malaya, Kuala Lumpur, Malaysia

Hiroaki Ikematsu Division of Advanced Gastroenterology and Endoscopy, The Institute of Medical Science, The University of Tokyo, Tokyo, Japan

Kenichiro Imai Division of Endoscopy, Shizuoka Cancer Center, Shizuoka, Japan

Haruhiro Inoue Digestive Diseases Center, Showa University Koto Toyosu Hospital, Tokyo, Japan

Kazuyuki Ishida Department of Diagnostic Pathology, Dokkyo Medical University, Tochigi, Japan

Mineo Iwatate Gastrointestinal Center, Sano Hospital, Kobe, Hyogo, Japan

Takao Kanemitsu Department of Endoscopy, Fukuoka University Chikushi Hospital, Fukuoka, Japan

Takashi Kanesaka Department of Gastrointestinal Oncology, Osaka International Cancer Institute, Osaka, Japan

Koichi Kano Department of Otorhinolaryngology, Head and Neck Surgery, Kitasato University School of Medicine, Sagamihara, Kanagawa, Japan

Hiroshi Kashida Department of Gastroenterology and Hepatology, Kindai University, Osaka-Sayama, Osaka, Japan

Chikatoshi Katada Department of Therapeutic Oncology, Kyoto University Graduate School of Medicine, Kyoto, Japan

Motohiko Kato Center for Diagnostic and Therapeutic Endoscopy, Keio University, School of Medicine, Tokyo, Japan

Takuji Kawamura Department of Gastroenterology, Kyoto Second Red Cross Hospital, Kyoto, Japan

Supakij Khomvilai Colorectal Surgery Division and Surgical Endoscopy, Department of Surgery, Chulalongkorn University, Bangkok, Thailand

Daisuke Kikuchi Department of Gastroenterology, Toranomon Hospital, Tokyo, Japan

Masaaki Kobayashi Department of Gastroenterology, Niigata Cancer Center Hospital, Niigata, Japan

Nozomu Kobayashi Endoscopy Division, National Cancer Center Hospital, Tokyo, Japan

Shin-ei Kudo Digestive Disease Center, Showa University Northern Yokohama Hospital, Yokohama, Kanagawa, Japan

Wataru Kurihara Department of Gastroenterology, Cancer Institute Hospital, Tokyo, Japan

Louis H. S. Lau Department of Medicine and Therapeutics, Faculty of Medicine, The Chinese University of Hong Kong, Shatin, NT, Hong Kong SAR, China

Ching-Tai Lee Department of Gastroenterology, E-Da Hospital and I-Shou University, Kaohsiung City, Taiwan

James Weiquan Li Department of Gastroenterology and Hepatology, Changi General Hospital, Singapore Health Services, Simei, Singapore

Noriko Matsuura Division of Research and Development for Minimally Invasive Treatment, Cancer Center, Keio University School of Medicine, Tokyo, Japan

Masashi Misawa Digestive Disease Center, Showa University Northern Yokohama Hospital, Yokohama, Kanagawa, Japan

Kurato Miyazaki Division of Gastroenterology and Hepatology, Department of Internal Medicine, Keio University, School of Medicine, Tokyo, Japan

Yasuhiko Mizuguchi Endoscopy Division, National Cancer Center Hospital, Tokyo, Japan

Makoto Naganuma Third Department of Internal Medicine, Kansai Medical University, Hirakata, Osaka, Japan

Shigenori Nagata Department of Diagnostic Pathology and Cytology, Osaka International Cancer Institute, Osaka, Japan

Shinji Nagata Department of Gastroenterology, Hiroshima City North Center Asa Citizens Hospital, Hiroshima, Japan

Keiichiro Nakajo Department of Gastroenterology and Endoscopy, National Cancer Center Hospital East, Kashiwa, Chiba, Japan

Shiro Oka Department of Gastroenterology, Hiroshima University Hospital, Hiroshima, Japan

Hiroyuki Ono Division of Endoscopy, Shizuoka Cancer Center, Shizuoka, Japan

Tsuneo Oyama Department of Endoscopy, Saku Central Hospital Advanced Care Center, Nagano, Japan

Rapat Pittayanon Division of Gastroenterology, Department of Medicine, Faculty of Medicine, Chulalongkorn University and King Chulalongkorn Memorial Hospital, The Thai Red Cross Society, Bangkok, Thailand

Yutaka Saito Endoscopy Division, National Cancer Center Hospital, Tokyo, Japan

Wataru Sano Gastrointestinal Center, Sano Hospital, Kobe, Hyogo, Japan

Yasushi Sano Gastrointestinal Center, Sano Hospital, Kobe, Hyogo, Japan

Kansai Medical University, Hirakata, Osaka, Japan

Stefan Seewald Centre of Gastroenterology, Hirslanden Clinic, Zurich, Switzerland

Kenjiro Shigita Department of Gastroenterology, Hiroshima City North Center Asa Citizens Hospital, Hiroshima, Japan

Yuto Shimamura Digestive Diseases Center, Showa University Koto Toyosu Hospital, Tokyo, Japan

Kensuke Shinmura Department of Gastroenterology and Endoscopy, National Cancer Center Hospital East, Kashiwa, Japan

Kazuo Shiotsuki Department of Gastroenterology (Endoscopy), Kanagawa Cancer Center, Yokohama, Kanagawa, Japan

Rajvinder Singh Department of Gastroenterology, The Lyell McEwin Hospital and University of Adelaide, Adelaide, SA, Australia

Naoki Sugimura Gastrointestinal Center, Sano Hospital, Kobe, Hyogo, Japan

Kazuki Sumiyama Department of Endoscopy, The Jikei University School of Medicine, Tokyo, Japan

Hisao Tajiri World Endoscopy Organization (WEO), The Jikei University School of Medicine, Tokyo, Japan

Kaoru Takabayashi Center for Diagnostic and Therapeutic Endoscopy, Keio University, Tokyo, Japan

Hiroyuki Takamaru Endoscopy Division, National Cancer Center Hospital, Tokyo, Japan

Masafumi Takatsuna Division of Gastroenterology and Hepatology, Nagaoka Red Cross Hospital, Niigata, Japan

Manabu Takeuchi Division of Gastroenterology and Hepatology, Nagaoka Red Cross Hospital, Niigata, Japan

Yoji Takeuchi Department of Gastroenterology and Hepatology, Gunma University Graduate School of Medicine, Maebashi, Gunma, Japan

Kohei Takizawa Department of Gastroenterology (Endoscopy), Kanagawa Cancer Center, Yokohama, Kanagawa, Japan

Naoto Tamai Department of Endoscopy, The Jikei University School of Medicine, Tokyo, Japan

Hidenori Tanaka Department of Gastroenterology, Hiroshima University Hospital, Hiroshima, Japan

Shinji Tanaka Omichi General Hospital, Onomichi, Hiroshima, Japan

Chin Kimg Tan Department of Gastroenterology and Hepatology, Changi General Hospital, Singapore, Singapore

Jin Lin Tan Department of Gastroenterology, The Lyell McEwin Hospital and University of Adelaide, Adelaide, Australia

Jun Liang Teh Department of Surgery, Ng Teng Fong General Hospital, National University Health System, Jurong East, Singapore

Akira Teramoto Department of Gastroenterology and Hepatology, Royal Brisbane and Women's Hospital, Herston, QLD, Australia

Kasenee Tiankanon Division of Gastroenterology, Department of Medicine, Faculty of Medicine, Chulalongkorn University and King Chulalongkorn Memorial Hospital, The Thai Red Cross Society, Bangkok, Thailand

Shigetsugu Tsuji Department of Gastroenterology, Ishikawa Prefectural Central Hospital, Kanazawa, Ishikawa, Japan

Kunihisa Uchita Department of Gastroenterology, Kochi Red Cross Hospital, Kochi, Japan

Noriya Uedo Department of Gastrointestinal Oncology, Osaka International Cancer Institute, Osaka, Japan

Hiroya Ueyama Department of Gastroenterology, Juntendo University School of Medicine, Tokyo, Japan

Takeshi Uozumi Endoscopy Division, National Cancer Center Hospital, Tokyo, Japan

Lai Mun Wang Department of Anatomical Pathology, Changi General Hospital, Singapore, Singapore

Wen-Lung Wang Department of Gastroenterology, E-Da Hospital and I-Shou University, Kaohsiung City, Taiwan

Tomonori Yano Department of Gastroenterology and Endoscopy, National Cancer Center Hospital East, Kashiwa, Chiba, Japan

Kenshi Yao Department of Endoscopy, Fukuoka University Chikushi Hospital, Chikushino, Japan

Hon Chi Yip Department of Surgery, Faculty of Medicine, The Chinese University of Hong Kong, Shatin, NT, Hong Kong SAR, China

Masao Yoshida Division of Endoscopy, Shizuoka Cancer Center, Shizuoka, Japan

Shigeaki Yoshida Aomori Prefectural Hospital Administration, Aomori, Japan

Edward Young Department of Gastroenterology, The Lyell McEwin Hospital and University of Adelaide, Adelaide, SA, Australia

Ni Yunbi Department of Anatomical and Cellular Pathology, Faculty of Medicine, The Chinese University of Hong Kong, Shatin, NT, Hong Kong SAR, China

Review-X1 System

Contents

EVIS X1 Endoscopy System

Akira Teramoto and Yasushi Sano

Contents

Y. Sano et al. (eds.), *Atlas of Advanced Endoscopy*, https://doi.org/10.1007/978-981-97-2732-2_1

1 Introduction

The history of camera-equipped endoscopes traces back to Olympus' prototype in 1950. An electronic endoscope capable of converting images into electrical signals and displaying them on a monitor was developed in Japan in 1986 [1]. Subsequently, models such as CV-240 (1997), CV-260, Lucera Spectrum (2006), and CV-290, Lucera Elite (2010) have evolved remarkably in terms of image quality and functionality. Notably, the recent introduction of the CV-1500, EVIS X1 (2020), marks a significant milestone after a decade and heralds further advancements (Fig. 1). Table 1 provides a summary of the distinctions between the previous and the latest models.

The X1 system essentially is a universal model that can be used with all types of scopes. Historically, Olympus endoscopes have been developed into two primary series due to differences in image quality demands. Frame-sequential endoscopy (Japan/UK/Hong Kong) involves a sequential image capture process employing alternating color channels, ultimately generating a full-color image, and potentially offering heightened resolution [2]. On the other hand, simultaneous endoscopy (Europe/USA) employs specialized sensors dedicated to individual color channels, allowing for the concurrent capture of images. This approach provides notable advantages, including enhanced color accuracy, diminished motion artifacts, and an overall elevation in image quality, albeit at a higher cost.

Consequently, limitations emerged in certain regions, including issues with relatively low image quality and the inability to use specific scopes such as magnifying endoscope and spiral motor power enteroscope. These problems have been resolved with the launch of the new system in 2020. In other words, it is believed that the era is changing, requiring endoscopists from all regions to be familiar with the imaging types and proficient in magnifying endoscopy diagnosis. This chapter provides a detailed description of the new features of the latest Olympus endoscopic platform, EVIS X1.

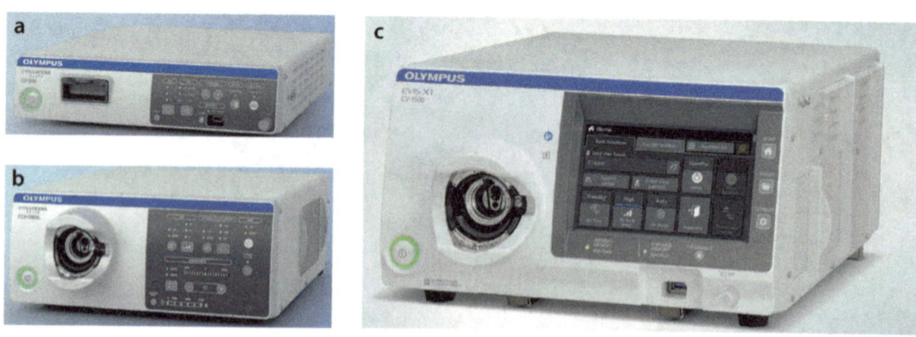

Fig. 1 **a** Lucera Elite (CV-290, video processor component). **b** CLV-290SL (light source component). **c** EVIS X1 (CV-1500, video processor with built-in LED light source)

Table 1 The main features of EVIS X1 and Lucera Elite/Exera III

	Lucera Elite/Exera III (CV-290/CV-190)	EVIS X1 (CV-1500)
Launch	2010	2020
Light source	Xenon	LED
Compatible monitor	HD	UHD 4K
Imaging method	Frame-sequential (CV-290)/ simultaneous (CV-190)	Frame-sequential (fast spin) and simultaneous
Image sensor compatibility	Monochromatic CCD	CMOS + CCD
Image mode	WLI, NBI (second generation) AFI, IRI	WLI, NBI (third generation) AFI, RDI
Digital manipulation	None	TXI, BAI-MAC
Component	Video processor only	Video processor with light source
Weight	9.4 kg + 18.5 kg (light source)	19.4 kg

LED light-emitting diode, *HD* high definition, *UHD* ultrahigh definition, *CCD* charge-coupled device, *CMOS* complementary metal oxide semiconductor, *WLI* white light imaging, *NBI* narrowband imaging, *AFI* autofluorescence imaging, *IRI* infrared imaging, *RDI* red dichromatic imaging, *TXI* texture and color enhancement imaging, *BAI-MAC* brightness adjustment imaging with maintenance of contrast

2 Features of EVIS X1

LED Light Source

The Olympus X1 introduces a new light-emitting diode (LED) light source, fundamentally altering the emission spectrum (Fig. 2). The LED light source integrates red, blue, yellow, violet, and a new amber color, alternately emitted and extinguished to create primary colors. These five LED light source offers a substantially extended lifespan compared to the previous xenon light source, which had a mere 500-hour lifespan. This enhancement not only prolongs the equipment's operational life but also contributes to energy conservation. LEDs provide consistent and stable light output, ensuring uniform illumination across the field of view. This uniformity is vital for accurate visualization during endoscopic procedures, eliminating the risk of underexposed or overexposed areas. The combination of amber, red, and green enables the application of a new red dichromatic imaging (RDI) function, while blue and violet combinations facilitate narrowband imaging (NBI).

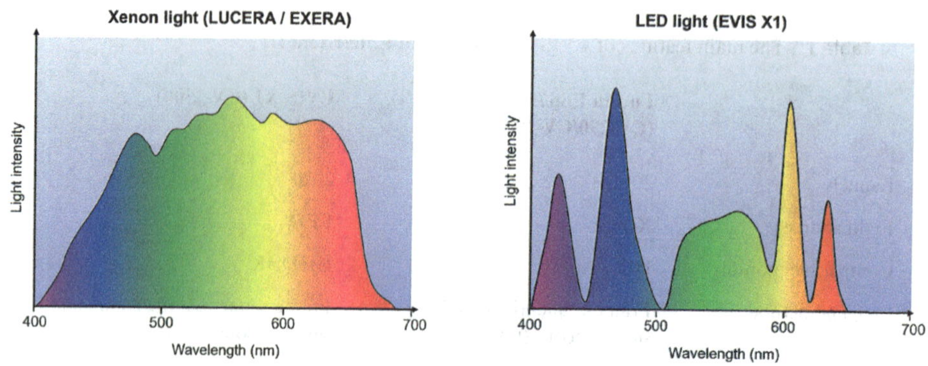

Fig. 2 Penetration length pattern of white light image generated by xenon and LED light sources. LED light consist of blue, violet, yellow amber, and red bulbs, offering more distinctiveness compared to xenon light

New Scopes

Flagship scopes for the X1 generation include GIF-EZ1500 and CF-EZ1500D models. The button-operated magnification function with dual focus from the previous GIF-HQ290 and CF-HQ290 models enabled magnification at the push of a single button, but as an optical zoom, it could only magnify up to 40×, making adequate evaluation difficult in certain cases. The new models have further enhanced functions, with magnification up to 100× and 90× for GIF and CF, respectively, and the GIF-EZ1500 allows the user to get as close as 3 mm to the subject and 1.5 mm in near-focus mode. In addition, the Expanded Depth of View (EDOF) function, which combines near and far resolution images to create a single image, provides clearer, widely focused image without demanding high-technical expertise. Moreover, the EZ-1500 scope is equipped with a newly developed color CMOS (complementary metal oxide semiconductor) sensor with high sensitivity, color reproduction, and low noise. As this scope has been widely marketed worldwide, it is predicted that magnified endoscopic diagnostics will become a standard in use. However, due to the seamless acquisition of magnified images with the new scope, there is a potential for debate regarding whether to use the classification of magnifying endoscopes or non-magnifying endoscopes, as the traditional clear distinction between non-magnified and magnified endoscopes may no longer apply.

Other models include the 1100 series, which is a non-magnifying scope with a color CCD (charge-coupled device) and is similar to the traditional Western models (Exera, 100 series), and the 1200 series, which uses a monochrome CCD and is similar to the traditional Japanese models (Lucera, 200 series) with a lever-type, high magnification function (GIF: 125×, CF: 135×, Fig. 3). Notably, the new 1200 series incorporates high-speed rotation among its frame-sequential methods, resulting in remarkably smooth and high-resolution performance compared to the previous generations. The GIF-1200 N model is a narrow-diameter scope but uses a new CMOS sensor, which reduces noise and produces high-definition image quality similar to conventional oral endoscopes.

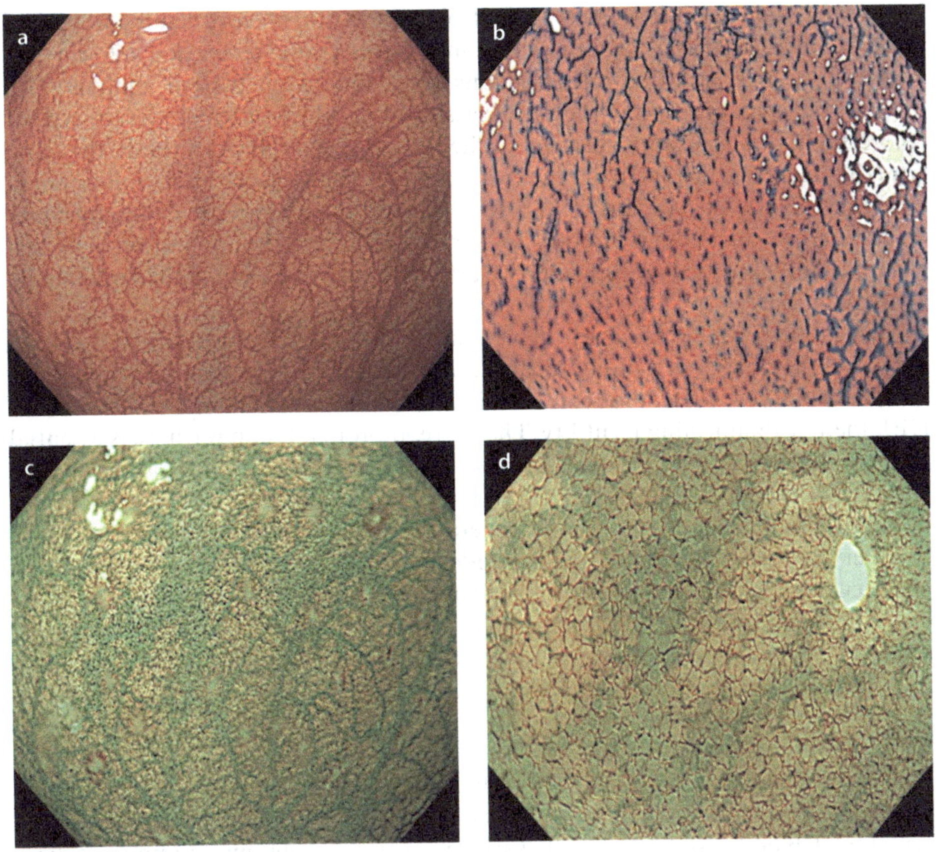

Fig. 3 Images of normal colonic mucosa captured using CF-1200XZ. **a** White light image (WLI) without magnification. **b** WLI with magnification (135×) after indigo carmine spraying showing the pit pattern. **c** Narrowband imaging (NBI) without magnification. **d** NBI with magnification (135×) showing the surface pattern

Image Enhancement

Narrowband Imaging (NBI)

NBI involves the transformation of regular xenon white light into blue and green wavelengths. This is achieved by introducing an NBI filter between the light source and the traditional red-green-blue filter [3]. This focused light carries wavelengths of 410 nm and 540 nm and is readily absorbed by hemoglobin, rendering dense hemoglobin vessels distinctly dark. This contrast enables a detail observation of capillary network patterns vital for diagnosing colorectal issues. Furthermore, the reduction in tissue reflection and scattering elevates the visibility of surface structures.

The implementation of NBI offers two primary benefits: Firstly, NBI is notably efficient and uncomplicated when compared to traditional dye-based chromoendoscopy. Given that crystal violet, a potentially carcinogenic substance, necessitates several minutes for staining, NBI enables instant observation of microstructures, devoid of potential risks. Secondly, magnifying NBI introduces a novel dimension, allowing the study of vascular network patterns alongside standard pit pattern analysis. This progress led to the creation of the Sano classification, which differentiates the type of colorectal polyps by examining their surface blood vessels [4]. As a result, the modern NBI diagnostic method combines both vascular and surface pattern information. When faced with low confidence in NBI diagnosis or when precise endoscopic prediction of invasion depth is essential for surgical decisions, the application of crystal violet staining provides supplementary clarity although there is a limited availability of the solution.

NBI, despite its advantages, has certain limitations. Particularly, its efficacy hinges on the quality of bowel preparation. In cases of inadequate bowel preparation, residual yellowish stool can be misleadingly represented as vivid red under NBI, thereby diminishing its utility. Therefore, thorough lesion washing is essential prior to undertaking detailed NBI assessment, regardless of preparation quality. Another drawback is that NBI often presents images darker than those produced by regular white light. The second-generation NBI in Lucera Elite/Exera III is brighter than the original NBI, and its effectiveness in detecting colorectal polyps has been demonstrated [5]. Due to the enhancements in LED light sources and sensors implemented in the X1, a higher level of brightness has been successfully attained. This increased brightness now provides sufficient illumination for routine practical use.

Red Dichromatic Imaging (RDI)

Red Dichromatic Imaging (RDI) is an advanced imaging technique that emphasizes red and brown colors using the newly developed amber LED light. RDI enhances the visibility of blood vessels and subtle tissue changes, contributing to improved diagnostic accuracy during endoscopic examinations [6].

Texture and Color Enhancement Imaging (TXI)

Olympus has introduced TXI in endoscopy, an innovative technique enhancing real-time visualization of tissue textures and colors. By utilizing advanced algorithms, TXI accentuates subtle patterns and color variations, aiding in the identification of tissue irregularities [7]. Using this feature can assist in the diagnosis of atrophic gastritis and is expected to improve detection rates for early gastric cancer and colorectal polyps. However, since its clinical data is not yet sufficient, future research is warranted.

Brightness Adjustment Imaging with Maintenance of Contrast (BAI-MAC)

BAI-MAC is another technology which involves digital adjustment of the brightness of endoscopic images to enhance visibility while preserving the contrast between different tissues. By iteratively fine-tuning the brightness and utilizing image enhancement algorithms, BAI-MAC helps endoscopists detect subtle abnormalities from a distance and make accurate diagnoses without compromising essential contrast information.

Application of Artificial Intelligence (AI)

Computer-aided tools for detecting polyps, using AI and deep-learning software (CADe), have garnered significant attention in recent times, with multiple trials displaying promising results. In 2020, Olympus introduced ENDO-AID (OIP-1), a new X1 compatible CADe software that subsequently demonstrated enhanced improved adenoma detection (52.9% vs. 37.4%, $p = 0.017$) [8]. This innovation not only streamlines the endoscopy process but also has the potential to simplify and improve efficiency for endoscopists by minimizing the need for excessive eye movements through its user-friendly presentation of lesions.

Limitations of EVIS X1

Given the financial constraints faced by many hospitals, obtaining the latest models might not be feasible within a reasonable timeframe. Consequently, older scope models are frequently still in operation, necessitating the utilization of the previous-generation light source alongside the new X1 to maintain compatibility. Furthermore, these aging models, when projected onto expansive 4 K screens, might encounter challenges in maintaining image stability. Particularly noteworthy is the variance in image settings between these older and newer generations. While this transition can be managed through the memory function, enabling users to switch between settings, it requires manual adjustment each time the scope generation changes. Additionally, the integration of touch panel interfaces, while enhancing interactivity, introduces an added layer of intricacy when making specific adjustments.

3 Conclusions

The combination of EZ-1500 and X1 provides stress-free practice with extremely high image quality, with high levels of sharpness, brightness, and frame rate that offer perfect white light and narrowband imaging in both upper and lower gastrointestinal tract. We anticipate that the new-generation endoscopy will offer clearly defined images and seamless motion capture, contributing to a comfortable working environment for all endoscopists. The combination of a high detection rate and magnifying endoscopy is expected to yield numerous benefits for patients worldwide.

References

1. Progress in medical devices for gastrointestinal endoscopy: Olympus inc. Dige Endosc. 2022;34:138–42.
2. Teramoto A, Hamada S, Ogino B, Yasuda I, Sano Y. Updates in narrow-band imaging for colorectal polyps: narrow-band imaging generations, detection, diagnosis, and artificial intelligence. Dig Endosc. 2023;35:453–70.
3. Gono K, Obi T, Yamaguchi M, et al. Appearance of enhanced tissue features in narrow-band endoscopic imaging. J Biomed Opt. 2004;9:568–77.

4. Sano Y, Ikematsu H, Fu KI, et al. Meshed capillary vessels by use of narrow-band imaging for differential diagnosis of small colorectal polyps. Gastrointest Endosc. 2009;69:278–83.
5. Atkinson NSS, Ket S, Bassett P, et al. Narrow-band imaging for detection of neoplasia at colonoscopy: a meta-analysis of data from individual patients in randomized controlled trials. Gastroenterology. 2019;157:462–71.
6. Fujimoto A, Saito Y, Abe S, et al. Clinical usefulness of red dichromatic imaging in hemostatic treatment during endoscopic submucosal dissection: first report from a multicenter, open-label, randomized controlled trial. Dig Endosc. 2022;34:379–90.
7. Sato T. Txi: texture and color enhancement imaging for endoscopic image enhancement. J Healthc Eng. 2021;2021:5518948.
8. Wong YT, Tai TF, Wong KF, et al. The study on artificial intelligence (AI) colonoscopy in affecting the rate of polyp detection in colonoscopy: a single Centre retrospective study. Surg Pract. 2022;26:115–9.

Texture and Color Enhancement Imaging (TXI)

Naoto Tamai and Kazuki Sumiyama

Contents

Y. Sano et al. (eds.), *Atlas of Advanced Endoscopy*, https://doi.org/10.1007/978-981-97-2732-2_2

1 Principles of TXI

Texture and color enhancement imaging (TXI) is an innovative image-enhanced endoscopy (IEE) developed by Olympus. Its core objective is to enhance the visibility of lesions by improving three vital image factors inherent to white light imaging (WLI): (1) brightness in dark areas; (2) texture, such as subtle surface changes; and (3) color, including slight color variations in the image. To achieve this, TXI applies image processing technology based on the retinex theory [1]. This theory involves decomposing images into two layers based on human visual system characteristics [2]. The first layer represents the scene's illumination light, while the second layer corresponds to the local contrast in brightness and color.

The flowchart in ◘ Fig. 1 outlines the principles of TXI, which comprises five processes. Initially, the input image was divided into base and detailed layers in the first step. The second step corrects brightness in dark areas using the base layer. Directly adjusting brightness via light source control is challenging, as it increases the brightness throughout the entire area, causing halation in the bright regions. However, manipulating the base layer, which represents the illumination light, enables partial brightness control in specific areas. In the third and fifth steps, texture and color are enhanced, respectively. The detailed layer captures local contrast and contains information on texture and subtle surface or color changes. Texture and color can be strengthened by enhancing the detailed layers. The corrected base and enhanced detailed layers were then combined in the fourth step to generate one of the TXI outputs. Additionally, in the fifth step, the color tone was significantly enhanced, particularly for distinguishing between red and white. The output from the fifth process represents one TXI output; the other TXI output is derived from the fourth process without color enhancement, resulting in a color tone close to that of WLI. TXI offers two modes: mode 1, the output from the fifth process with color enhancement, and mode 2, the output from the fourth process without color enhancement, close to the WLI color tone [3].

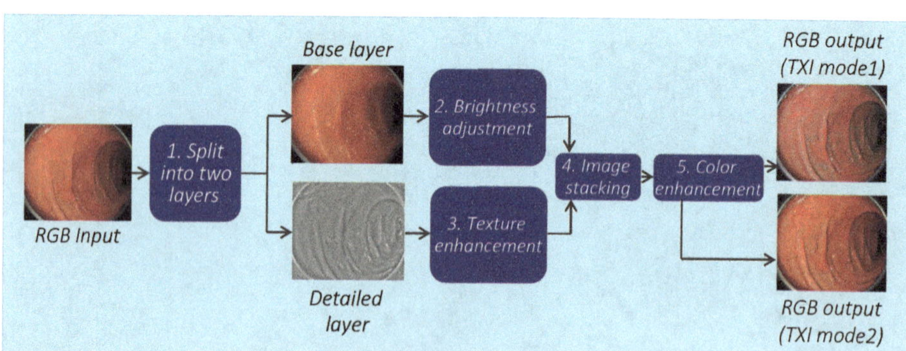

◘ **Fig. 1** Principles of texture and color enhancement imaging. *TXI* texture and color enhancement imaging

◘ Figure 2 illustrates the features of the TXI. In TXI modes 1 and 2, gaps between the nodules were more distinctly depicted than in WLI and narrowband imaging (NBI). Furthermore, owing to the emphasis on brightness of the distal portion, the appendix orifice can be more clearly observed with TXI than with WLI. In TXI mode 1, pale areas appear even paler, and red areas appear redder, indicating a stronger contrast in color tone compared with white light. Conversely, in TXI mode 2, color tones are not emphasized, allowing for observation more similar to WLI conditions. Both TXI modes 1 and 2 offer improved visualization of vessels in the mucosa, particularly where residual liquid has accumulated. This suggests that TXI is less affected by the residual liquid than NBI. The advantage of TXI in improving visibility is also evident in non-granular laterally spreading tumors, which are often difficult to detect using WLI (◘ Fig. 3).

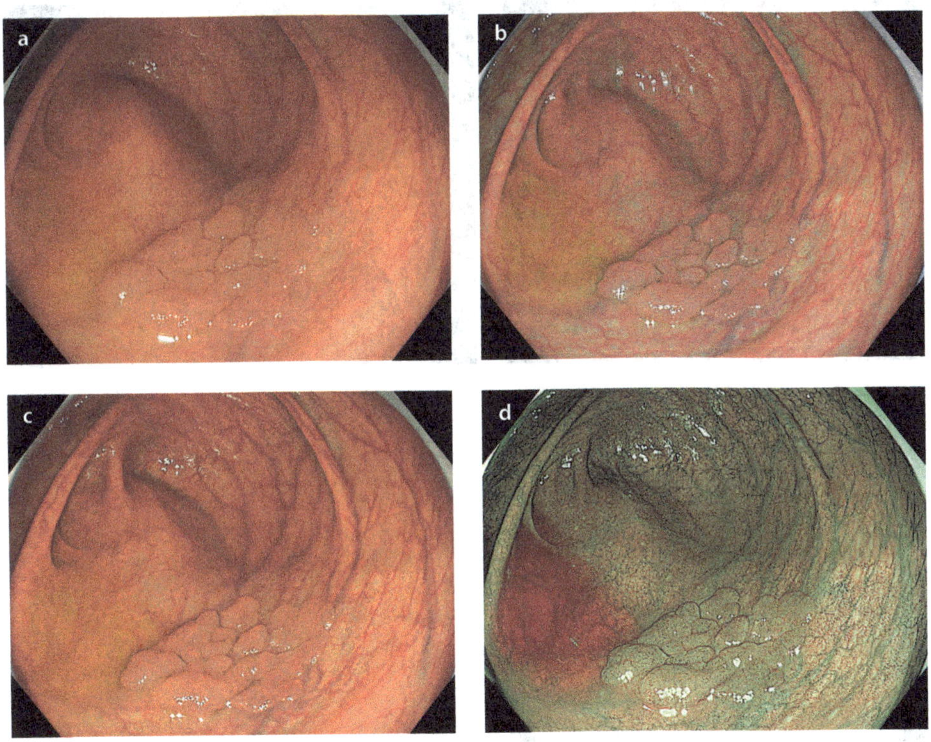

◘ **Fig. 2** High-grade dysplasia, IIa (LST-G homogenous) lesion located in the cecum. **a** WLI, **b** TXI mode 1, **c** TXI mode 2, **d** NBI. *LST-G* laterally spreading granular-type tumor, *TXI* texture and color enhancement imaging, *NBI* narrowband imaging

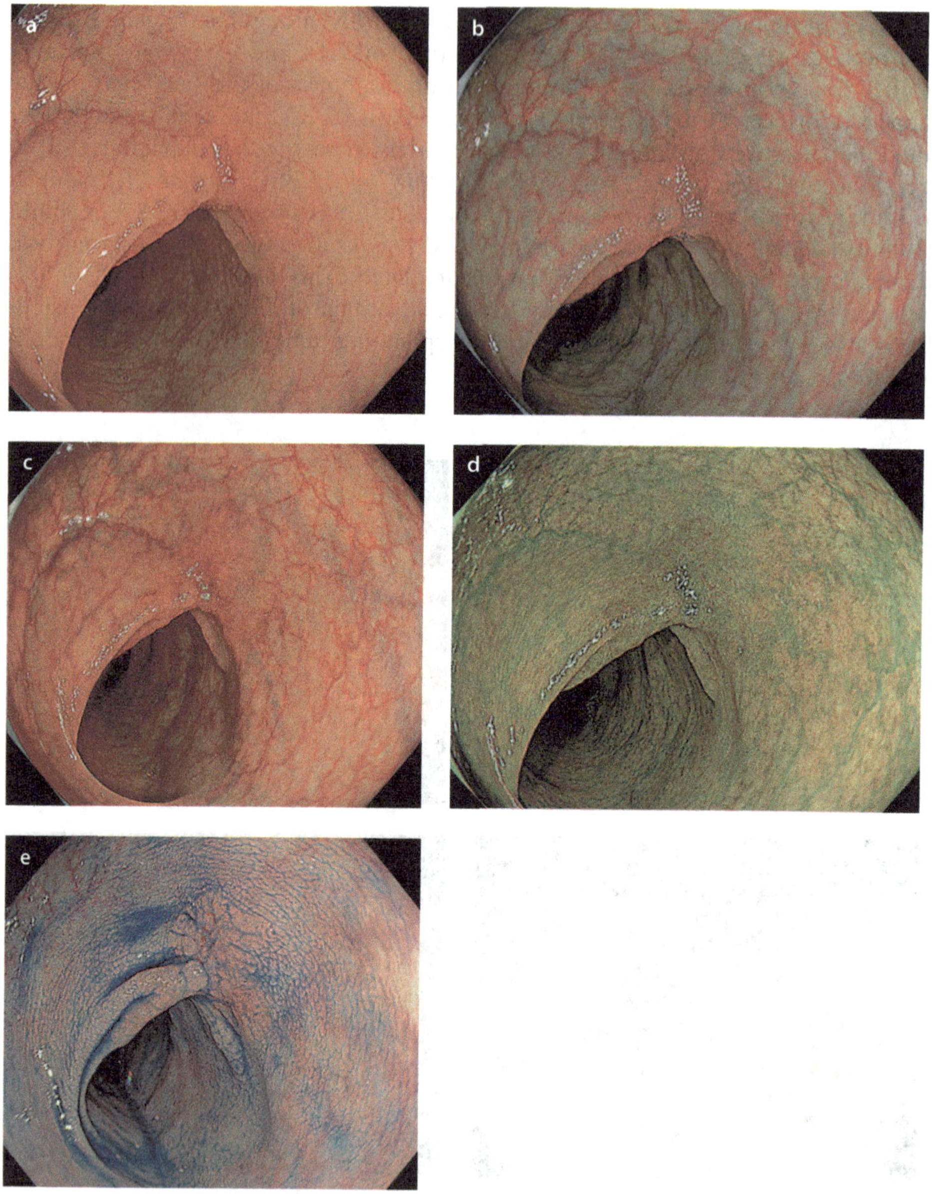

Fig. 3 High-grade dysplasia, IIa (LST-NG flat-elevated) lesion located in the sigmoid colon. **a** WLI, **b** TXI mode 1, **c** TXI mode 2, **d** NBI, **e** chromoendoscopy with indigo carmine. *LST-NG* laterally spreading non-granular-type tumor, *TXI* texture and color enhancement imaging, *NBI* narrowband imaging

2 Clinical Utility of TXI

Upper GI and Pancreaticobiliary

Studies in upper gastrointestinal endoscopy have demonstrated TXI's efficacy in enhancing the visibility of esophageal squamous cell carcinoma [4], early gastric cancer [5], and superficial nonampullary duodenal tumors [6]. Furthermore, reports indicate that TXI enhances the effectiveness of assessing *Helicobacter pylori* gastritis [7] and Barrett's esophagus [8].

In the field of pancreaticobiliary, case reports suggest that TXI improves the recognition of bile and pancreatic duct orifices [9, 10].

Colonoscopy

Regarding colonoscopy, research has revealed the effectiveness of TXI in providing better visualization of colorectal lesions [11] and assessing mucosal healing in patients with ulcerative colitis. Additionally [12], two studies revealed TXI's efficacy in improving detectability of colorectal lesions. Sakamoto et al. reported that TXI improved the mean number of adenomas detected per procedure (TXI, 1.5, vs. WLI, 1.0) and adenoma detection rate (ADR) (TXI, 58.2%, vs. WLI, 46.8%) compared to WLI in a retrospective multicenter observational study [13]. Antonelli et al. conducted an international multicenter randomized trial comparing ADR between TXI and high-definition WLI groups in screening, surveillance, and diagnostic colonoscopy. They found that the TXI group had a significantly higher ADR than the WLI group (TXI, 58.9%, vs. WLI, 42.7%). Significant differences were observed for both ≤5-mm and 6–9-mm adenomas [14].

3 Future Perspective of TXI

In endoscopy, where detection and diagnosis of lesions occur on a two-dimensional monitor, TXI is an innovative IEE that facilitates the adjustment of brightness and emphasizes surface irregularities and color differences in endoscopic images. Although its effectiveness in improving the visibility of neoplastic lesions in the upper gastrointestinal tract is evident, further investigations are required to determine its potential in enhancing the detectability of neoplastic lesions. Regarding colonoscopy, TXI's role in improving detectability of colorectal lesion has been established, and it may provide better detection of colorectal neoplasia, even in patients with poor bowel preparation. However, additional evidence is required to understand the specific conditions under which TXI excels.

The combination of TXI with chromoendoscopy or its integration with NBI has the potential to enhance endoscopic diagnostic accuracy and detectability of neoplastic lesions in the future. As evidence is accumulated, TXI is expected to significantly contribute to routine clinical practice.

References

1. Land EH, McCann JJ. Lightness and retinex theory. J Opt Soc Am. 1971 Jan;61(1):1–11.
2. Meylan L, Süsstrunk S. High dynamic range image rendering with a Retinex-based adaptive filter. IEEE Trans Image Process. 2006;15(9):2820–30.
3. Sato T. TXI: texture and color enhancement imaging for endoscopic image enhancement. J Healthc Eng. 2021;2021:5518948.
4. Dobashi A, Ono S, Furuhashi H, Futakuchi T, Tamai N, Yamauchi T, Suka M, Sumiyama K. Texture and color enhancement imaging increases color changes and improves visibility for squamous cell carcinoma suspicious lesions in the pharynx and esophagus. Diagnostics (Basel). 2021;11(11):1971.
5. Abe S, Yamazaki T, Hisada IT, Makiguchi ME, Yoshinaga S, Sato T, Nonaka S, Suzuki H, Oda I, Saito Y. Visibility of early gastric cancer in texture and color enhancement imaging. DEN Open. 2021;2(1):e46.
6. Okimoto K, Matsumura T, Maruoka D, Kurosugi A, Shiratori W, Nagashima A, Ishikawa T, Kaneko T, Kanayama K, Akizue N, Ohta Y, Taida T, Saito K, Kato J, Kato N. Magnified endoscopy with texture and color enhanced imaging with indigo carmine for superficial nonampullary duodenal tumor: a pilot study. Sci Rep. 2022;12(1):10381.
7. Kitagawa Y, Koga K, Ishigaki A, Nishii R, Sugita O, Suzuki T. Endoscopic diagnosis of helicobacter pylori gastritis using white light imaging and texture and color enhancement imaging. Endosc Int Open. 2023;11(2):E136–41.
8. Ikeda A, Takeda T, Ueyama H, Uemura Y, Iwano T, Yamamoto M, Uchida R, Utsunomiya H, Oki S, Suzuki N, Abe D, Yatagai N, Akazawa Y, Ueda K, Asaoka D, Shibuya T, Hojo M, Nojiri S, Nagahara A. Comparison of texture and color enhancement imaging with white light imaging in 52 patients with short-segment Barrett's esophagus. Med Sci Monit. 2023;29:e940249.
9. Tanisaka Y, Mizuide M, Fujita A, Shiomi R, Shin T, Hirata D, Ryozawa S. Texture and color enhancement imaging facilitates the identification of pancreatic and bile duct orifices after endoscopic papillectomy. Endoscopy. 2023;55(S 01):E22–3.
10. Toyonaga H, Kin T, Hayashi T, Takahashi K, Katanuma A. The application of texture and color enhancement imaging in transpapillary biliary cannulation. Endoscopy. 2022;54(S 02):E786–7.
11. Tamai N, Horiuchi H, Matsui H, Furuhashi H, Kamba S, Dobashi A, Sumiyama K. Visibility evaluation of colorectal lesion using texture and color enhancement imaging with video. DEN Open. 2022;2(1):e90.
12. Hayashi Y, Takabayashi K, Kato M, Tojo A, Aoki Y, Hagihara Y, Yoshida K, Yoshimatsu Y, Kiyohara H, Sugimoto S, Nanki K, Mikami Y, Sujino T, Mutaguchi M, Kawaguchi T, Hosoe N, Yahagi N, Ogata H, Kanai T. Usefulness of texture and color enhancement imaging in assessing mucosal healing in patients with ulcerative colitis. Gastrointest Endosc. 2023;97(4):759–766.e1.
13. Sakamoto T, Ikematsu H, Tamai N, Mizuguchi Y, Takamaru H, Murano T, Shinmura K, Sasabe M, Furuhashi H, Sumiyama K, Saito Y. Detection of colorectal adenomas with texture and color enhancement imaging: multicenter observational study. Dig Endosc. 2023;35(4):529–37.
14. Antonelli G, Bevivino G, Pecere S, Ebigbo A, Cereatti F, Akizue N, Di Fonzo M, Coppola M, Barbaro F, Sharma P, Walter BM, Caruso A, Okimoto K, Antenucci C, Matsumura T, Zerboni G, Grossi C, Meinikheim M, Papparella LG, Correale L, Costamagna G, Repici A, Spada C, Messmann H, Hassan C, Iacopini F. Texture and colour enhancing imaging versus high-definition white light endoscopy for detection of colorectal neoplasia: a randomised trial. Endoscopy. 2023;55(12):1072–80. https://doi.org/10.1055/a-2129-7254.

Red Dichromatic Imaging (RDI)

Kurato Miyazaki and Motohiko Kato

Contents

Y. Sano et al. (eds.), *Atlas of Advanced Endoscopy*, https://doi.org/10.1007/978-981-97-2732-2_3

1 Introduction

RDI (red dichromatic imaging) is a kind of novel image enhancement endoscopy technology included in EVIS X1 system recently launched by Olympus. It is an image enhancement technique that forms contrast of deep tissue and blood vessels and is expected to be used more widely in the future. In this chapter, we will describe the principle of RDI and its clinical applications.

2 Principle of RDI

RDI utilizes long wavelength light that penetrates into deeply and visualize blood vessels and provides tissue contrast by irradiating three types of narrowband light (red, amber, and green) at different wavelengths (630 nm, 600 nm, and 540 nm, respectively) [1]. Deep and superficial blood vessels are differentiated by utilizing the differences in the absorption and reflectance of hemoglobin (◘ Fig. 1):

1. Red (620–640 nm): Reaches deep into the mucosa at approximately 1.0 ~ 1.5 mm and displays deep blood vessels and mucosa at a constant brightness.
2. Amber (595–610 nm): Reaches deep into the mucosa at approximately 1.0 ~ 1.5 mm and displays deep blood vessels.
3. Green (520–550 nm): Does not reach deep into the mucosa and displays superficial capillaries.

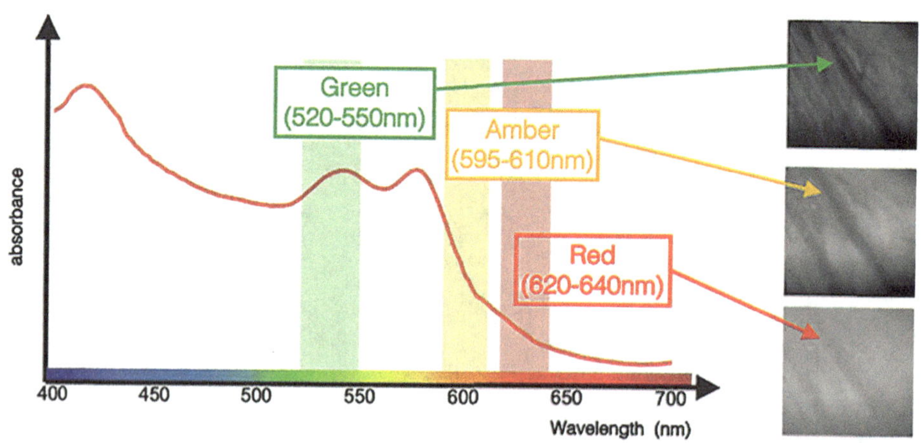

◘ **Fig. 1** Principle of RDI (courtesy by Olympus Co.). Three types of light with different wavelengths are irradiated to form a contrast of deep tissue and blood vessels

3 Three Modes in RDI

By one touch of a button on the scope, it is possible to switch from white light mode to RDI mode. RDI has three different modes, modes 1, 2, and 3 (◘ Fig. 2). These modes can be switched in sequence by pressing a switch on the endoscope. By switching modes, it is possible to change the level of deep vessel enhancement. The features of each mode are described below.

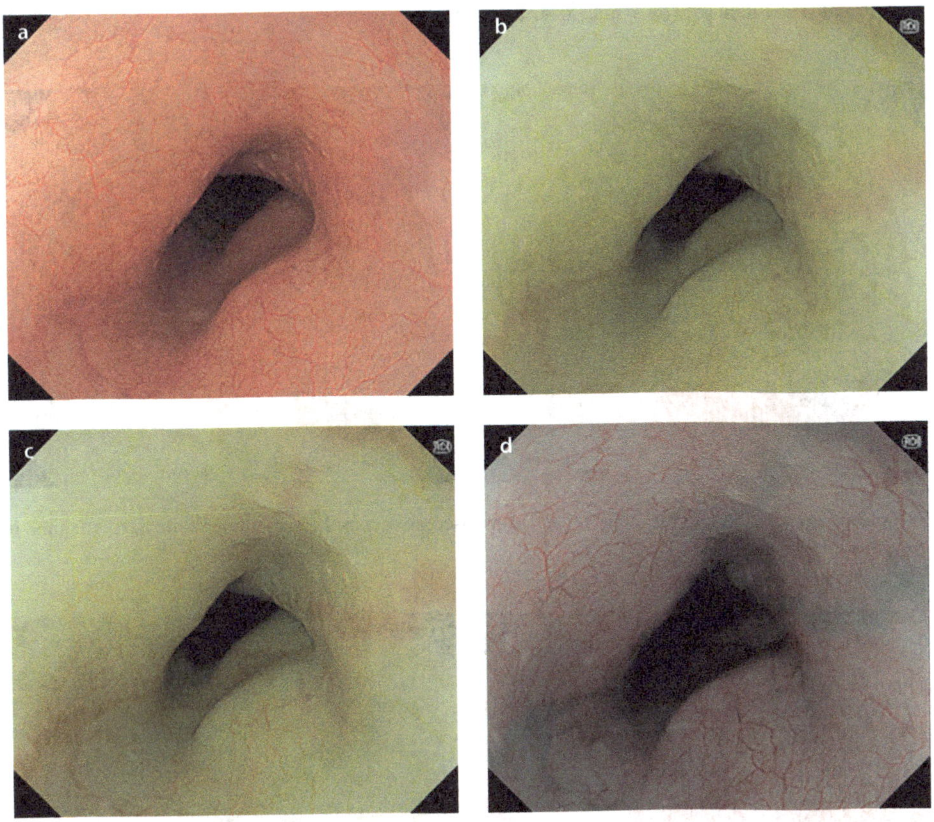

◘ **Fig. 2** Figure of the esophagus in each mode of RDI. **a** White light imaging. **b** RDI mode 1. **c** RDI mode 2. **d** RDI mode 3

Mode 1

This mode is mainly aiming for identification of bleeding points. In cases of active bleeding, it is often difficult to continue the hemostasis because the pooled blood disturbs visualization. Potential usefulness of RDI in ESD by the improvement of identification of bleeding point has been reported [2–4]. A multicenter randomized controlled trial revealed significant reduction of endoscopists' psychological stress during hemostasis even though it failed to show the shortening of hemostasis time [5]. There is also a case report of successful hemostasis for active bleeding due to peptic gastric ulcer [6] (Figs. 3, 4, and 5).

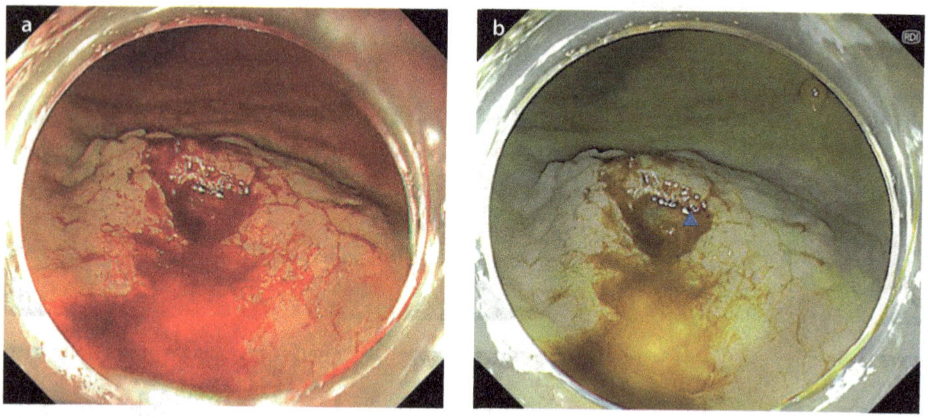

Fig. 3 Figure of oozing bleeding in gastric endoscopic mucosal resection. **a** White light imaging. **b** RDI mode 1

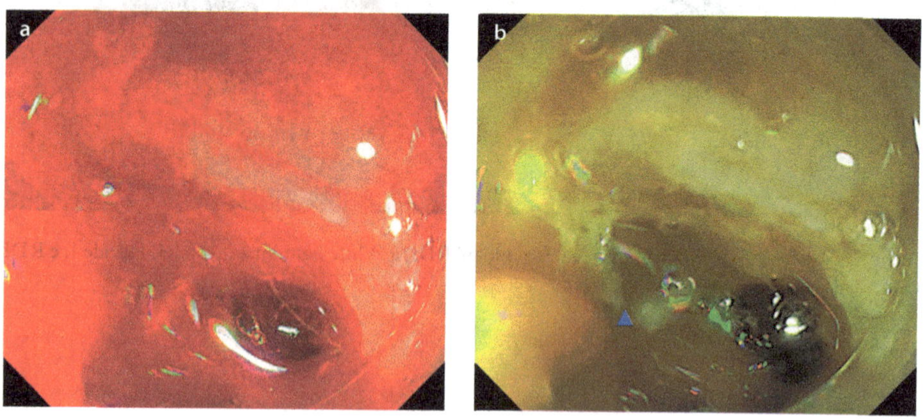

Fig. 4 Figure of spurting bleeding in duodenal ulcer. **a** White light imaging. **b** RDI mode 1

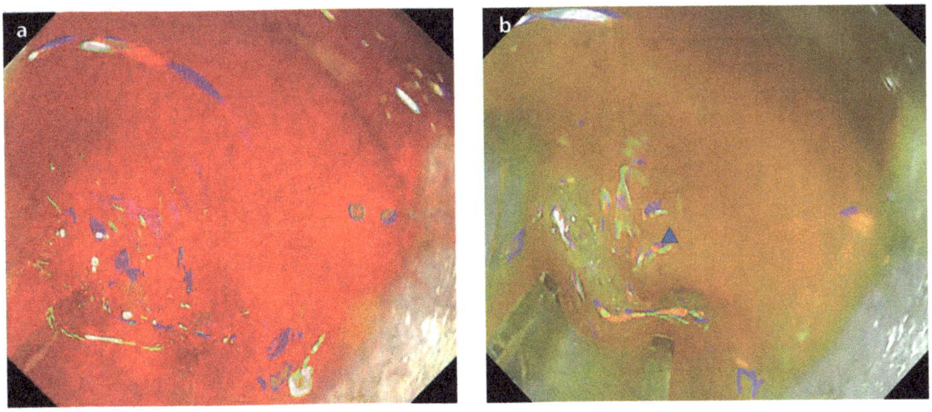

Fig. 5 Figure of spurting bleeding during duodenal endoscopic submucosal dissection. **a** White light imaging. **b** RDI mode 1

Mode 2

This mode is developed aiming improvement of visualization of deep blood vessels. Compared to mode 1, mode 2 enhances redness of blood vessels and improves the visibility of blood vessels in deep tissues. This is especially useful for initial submucosal injection during endoscopic submucosal dissection (ESD). Normally, when performing injection during ESD, blind injection cannot be avoided, because we cannot directly observe blood vessels below the mucosa. However, by identifying deep blood vessels under RDI mode 2 and avoiding unnecessary injury of those vessels, it prevents bleeding or the submucosal hematoma formation (Figs. 6 and 7) [7]. Similarly, this mode is also reported to be useful for predicting the depth of esophageal varices and confirming red color signs [8].

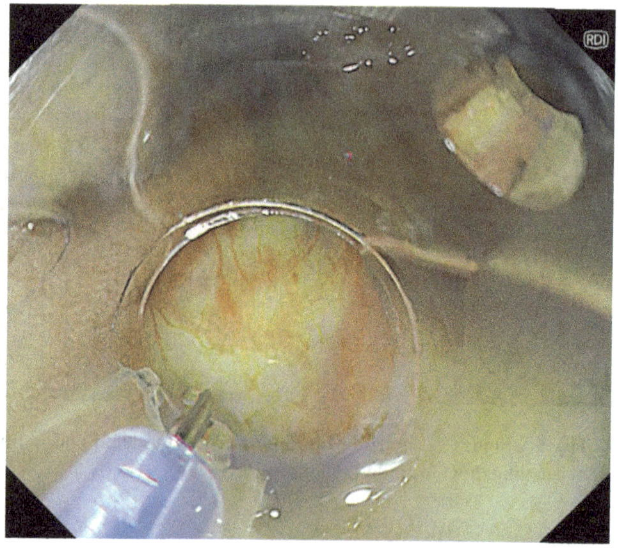

Fig. 6 Scene of injection using RDI mode 2 in rectal ESD with abundant hemorrhoids, avoiding the blood vessels

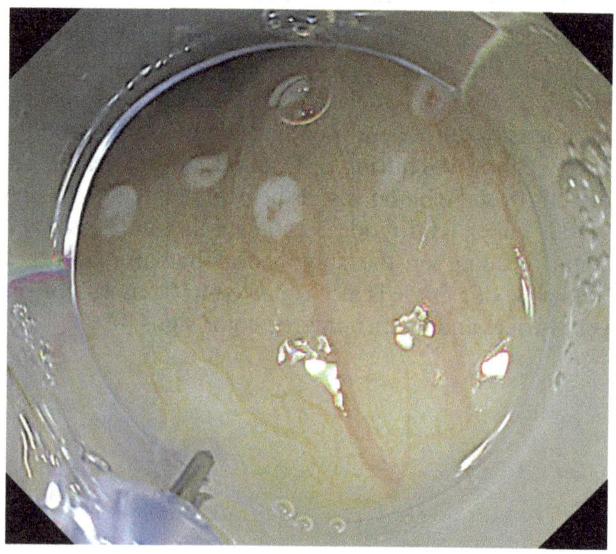

Fig. 7 Scene of injection using RDI mode 2 in esophageal ESD, avoiding the blood vessels

Mode 3

In this mode, deep blood vessels are displayed in green and superficial blood vessels in red. Images obtained by this mode are similar to those by narrowband imaging; however, bile in the digestive tract lumen does not turn red, facilitating mucosal observation in colonoscopy with poor preparation. It is also reported that this mode enables to visualize subtle inflammatory activity that is unable to find in white light imaging in patients with ulcerative colitis (Fig. 8) [9].

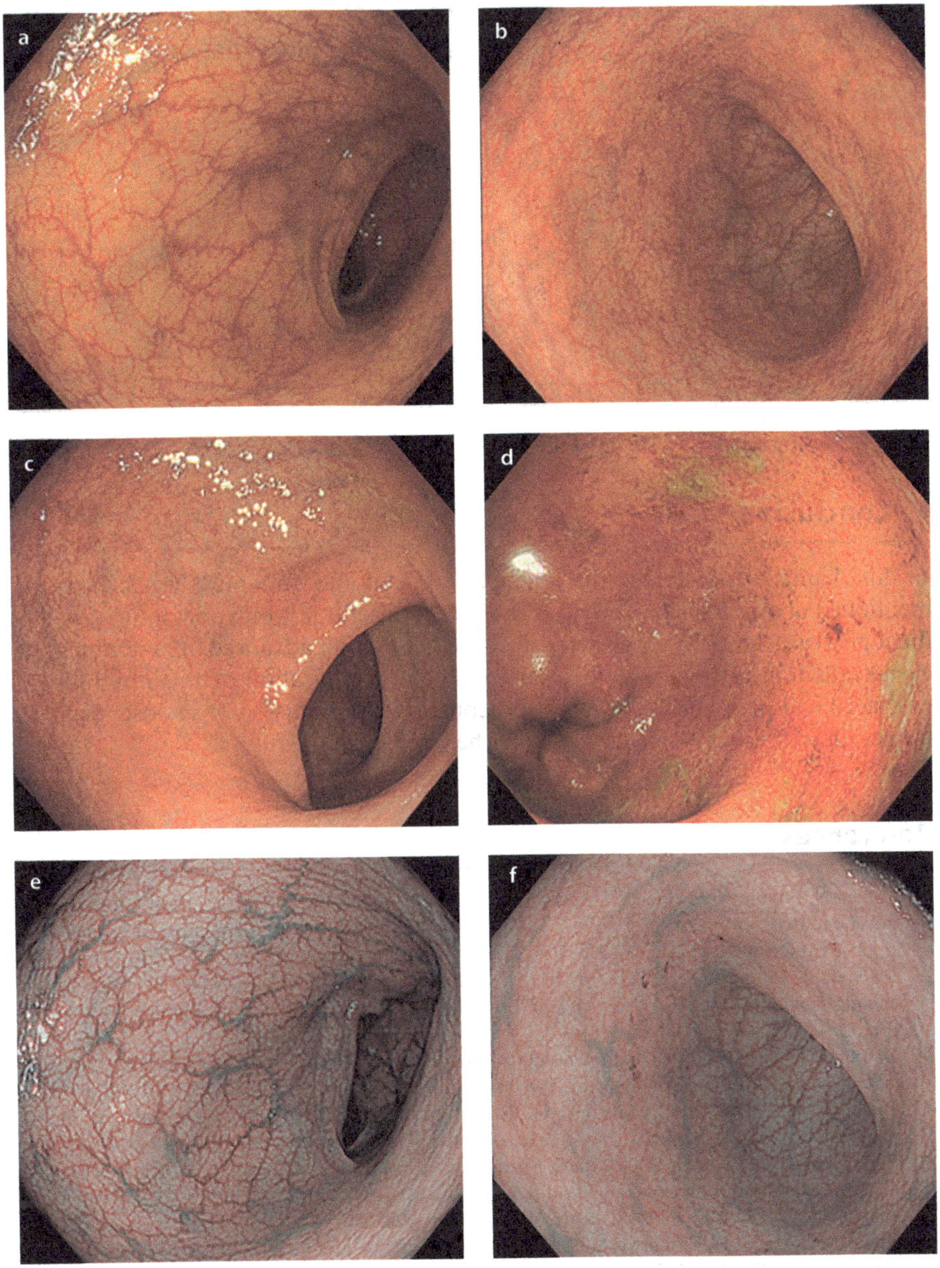

Fig. 8 White light imaging versus RDI mode 3 in cases with ulcerative colitis. **a** White light imaging of case with Mayo score 0. **b** RDI mode 3 of case with Mayo score 0. **c** White light imaging of case with Mayo score 1. **d** RDI mode 3 of case with Mayo score 1. **e** White light imaging of case with Mayo score 2. **f** RDI mode 3 of case with Mayo score 2. **g** White light imaging of case with Mayo score 3. **h** RDI mode 3 of case with Mayo score 3

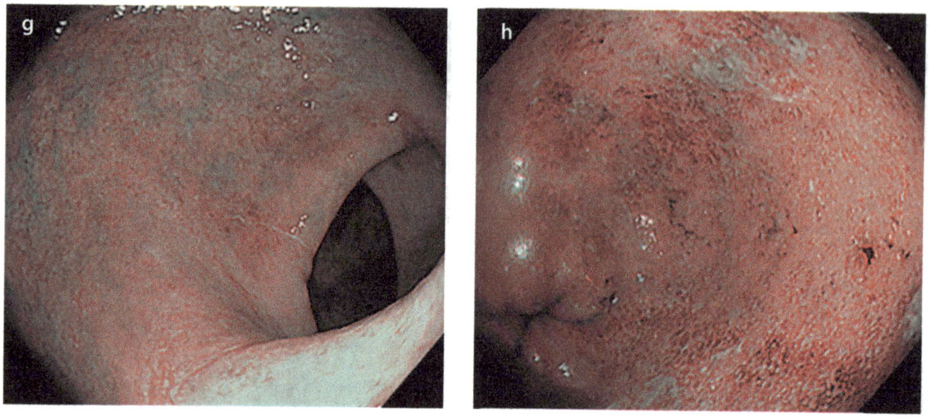

Fig. 8 (continued)

4 Conclusion

In this chapter, we described the principles of RDI and its clinical applications. As mentioned above, RDI has been reported to be useful not only for initial development purpose, the endoscopic hemostasis, but also other advanced interventions and even evaluation of inflammation activity. Further application of this novel image-enhanced endoscopy technology would contribute progression of endoscopic techniques for wider and better provision for health care in the future.

References

1. Yahagi N, Fujimoto A, Horii J, Uraoka T, Shimoda M, Takabayashi K, et al. Dual red imaging: a novel endoscopic imaging technology visualizing thick blood vessels in the gastrointestinal wall. Endosc Int Open. 2019;7(12):E1632–E5.
2. Maehata T, Fujimoto A, Uraoka T, Kato M, Horii J, Sasaki M, et al. Efficacy of a new image-enhancement technique for achieving hemostasis in endoscopic submucosal dissection. Gastrointest Endosc. 2020;92(3):667–74.
3. Ninomiya Y, Oka S, Tanaka S, Hirano D, Sumimoto K, Tamaru Y, et al. Clinical impact of dual red imaging in colorectal endoscopic submucosal dissection: a pilot study. Ther Adv Gastroenterol. 2016;9(5):679–83.
4. Tanaka H, Oka S, Tanaka S, Matsumoto K, Boda K, Yamashita K, et al. Dual red imaging maintains clear visibility during colorectal endoscopic submucosal dissection. Dig Dis Sci. 2019;64(1):224–31.
5. Fujimoto A, Saito Y, Abe S, Hoteya S, Nomura K, Yasuda H, et al. Haemostasis treatment using dual red imaging during endoscopic submucosal dissection: a multicentre, open-label, randomised controlled trial. BMJ Open Gastroenterol. 2019;6(1):e000275.
6. Kubosawa Y, Mori H, Fujimoto A. Utility of dual red imaging for endoscopic hemostasis of gastric ulcer bleeding. Dig Dis. 2020;38(4):352–4.
7. Miyazaki K, Kato M, Sasaki M, Iwata K, Masunaga T, Kubosawa Y, et al. Red dichromatic imaging reduces bleeding and hematoma during submucosal injection in esophageal endoscopic submucosal dissection. Surg Endosc. 2022;36(11):8076–85.
8. Furuichi Y, Gotoda T, Moriyasu F, Ogawa S, Kasai Y, Takeuchi H, et al. Dual red imaging (novel advanced endoscopy) can increase visibility and can predict the depth in diagnosing esophageal varices. J Gastroenterol. 2017;52(5):568–76.
9. Naganuma M, Yahagi N, Bessho R, Ohno K, Arai M, Mutaguchi M, et al. Evaluation of the severity of ulcerative colitis using endoscopic dual red imaging targeting deep vessels. Endosc Int Open. 2017;5(1):E76–82.

Artificial Intelligence (AI) in Colonoscopy

Masashi Misawa and Shin-ei Kudo

Contents

Y. Sano et al. (eds.), *Atlas of Advanced Endoscopy*, https://doi.org/10.1007/978-981-97-2732-2_4

1 Introduction

Artificial Intelligence (AI) has ushered in an era of exciting new technology that is evident in numerous domains, including healthcare. One significant focus has been the integration of AI in medical imaging. In particular, systems referred to as computer-aided detection and diagnosis (CADe and CADx) have demonstrated considerable promise in the realm of gastrointestinal endoscopy. These systems employ AI to aid endoscopists in identifying abnormal regions and differentiating detected abnormalities, and this may contribute to improved patient outcomes.

One specific procedure in which these systems could prove highly beneficial is colonoscopy, a crucial tool for colorectal cancer screening. The practice of colonoscopy presents known limitations. Approximately one-fourth of colorectal neoplasms are missed during a single colonoscopy procedure [1], and the differentiation between neoplasms and nonneoplasms frequently falls below expected thresholds [2]. Several factors can influence the quality of colonoscopies, including the physician's experience and the quality of bowel preparation, among others. AI-assisted systems employing deep learning may enhance the quality of the procedure by facilitating the detection and diagnosis of colorectal polyps.

While these systems appear to offer numerous benefits, it is imperative to rigorously analyze their performance in real-world settings. In this section, we endeavor to provide an overview of current research trends in AI applications for colonoscopy. We will also investigate how these systems have enhanced clinical outcomes and explore emerging concepts such as computer-aided quality (CAQ) improvement. By examining these topics, we aim to contribute to the growing body of knowledge regarding the utilization of AI in colonoscopy and suggest potential future directions.

2 Research Trends in CADe

In the field of colonoscopy, overlooking lesions, such as polyps, has long posed a significant challenge [1]; consequently, there was a desire for the development of CADe systems to detect polyps. Since the 2000s, various image feature parameters, such as edge detection and texture analysis, have been examined for integration into machine learning processes [3], primarily within the domain of information engineering for CADe of polyps. However, the accuracy of these CADe systems never stably exceeded 90%, with no instances of successful real-time detection assistance due to computational limitations.

Nevertheless, the advent of deep learning in the 2010s revolutionized the situation, with accuracy levels exceeding 90% and the emergence of the potential for real-time detection. In 2018, Misawa [4], Urban [5], and Wang [6] sequentially reported in medical journals the use of deep learning for real-time CADe systems. This highlighted the possible clinical applications of CADe in colonoscopy and spurred further fervent research and development. As of July 2023, in Japan, commercial distribution has commenced for products such as EndoBRAIN®-EYE (developed by Showa University and Nagoya University), Wise Vision (developed by NEC

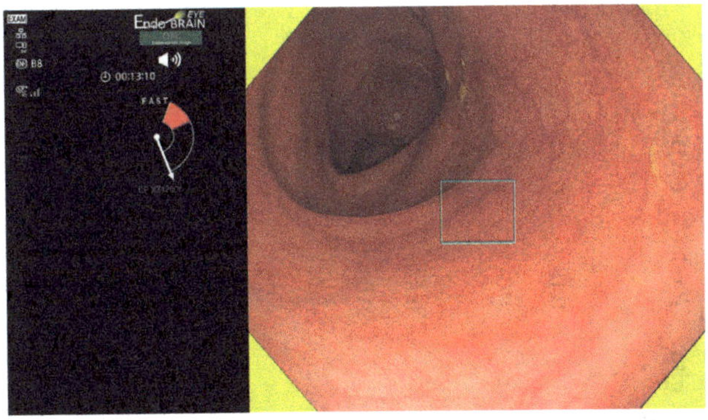

Fig. 1 An example of the computer-aided detection (CADe) in real-time clinical practice. CADe (EndoBRAIN-EYE) detected flat-type lesion by blur rectangle

Corporation and the National Cancer Center), CAD EYE by Fujifilm Corporation, and EIRL Colon Polyp by L-pixel, all of which have received approval from the Japanese regulatory authorities. Figure 1 shows an example of the clinical use of the EndoBRAIN-EYE. Looking around the world, numerous other products have been released. Within the scope of our research, several products have been launched on the market as CADe systems, including GI Genius (Medtronic), DISCOVERY (Pentax), Endo-AID (Olympus), MAGENTIQ-COLO (Magentiq EYE), CADDIE (Ordin Vision), EndoScreener (Wision AI), and Skout (ITERATIVE). Among these, GI Genius, EndoScreener, and Skout have become commercially available following approval from the US FDA.

To what extent do these CADe systems improve clinical outcomes? Table 1 summarizes the reported randomized controlled trials (RCTs) of CADe. In this context, RCTs refer to a study design wherein one group undergoes standard colonoscopy, while another is subjected to colonoscopy with CADe, implying a non-back-to-back comparison. With regard to the adenoma detection rate (ADR), in all but two trials, the utilization of CADe has been observed to significantly increase the ADR. A meta-analysis of CADe has demonstrated that its application can enhance the ADR by approximately 10% [18]. However, this meta-analysis also indicated that the polyps with increased detection rates due to CADe were predominantly diminutive adenomas (under 5 mm in size) and CADe did not contribute to the detection of advanced adenomas.

However, a recent large-scale RCT conducted by Xu et al. revealed a slight but statistically significant improvement in the detection of advanced adenomas within the CADe group [11]. These findings suggest that it may be premature to conclude that CADe is solely effective for diminutive adenomas. Furthermore, there may be potential biases in these studies, including but not limited to CADe. An example is the Hawthorne effect, where the examiner in the intervention group alters their behavior to improve the outcome. This bias may be present in reports from research groups involved in AI development or those with conflicts of interest with AI development companies. Consequently, large-scale prospective clinical studies, independent of AI developers, are deemed necessary to accurately ascertain the value of CADe.

Table 1 Reported randomized controlled trial of computer-aided detection (CADe)

Author	Year	Patients *n*	ADR without CADe (%)	ADR with CADe (%)	*P* value	AADR without CADe (%)	AADR with CADe (%)	*P* value	CADe system
Karsenti D et al. [7]	2023	2592	33.7	37.5	0.051	7.6	9.3	0.18	GI Genius
Nakashima H et al. [8]	2023	415	47.6	59.4	0.018	7.2	7.7	1	CAD EYE
Ahmad A et al. [9]	2023	614	65	71.4	0.09	–	–	–	CAD EYE
Gimeno-García A et al. [10]	2023	370	40.8	54.8	0.01	12.1	11.6	0.89	Endo-AID
Xu H et al. [11]	2023	3059	32.4	39.9	0.001	4.4	6.6	0.041	Original
Shaukat A et al. [12]	2022	1440	43.9	47.8	0.065	–	–	–	SKOUT
Repici A et al. [13]	2020	685	40.4	54.8	0.04	8	9.8	0.26	GI Genius
Gong D et al. [14]	2020	704	8	16	0.001	–	–	–	ENDOAN-GEL
Wang P et al. [15]	2020	1046	28	34	0.03	4	2	0.66	EndoScreener
Liu WN et al. [16]	2020	1026	23.89	39.1	<0.001	6.45	2.88	0.821	Original CADe
Wang P et al. [17]	2019	1058	20.3	29.1	<0.001	5.95	3.41	0.803	Original CADe

ADR adenoma detection rate, *AADR* advanced adenoma detection rate, *CADe* computer-aided detection

3 Trends in the Research of CADx

In routine clinical practice, the ability to differentiate whether a detected lesion is neoplastic and, if so, whether it is benign or malignant, along with the determination of suitable surgical or endoscopic treatment, requires a high accuracy. However, achieving an accurate differential diagnosis can be challenging without substantial training. Therefore, support from CADx may prove valuable in enhancing diagnostic precision within this field.

Research on CADx within the field of colonoscopy has been active since the 2010s. Multiple research groups have reported studies targeting vessels and surface patterns obtained from magnifying narrow-band imaging (NBI). In this research field, a group from Hiroshima University has been a pioneer. In a trial conducted by Kominami et al., CADx was employed to analyze 118 lesions in 41 patients, achieving a sensitivity and specificity of 93.3% for neoplasms, as reported in 2016 [19]. Subsequently, researchers Chen [20] and Byrne [21] reported the utilization of deep learning in CADx, with both studies achieving over 90% sensitivity in differentiating neoplasms from nonneoplasms. Mori et al. conducted a prospective study using EndoBRAIN, a CADx system designed for ultra-magnifying endoscopy, and reported that neoplasms could be differentiated with a sensitivity of 92.7% and specificity of 89.8%. This study underscores the potential utility of CADx in real-world clinical practice [22].

EndoBRAIN has been reevaluated in international multicenter studies in Japan, Norway, and the UK. Specifically, this study only included trainee endoscopists, who are expected to benefit most from AI. While the use of CADx did not lead to a significant increase in sensitivity (88.4% vs. 90.4%), it did result in a significant improvement in the proportion of high-confidence diagnoses (74.2% vs. 92.6%) [23]. In 2021, Fujifilm Corporation launched CAD EYE, a system capable of differentiating between neoplasms and nonneoplasms. CADx systems that classify neoplasms and nonneoplasms are gradually becoming integrated into daily clinical practice. ◻ Table 2 lists prospective studies using CADx, as obtained by searching PubMed. As of July 2023, no RCT has been conducted, leading to a lower level of evidence. However, reports indicate relatively high diagnostic accuracy in real-world clinical practice. It should be noted that many of these reports emanate from AI developers or groups with conflicts of interest, warranting caution regarding the potential for bias.

Table 2 Reported prospective studies targeting on computer-aided diagnosis (CADx)

Authors	Year	Number of the polyps	Sensitivity	Specificity	Accuracy	CADx system	Study setting
Houwen B et al. [24]	2023	423	89	38	79	Original	AI-alone (including SSL)
Hassan C et al. [25]	2023	319	81.8-86.4	92.4–94.0	–	CAD EYE, GI Genius	AI-assisted
Weiquan Li J et al. [26]	2023	661	61.8	–	71.6	CAD EYE	AI-assisted
Minegishi Y et al. [27]	2022	465	94.4	62.5	85.5	EndoBRAIN-X	AI-assisted (including SSL)
Rondonotti E et al. [28]	2023	596	88.6	88.1	88.4	CAD EYE	AI-assisted
Hassan C et al. [29]	2022	544	82	93.2	91.8	GI Genius	AI-alone
Barua I et al. [23]	2022	892	90.4	85.9	–	EndoBRAIN	AI-assisted
Mori Y et al. [22]	2018	466	92.7	89.9	–	EndoBRAIN	AI-alone

AI-alone: the diagnostic performances were measured by standalone AI
AI-assisted: the diagnostic performances were measured according to endoscopists' diagnosis with reference to the CADx
SSL sessile serrated lesion

4 Other Developments in CADx

Early versions of CADx were primarily designed to assist in differentiating between neoplasms and nonneoplasms. However, as AI development technology has advanced, new targets for CADx are being eagerly investigated. These include the depth prediction of early colorectal cancer and differentiation between sessile serrated lesions (SSLs) and hyperplastic polyps. In real-world clinical practice, SSLs recently treated as a type of neoplastic polyp are gaining attention. Diagnosing the depth of a cancer is acknowledged to be challenging. Consequently, there is an increasing focus on more sophisticated CADx systems, moving beyond simple differentiation between neoplasms and nonneoplasms.

First, research on AI for the diagnosis of cancer depth has taken precedence, given its greater clinical importance. Tamai et al. developed a CADx for diagnosing invasive cancers (requiring surgical resection) using magnifying NBI images, reporting a sensitivity of 83.9% and specificity of 82.6% [30]. Takeda et al. also reported a correct diagnosis rate of 94.1% for invasive cancers using CADx for endocytoscopy (EndoBRAIN-Plus) [31]. ◘ Figure 2 shows an early-stage colorectal cancer (T1) that diagnosed using the EndoBRAIN-Plus. Tokunaga et al. created a more versatile and practical CADx system, capable of determining whether the target lesion can be treated endoscopically, based on white light images [32]. This system was retrospectively verified and reported an accuracy of 90.3%. Okamoto et al. developed a CADx to identify JNET classification types 1, 2A, 2B, and 3, achieving an accuracy of 94.1% for JNET type 3 [33].

Minegishi et al. conducted a prospective study using a CADx system capable of identifying SSL. According to their study, the sensitivity for neoplastic lesions, including SSL, was 94.4%, while the specificity was 62.5% [27]. This CADx, named EndoBAIN-X, obtained approval from the Japanese regulatory body. Houwen et al. are developing a similar CADx system and conducting prospective clinical studies, and the sensitivity of their CADx system for tumorous lesions, including SSL, was 89%, while the specificity was 38% [24]. Consequently, clinical studies of CADx systems treating SSL as neoplasms are being gradually reported.

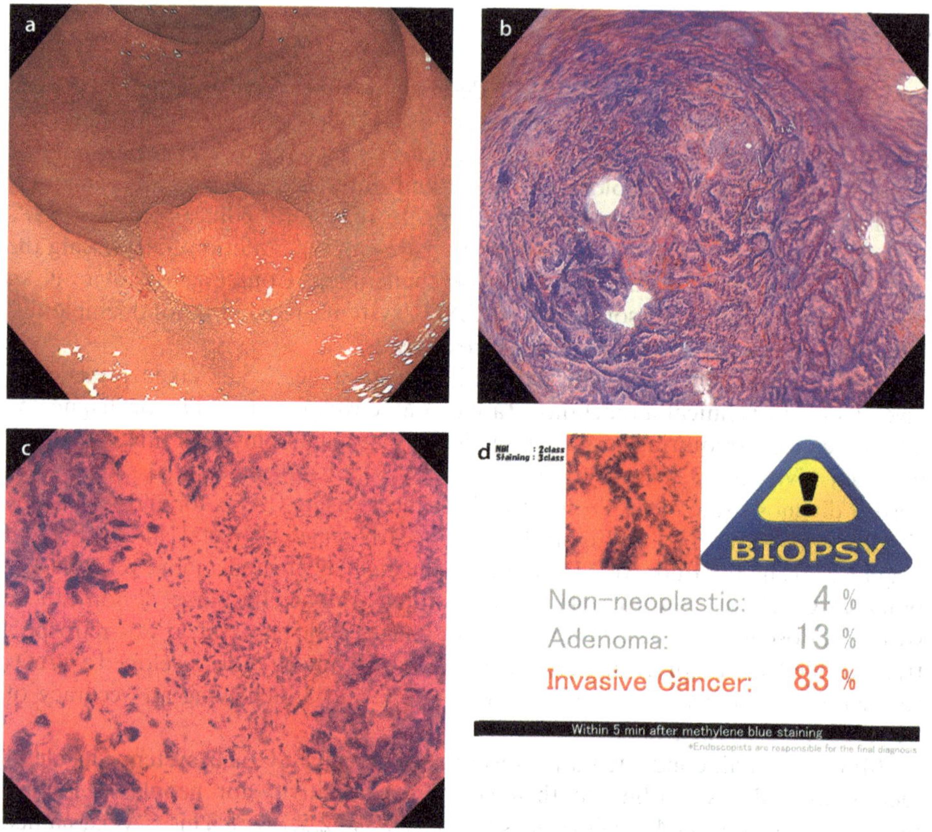

Fig. 2 A case of T1 colorectal cancer diagnosed using computer-aided diagnosis (CADx). **a** White light endoscopy showed a 10-mm-diameter 0-Is+IIc lesion. **b** Magnifying endoscopy after crystal violet staining showed V_I high-grade pit pattern. **c** Endocytoscopy after methylene blue staining. Lumens (opening of the crypt) were indistinct and enlarged; irregular-shaped stained cell nuclei were observed. **d** The prediction of CADx (EndoBRAIN-Plus) was invasive cancer. This lesion surgically resected and the pathological diagnosis was deep submucosal cancer (T1b)

5 Other Applications of AI

It has been observed that the usage of AI alone may not adequately enhance clinical outcomes. Recent negative studies evaluating the performance of CADe have been reported [34]. This has led to the emergence of, and growing acceptance for, a novel concept known as CAQ. The ability of CADe to detect lesions is limited to instances in which lesions, such as polyps or cancers, are visually depicted in endoscopic monitor. For instance, if polyps or cancers are located in blind spots, such as folds, residual feces, or behind flexures, CADe is found to be ineffective. This limitation arises because CADe, being solely an image analysis technology, is only effective for lesions that are visualized on the monitor.

It is hypothesized that the development of a system to guide colonoscopy with fewer blind spots could work in conjunction with CADe to further reduce missed lesions. In this context, Yao and his team developed a system that monitors the withdrawal speed of endoscope. The system issues real-time alarm if the withdrawal is excessively fast. The use of this system significantly improved the lesion detection rate, as reported by Yao et al. [35]. The monitoring of withdrawal speed likely encouraged more careful observation of the colonic mucosa, thus altering endoscopists' practices.

6 Conclusions

The introduction of deep learning has democratized AI technology, enabling the rapid development of high-accuracy AI systems. This breakthrough has accelerated the clinical implementation of AI, making it an integral part of many medical procedures, including endoscopy.

Every day, numerous new studies on endoscopic AI are reported, adding to our understanding of AI's potential and limitations in this field. Some AI systems, including EndoBRAIN, have already received regulatory approval and are being implemented in clinical settings.

However, while AI has shown tremendous potential in enhancing detection and diagnostic capabilities in colonoscopy, it is essential to remember that the long-term benefits of AI usage, such as reducing cancer incidence and mortality, remain unproven. Therefore, moving forward, the focus should be on conducting large-scale prospective studies to verify these benefits and assess the overall utility of AI in endoscopy. The exploration of AI's role in medicine is only beginning, and the horizon promises many more exciting discoveries.

References

1. van Rijn JC, Reitsma JB, Stoker J, et al. Polyp miss rate determined by tandem colonoscopy: a systematic review. Am J Gastroenterol. 2006;101:343–50.
2. Rees CJ, Rajasekhar PT, Wilson A, et al. Narrow band imaging optical diagnosis of small colorectal polyps in routine clinical practice: the detect inspect characterise resect and DISCARD 2 (DISCARD 2) study. Gut. 2017;66:887–95.
3. Bernal J, Sánchez J, Vilarino F. Towards automatic polyp detection with a polyp appearance model. Pattern Recogn. 2012;45:3166–82.
4. Misawa M, Kudo SE, Mori Y, et al. Artificial intelligence-assisted polyp detection for colonoscopy: initial experience. Gastroenterology. 2018;154(2027-2029):e3.
5. Urban G, Tripathi P, Alkayali T, et al. Deep learning localizes and identifies polyps in real time with 96% accuracy in screening colonoscopy. Gastroenterology. 2018;155(1069-1078):e8.
6. Wang P, Xiao X, Glissen Brown JR, et al. Development and validation of a deep-learning algorithm for the detection of polyps during colonoscopy. Nat Biomed Eng. 2018;2:741–8.
7. Karsenti D, Tharsis G, Perrot B, et al. Effect of real-time computer-aided detection of colorectal adenoma in routine colonoscopy (COLO-GENIUS): a single-Centre randomised controlled trial. Lancet Gastroenterol Hepatol. 2023;8:726–34.
8. Nakashima H, Kitazawa N, Fukuyama C, et al. Clinical evaluation of computer-aided colorectal neoplasia detection using a novel endoscopic artificial intelligence: a single-center randomized controlled trial. Digestion. 2023;104(3):193–201.
9. Ahmad A, Wilson A, Haycock A, et al. Evaluation of a real-time computer-aided polyp detection system during screening colonoscopy: AI-DETECT study. Endoscopy. 2023;55:313–9.

10. Gimeno-Garcia AZ, Hernandez Negrin D, Hernandez A, et al. Usefulness of a novel computer-aided detection system for colorectal neoplasia: a randomized controlled trial. Gastrointest Endosc. 2023;97(528-536):e1.
11. Xu H, Tang RSY, Lam TYT, et al. Artificial intelligence-assisted colonoscopy for colorectal cancer screening: a multicenter randomized controlled trial. Clin Gastroenterol Hepatol. 2023;21(337-346):e3.
12. Shaukat A, Lichtenstein DR, Somers SC, et al. Computer-aided detection improves adenomas per colonoscopy for screening and surveillance colonoscopy: a randomized trial. Gastroenterology. 2022;163:732–41.
13. Repici A, Badalamenti M, Maselli R, et al. Efficacy of real-time computer-aided detection of colorectal neoplasia in a randomized trial. Gastroenterology. 2020;159(512-520):e7.
14. Gong D, Wu L, Zhang J, et al. Detection of colorectal adenomas with a real-time computer-aided system (ENDOANGEL): a randomised controlled study. Lancet Gastroenterol Hepatol. 2020;5:352–61.
15. Wang P, Liu X, Berzin TM, et al. Effect of a deep-learning computer-aided detection system on adenoma detection during colonoscopy (CADe-DB trial): a double-blind randomised study. Lancet Gastroenterol Hepatol. 2020;5:343–51.
16. Liu WN, Zhang YY, Bian XQ, et al. Study on detection rate of polyps and adenomas in artificial-intelligence-aided colonoscopy. Saudi J Gastroenterol. 2020;26:13–9.
17. Wang P, Berzin TM, Glissen Brown JR, et al. Real-time automatic detection system increases colonoscopic polyp and adenoma detection rates: a prospective randomised controlled study. Gut. 2019;68:1813–9.
18. Barua I, Vinsard DG, Jodal HC, et al. Artificial intelligence for polyp detection during colonoscopy: a systematic review and meta-analysis. Endoscopy. 2021;53:277–84.
19. Kominami Y, Yoshida S, Tanaka S, et al. Computer-aided diagnosis of colorectal polyp histology by using a real-time image recognition system and narrow-band imaging magnifying colonoscopy. Gastrointest Endosc. 2016;83:643–9.
20. Chen PJ, Lin MC, Lai MJ, et al. Accurate classification of diminutive colorectal polyps using computer-aided analysis. Gastroenterology. 2018;154:568–75.
21. Byrne MF, Chapados N, Soudan F, et al. Real-time differentiation of adenomatous and hyperplastic diminutive colorectal polyps during analysis of unaltered videos of standard colonoscopy using a deep learning model. Gut. 2017;68(1):94–100.
22. Mori Y, Kudo SE, Misawa M, et al. Real-time use of artificial intelligence in identification of diminutive polyps during colonoscopy: a prospective study. Ann Intern Med. 2018;169:357–66.
23. Barua I, Wieszczy P, Kudo S-e, et al. Real-time artificial intelligence–based optical diagnosis of neoplastic polyps during colonoscopy. NEJM Evid. 2022;1:EVIDoa2200003.
24. Houwen B, Hazewinkel Y, Giotis I, et al. Computer-aided diagnosis for optical diagnosis of diminutive colorectal polyps including sessile serrated lesions: a real-time comparison with screening endoscopists. Endoscopy. 2023;55(8):756–65.
25. Hassan C, Sharma P, Mori Y, et al. Comparative performance of artificial intelligence optical diagnosis Systems for Leaving in situ colorectal polyps. Gastroenterology. 2023;164(467-469):e4.
26. Weiquan Li J, Ho Wu CC, Jie Lee JW, et al. Real-world validation of a computer-aided diagnosis system for prediction of polyp histology in colonoscopy: a prospective multicenter study. Am J Gastroenterol. 2023;118(8):1353–64.
27. Minegishi Y, Kudo SE, Miyata Y, et al. Comprehensive diagnostic performance of real-time characterization of colorectal lesions using an artificial intelligence-assisted system: a prospective study. Gastroenterology. 2022;163(323-325):e3.
28. Rondonotti E, Hassan C, Tamanini G, et al. Artificial intelligence-assisted optical diagnosis for the resect-and-discard strategy in clinical practice: the artificial intelligence BLI characterization (ABC) study. Endoscopy. 2023;55:14–22.
29. Hassan C, Balsamo G, Lorenzetti R, et al. Artificial intelligence allows leaving-in-situ colorectal polyps. Clin Gastroenterol Hepatol. 2022;20(2505-2513):e4.
30. Tamai N, Saito Y, Sakamoto T, et al. Effectiveness of computer-aided diagnosis of colorectal lesions using novel software for magnifying narrow-band imaging: a pilot study. Endosc Int Open. 2017;5:E690–4.

31. Takeda K, Kudo SE, Mori Y, et al. Accuracy of diagnosing invasive colorectal cancer using computer-aided endocytoscopy. Endoscopy. 2017;49:798–802.
32. Tokunaga M, Matsumura T, Nankinzan R, et al. Computer-aided diagnosis system using only white-light endoscopy for the prediction of invasion depth in colorectal cancer. Gastrointest Endosc. 2021;93:647–53.
33. Okamoto Y, Yoshida S, Izakura S, et al. Development of multi-class computer-aided diagnostic systems using the NICE/JNET classifications for colorectal lesions. J Gastroenterol Hepatol. 2022;37:104–10.
34. Levy I, Bruckmayer L, Klang E, et al. Artificial intelligence-aided colonoscopy does not increase adenoma detection rate in routine clinical practice. Am J Gastroenterol. 2022;117:1871–3.
35. Yao L, Zhang L, Liu J, et al. Effect of an artificial intelligence-based quality improvement system on efficacy of a computer-aided detection system in colonoscopy: a four-group parallel study. Endoscopy. 2022;54:757–68.

Classification

Contents

Classification of Endoscopic Imaging

Hisao Tajiri

Content

Y. Sano et al. (eds.), *Atlas of Advanced Endoscopy*, https://doi.org/10.1007/978-981-97-2732-2_5

Gastrointestinal endoscopy began with the introduction of the gastroscope, which prompted active clinical use of endoscopes in subsequent years, and thereafter various fiberscopes were also applied clinically. In 1984, electronic videoendoscopes were also developed. Unlike fiberscopes, which directly detect light signals, electronic videoendoscopes convert electronic signals into images via semiconductor elements and allow various forms of electronic image processing and analysis.

Between the late 1980s and early 1990s, research on image processing and analysis was conducted mainly by the National Cancer Center Hospital East and Olympus Group, which led to basic experiments on narrow band light imaging starting in 1994. As a result, a patent application was made in 1999, and a narrow band imaging (NBI) device (Olympus Co., Tokyo, Japan) was introduced commercially in 2006. Since the early 2000s, it has been demonstrated by a number of researchers that NBI is useful for early diagnosis of cancers of the oropharynx, hypopharynx, esophagus, stomach, and large intestine. A succession of similar techniques was later made public, and interpretation of the term "special light observation" began to differ among academic societies and research organizations. In view of this problem and the need to establish internationally applicable terminology for endoscopy, we proposed an object-oriented classification for endoscopic imaging in 2008. Basically, the concept was that endoscopic imaging can be divided into five categories: (1) conventional endoscopy (white light endoscopy (WLE)), (2) image-enhanced endoscopy (IEE), (3) magnified endoscopy, (4) microscopic endoscopy, and (5) tomographic endoscopy [1].

IEE is subdivided into optical, digital, optical-digital, and chromoendoscopy methods.

While NBI has spread and contributed to the standardization of diagnosis on a global level, our colleagues have worked tirelessly to further improve the quality of endoscopy. Over the last 15 years, owing to the development and worldwide spread of NBI, international classifications have been introduced in the field of gastric cancer, Barrett's esophagus, and colorectal neoplasia.

Since the introduction of this classification, several IEE techniques have become commercially available through advances in endoscopy technology, including blue light imaging (BLI), linked color imaging (LCI), red dichromatic imaging (RDI), and texture and color enhancement imaging (TXI). Therefore, a revised version including those techniques is presented in ◘ Fig. 1.

The LASEREO system (FUJIFILM Co., Tokyo, Japan) based on laser endoscopy was developed for advanced IEE, including WLI, BLI, and LCI. LCI using short-wavelength narrow band laser light combined with white laser light is a new technique that enhances differences in red coloration through digital processing [2]. This enables LCI to visualize red lesions better by enhancing their intensity relative to whitish lesions, which appear whiter. Recently, a controlled, multicenter trial with randomization using minimization has reported that LCI is more effective than WLI for detecting neoplastic lesions in the pharynx, esophagus, and stomach [3].

RDI is a type of novel IEE technology included in the EVIS X1 system launched by Olympus Co. in 2020. It is an imaging technique that enhances the contrast of deep tissue and blood vessels [4]. RDI has been reported to be useful for not only its initial development purpose, i.e., endoscopic hemostasis, but also other advanced interventions and even evaluation of inflammation activity. On the other hand, TXI developed by Olympus Co. is designed to enhance three image factors in WLI—texture, brightness, and color—to clearly define subtle tissue differences [5]. TXI is an

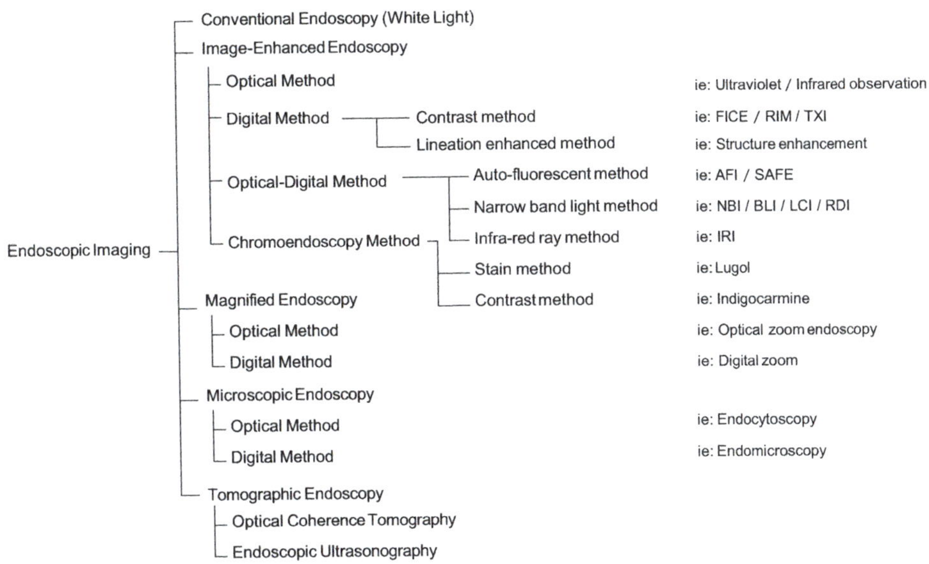

Fig. 1 Classification of endoscopic imaging

innovative IEE that facilitates the adjustment of brightness and emphasizes surface irregularities and color differences in endoscopic images. As further data are accumulated, these new IEE technologies are expected to make a significant contribution to clinical practice.

References

1. Tajiri H, Niwa H. Proposal for a consensus terminology in endoscopy: how should different endoscopic imaging techniques be grouped and defined? Endoscopy. 2008;40:775–8.
2. Fukuda H, Miura Y, Hayashi Y, et al. Linked color imaging focused on neoplasm detection in the technology facilitates early detection of flat gastric cancers. Clin J Gastroenterol. 2015;8:385–9.
3. Ono S, Kawada K, Dohi O, et al. Linked color imaging focused on neoplasm detection in the upper gastrointestinal tract. A randomized trial. Ann Intern Med. 2021;174:18–24.
4. Yahagi N, Fujimoto A, Horii J, et al. Dual red imaging: a novel endoscopic imaging technology visualizing thick blood vessels in the gastrointestinal wall. Endosc Int Open. 2019;7(12):E1632–5.
5. Sato T. TXI: texture and color enhancement imaging for endoscopic image enhancement. J Healthcare Eng. 2021;2021:5518948.

Paris Classification

Mineo Iwatate and Santa Hattori

Contents

Y. Sano et al. (eds.), *Atlas of Advanced Endoscopy*, https://doi.org/10.1007/978-981-97-2732-2_6

1 The Paris Endoscopic Classification for Superficial GI Lesions

In 2002, an international group of endoscopists, surgeons, and pathologists gathered in Paris to develop the endoscopic morphological classification for "superficial" neoplastic lesions in the esophagus, stomach, and colon [1]. A neoplastic lesion is considered "superficial" when its endoscopic appearance suggests that it has not penetrated the digestive wall beyond the submucosa. In Japan, neoplastic stomach lesions that appear "superficial" during endoscopy are classified as "type 0" subtypes. The term "type 0" was chosen to distinguish the classification of "superficial" lesions from the Borrmann classification, which was proposed in 1926 for "advanced" gastric tumors and included types 1 to 4. The endoscopic morphology of superficial lesions is associated with invasive potential and can help determine treatment strategies [2].

2 Definition of Endoscopic Morphology (Fig. 1)

The classification of type 0 lesions is based on the distinction between polypoid (type 0-I); non-polypoid, nonexcavated (type 0-II); and non-polypoid, excavated (type 0-III) lesions. In addition, type 0-II lesions are further divided into three types, including slightly elevated (type 0-IIa), flat (type 0-IIb), and slightly depressed (type 0-IIc). Type 0-I is subdivided into pedunculated (Ip) and sessile (Is). The differentiation between type 0-I and type 0-IIa is based on the height of the closed cups of biopsy forceps (2.5 mm). Lesions protruding above the level of the closed cups of a biopsy forceps are classified as type 0-I; those below this level are classified as type 0-IIa.

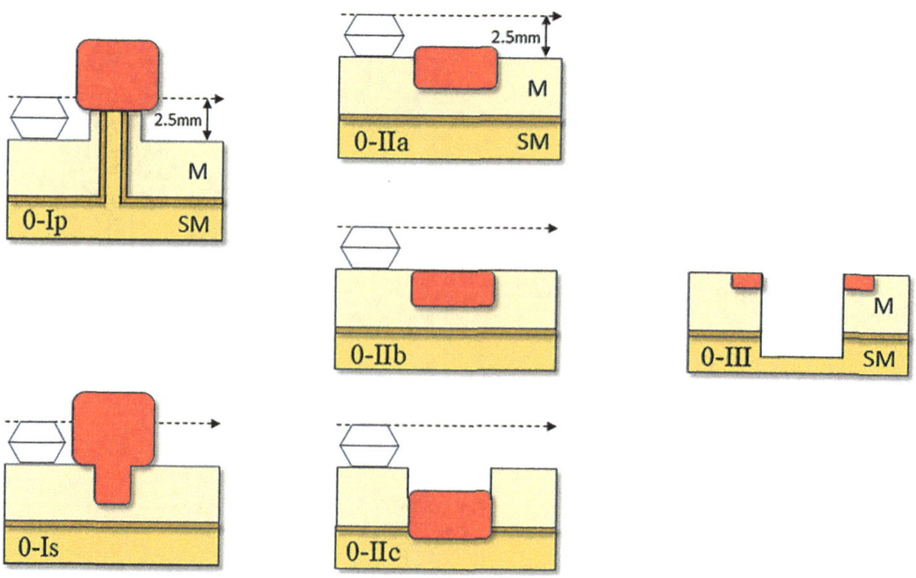

Fig. 1 Definition of endoscopic morphology: polypoid (Ip and Is), non-polypoid (IIa, IIb, and IIc), non-polypoid and excavated (III)

Mixed types associate two distinct types of morphology. The most common pattern of mixed types consists of an elevation (IIa) and a depression (IIc). An elevated lesion with a central depression at its top is classified as type 0-IIa + IIc. In comparison, a depressed lesion with elevated borders or a central elevation is classified as type 0-IIc + IIa. Other mixed types include type 0-IIc + Is, 0-Is+IIc, 0-IIc + III, and 0-III + IIc. However, the exact placement of this mixed-type classification requires a precise evaluation of the morphology, and there is room for interobserver disagreement because the relative surface of each type is not the only factor relevant to prognosis.

3 Endoscopic Staging

The morphology of a type 0 neoplastic lesion can predict the depth of invasion into the digestive tract wall, providing an "endoscopic staging." Therefore, endoscopic descriptive morphology can help treatment decisions involving endoscopic resection or surgery. The primary role of endoscopic staging is to predict the risk of submucosal invasion and the associated risk of nodal metastases. For a type 0-I lesion, the size is a reliable predictive criterion. The risk of submucosal invasion increases as the size is larger. On the other hand, with type 0-II lesions, the morphologic subtypes have greater importance. Submucosal invasion is more frequent in depressed lesions (IIc) comparing the same size [3]. Mixed-type lesions are more likely to invade into submucosa at a higher proportion than non-mixed-type lesions. As a rule, type 0-IIa + IIc lesions have a poorer prognosis, with a risk of deep invasion in the submucosa. Because morphological endoscopic staging is not accurate enough, following advanced diagnoses such as chromoendoscopy, image-enhanced endoscopy, and magnifying endoscopy is required for better staging.

References

1. The Paris endoscopic classification of superficial neoplastic lesions: Esophagus, stomach, and colon: November 30 to December 1, 2002. Gastrointest Endosc. 2003;58:S3–43.
2. Sano Y, Iwadate M. The importance of the macroscopic classification of colorectal neoplasms. Gastrointest Endosc Clin N Am. 2010;20:461–9.
3. Se K, Lambert R, Allen JI, et al. Nonpolypoid neoplastic lesions of the colorectal mucosa. Gastrointest Endosc. 2008;68:S3–47.

Esophagus: The Japan Esophageal Society (JES) Classification

Kenichi Goda and Tsuneo Oyama

Contents

Y. Sano et al. (eds.), *Atlas of Advanced Endoscopy*, https://doi.org/10.1007/978-981-97-2732-2_7

1 Squamous Cell Carcinoma

Predicting invasion depth of superficial esophageal squamous cell carcinoma is crucial in determining the precise indication for endoscopic resection (ER) because the rate of lymph node metastasis increases in proportion to the invasion depth of the carcinoma. Previous studies have shown a close relationship between microvascular patterns observed by magnifying endoscopy and invasion depth of the superficial carcinoma. Although there were two major classifications, Inoue and Arima, the Japan Esophageal Society (JES) integrated the two classifications and simplified it and developed a new magnifying endoscopic classification for the characterization and predicting invasion depth of superficial esophageal squamous cell carcinomas (SESCCs). This is essential for developing a treatment strategy for SESCC, in particular the indication for ER. Therefore, in this classification, morphological types of microvessels are classified into two categories of noncancerous [**type A:** normal epithelium, inflammation, and **squamous** intraepithelial neoplasia (SIN)] and cancerous (**Type B**: SCC) lesions. The cancerous types of microvessels corresponding to SESCCs are subclassified into three groups based upon an indication for ER as follows: an absolute indication type (**Type B1**: T1a-EP or T1a-LPM), a relative indication type [**Type B2**: T1a-MM or **T1b**-SM1(tumor invades the submucosa to a depth of 200 μm or less from the muscularis mucosa)], and a contraindication type [**Type B3: T1b-SM2** (tumor invades the submucosa to a depth more than 200 μm)]. Diagnostic criteria of the JES classification are based on the degree of microvascular irregularity in the target lesion observed by magnifying endoscopy. Intrapapillary capillary loops (IPCL) are a basic unit of microvasculature in the stratified squamous epithelial layer. The microvascular irregularity is evaluated for the presence or absence of each of the following morphological factors: weaving (i.e., tortuosity), dilatation, irregular caliber, and different shape (i.e., various shapes). Microvessels are classified as type A if they have three or fewer factors (i.e., without severe abnormality; ◘ Fig. 1a) and type B if they have all four (i.e., with severe abnormality). Type B is then subclassified into B1, B2, and B3 (◘ Fig. 1b–d, respectively) based on the running pattern or degree of dilatation of severely irregular microvessels. The definitions and schemas of type A and B vessels and predicted histology of invasion depth by type B vessels are summarized in ◘ Table 1. A large scale validation study showed high overall accuracy (90.5%) of type B vessels of the JES classification. The most important auxiliary criterion in the JES classification is avascular area (AVA). AVA is defined as a low or no vascularity area surrounded by all subtypes of type B microvessels including B1 vessels. Diameters of AVA are positively correlated with sizes of histological cancer nest and the histological invasion depth of SESCC. Small (<0.5 mm), middle (0.5 ≤< 3 mm), and large (≤3 mm) AVA are suggestive of T1a-EP/LPM, T1a-MM/T1b-SM1, and T1b-SM2, respectively. A key point to note is that any types of AVA (small, middle, and large) surrounded by B1 vessels are suggestive of T1a-EP or T1a-LPM SCC.

Table 1 The JES classification

Type of vessels	Schema	Definitions		Invasion depth	Histology
A	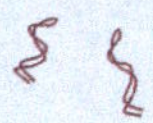	Normal IPCL or abnormal microvessels without severe irregularity		No invasion	Normal epithelium, inflammation, and LGIN
B1		Abnormal microvessels with severe irregularity or highly dilated abnormal vessels.	Type B vessels with a loop-like formation	T1a-EP or T1a-LPM	HGIN and invasive SCC
B2	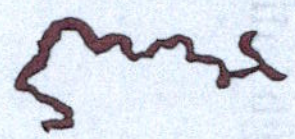		Type B vessels without a loop-like formation	T1a-MM or T1b-SM1	
B3			Highly dilated vessels which calibers appear to be more than three times that of B2 vessels	T1b-SM2	

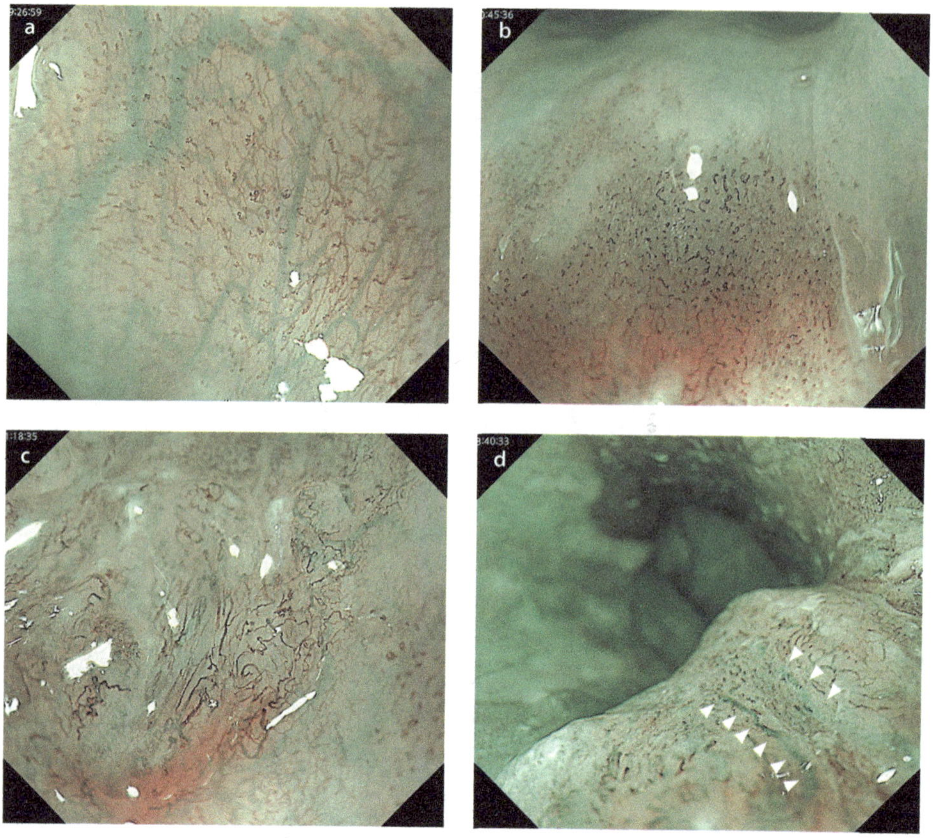

Fig. 1 **a** Type A microvessels, **b** Type B1 microvessels, **c** Type B2 microvessels, **d** Type B3 microvessels

A meta-analysis comparing the Inoue, Arima, and JES classifications in the diagnostic accuracy and staging of SESCC showed that diagnostic accuracy values of the three classifications for epithelial/lamina propria mucosa (EP/LPM) tumor staging were all high (Inoue, 87.2%; Arima, 98.7%; and JES, 86.7%), but for MM or SM1, the diagnostic accuracy of the JES was significantly higher than the Inoue classification (75.5% vs. 58.7%, $P < 0.05$), and for EP/LPM and SM2 or deeper, the diagnostic accuracy of the Arima classification was significantly higher than that of the other two ($P < 0.01$). However, diagnostic reproducibility (κ value) of the Arima classification was lower than that of the JES classification. **The meta-analysis suggested that** the JES classification will have high validity and reliability in the diagnosis of SESCC.

2 Barrett's Esophagus (BE) and BE-Related Neoplasia

Studies showed an excellent prognosis among patients with superficial BE-related neoplasms (SBERN), including dysplasia and esophageal adenocarcinoma (EAC) confined to the submucosal layer. Early detection is the primary key factor for a favorable prognosis among patients with EAC. SBERN, especially the flat macroscopic type, is often difficult to detect by standard white-light endoscopy. Current guidelines in the West recommend endoscopic surveillance for patients with BE using random four-quadrant biopsies, in which samples are obtained based on the Seattle protocol to detect SBERN. However, the Seattle protocol is labor-intensive, time-consuming, expensive, and additionally associated with a high risk of sampling error.

A recent meta-analysis suggested that acetic acid enhancement, narrowband imaging (NBI), and endoscopy based confocal laser endomicroscopy are promising techniques for targeted biopsy that could be employed to eliminate the need for random biopsies. Several classifications using NBI magnifying endoscopy (NBI-M) have been proposed, thereby suggesting the high utility of the classifications for the diagnosis of SBERN. However, none of them have been widely accepted; thus, recent studies attempted to integrate and simplify the magnifying endoscopic classifications.

The Japan Esophageal Society Barrett's Esophagus (JES-BE) working group has developed and proposed a new magnifying endoscopic classification system, including a diagnostic flowchart (◘ Fig. 2), to identify SBERN, namely, the JES-BE classification. The classification has three characteristics, simplified, an easily understood classification by incorporating well-known diagnostic criteria for early gastric cancer and modified criteria for a flat pattern.

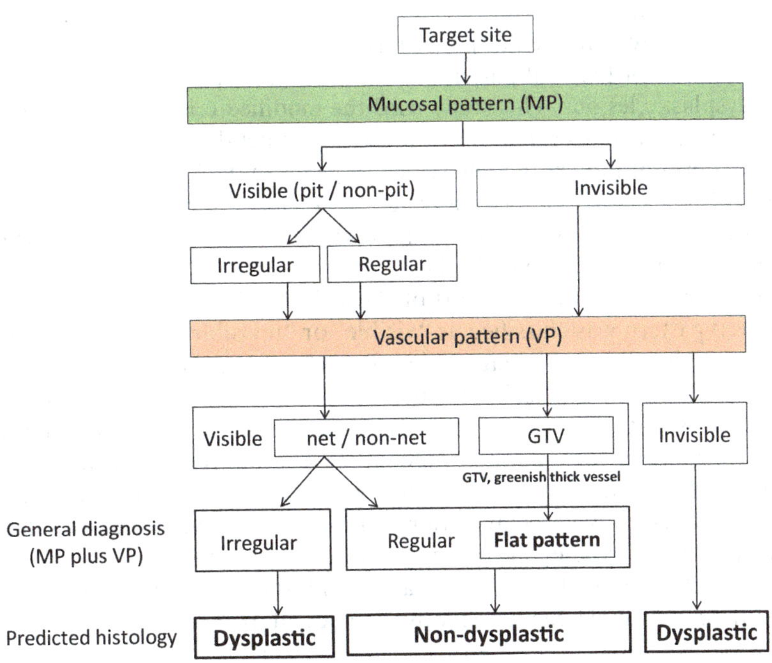

◘ **Fig. 2** Diagnostic flowchart using the JES-BE classification

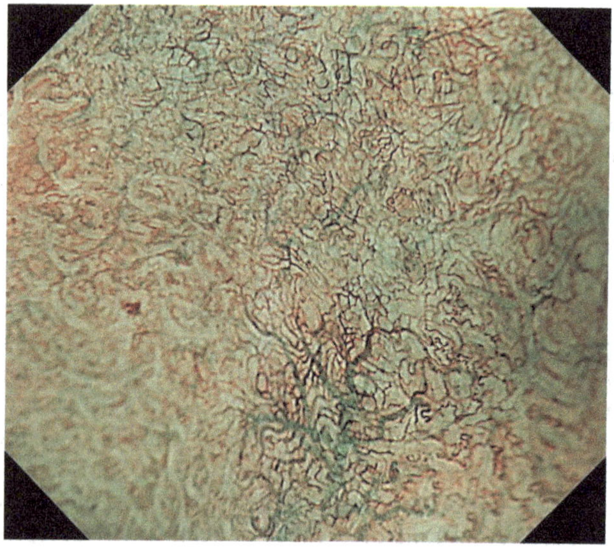

Fig. 3 Flat pattern corresponding to non-dysplastic histology (intestinal metaplasia)

The flat pattern is a unique mucosal surface in BE showing an invisible/absent mucosal pattern, corresponding not to dysplastic but to non-dysplastic histology. The flat pattern was originally defined as invisible mucosal pattern (i.e., absence of pits and villi) with normal-appearing, long branching vessels. The flat pattern mimics an absent micro-surface (mucosal) pattern, which is significantly suggestive of early gastric cancer. As it was difficult for clinicians working in areas with a high incidence of gastric cancer to rate the flat pattern as non-dysplastic, recent studies proposed modified criteria for the "flat pattern." This was to enhance the diagnostic accuracy for non-dysplastic lesions. This study used the modified criteria; according to these criteria, an invisible mucosal pattern without a distinct demarcation line and visible vascular patterns of long branching vessels or greenish thick vessels (GTV) was rated as "regular." Thus, a flat pattern was rated as "regular" and the predicted histology as "non-dysplastic." Figure 3 shows representative NBI-M images of the flat pattern according to the modified criteria.

By using the diagnostic flowchart of the JES-BE classification (Fig. 1), first, the mucosal pattern was classified as "visible" or "invisible" and rated as "regular" or "irregular" based on the diagnostic criteria for irregularity, as reported previously. The "invisible" mucosal pattern cannot be rated. Second, the vascular pattern was classified as "visible" or "invisible." The "visible" vascular pattern included normal-appearing, long branching vessels and GTV previously reported and after-mentioned. General diagnosis was rated as "regular" or "irregular" based on mucosal plus vascular patterns. Finally, histology ("non-dysplastic" vs. "dysplastic") was predicted according to the general diagnosis. "Dysplastic" corresponds to SBERN, including low-grade dysplasia (LGD), high-grade dysplasia (HGD), and superficial adenocarcinoma. Representative NBI-M images are shown in Fig. 4.

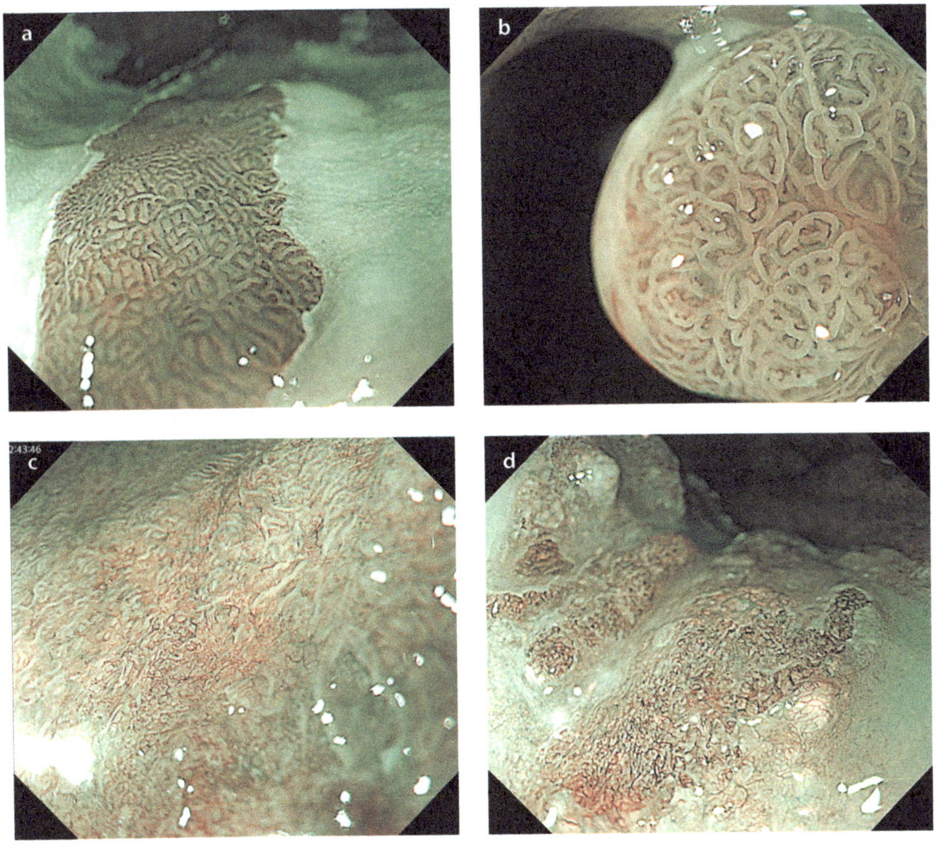

Fig. 4 **a** Regular pattern (MP, visible (pit), regular; VP, visible (net), regular) corresponding to non-dysplastic histology (oxyntocardiac glands without intestinal metaplasia). **b** Regular pattern (MP, visible (non-pit); VP, visible (non-net), regular) corresponding to non-dysplastic histology (cardiac-type glands with foveolar hyperplasia and intestinal metaplasia). **c** Irregular pattern (MP, invisible; VP, visible (**non-net and net**), irregular) corresponding to dysplastic histology (well-differentiated adenocarcinoma with low-grade atypia (noninvasive). **d** Irregular pattern (MP, invisible: VP, visible (**non-net and net**), irregular) corresponding to dysplastic histology (invasive differentiated adenocarcinoma (with submucosal invasion)

A nationwide multicenter study demonstrated the sensitivity and specificity for 10 reviewers (5 experts and 5 non-experts) were 87% and 97%, respectively. Overall accuracy, positive predictive value, and negative predictive value were 91%, 98%, and 83%, respectively. The overall strength of inter-observer and intra-observer agreements for dysplastic histology prediction was $\kappa = 0.77$ and $\kappa = 0.83$, respectively. No significant difference in diagnostic accuracy and reproducibility between experts and non-experts was found. The JES-BE classification system, including the diagnostic flowchart for predicting dysplastic BE, is acceptable and reliable, regardless of the clinician's experience level.

Esophagus: The BING Classification

Jin Lin Tan and Rajvinder Singh

Contents

Y. Sano et al. (eds.), *Atlas of Advanced Endoscopy*, https://doi.org/10.1007/978-981-97-2732-2_8

1 Background

Barrett's esophagus (BE) is a metaplastic alteration of the normal esophagus epithelium, predisposing to the development of esophageal adenocarcinoma (EAC), of which the 5-year overall survival rate is a dismal 17% [1, 2]. International guidelines have recommended endoscopic surveillance for patients with BE to detect early dysplasia [3, 4]. Endoscopic surveillance is typically performed with high-definition white light imaging (WLI) with quadrantic biopsies taken as per the Seattle protocol [5]. However, early Barrett's dysplasia are easily missed due to subtle mucosal change [6, 7].

To enhance the accuracy of BE surveillance, narrowband imaging (NBI), along with other advanced imaging tools, is commonly employed in clinical practice [8]. NBI enables improved visualization of both mucosal and vascular patterns within the BE segment, thereby enhancing the detection of dysplasia [9]. Several classification systems, including the Kansas Classification, Amsterdam Classification, Nottingham Classification, and Asia Pacific Barrett's Consortium Classification, have been developed to predict high-grade dysplasia and esophageal adenocarcinoma based on mucosal characteristics [10–13]. However, these classifications originated from single-center studies, and subsequent validation studies have indicated less than optimal accuracy and interobserver/intraobserver agreements [14, 15]. Given the challenges in detecting dysplasia within Barrett's esophagus, an accurate and reproducible classification system is essential.

2 BING Classification

The Barrett's International NBI Group (BING) was convened to develop and validate a prospective, consensus-driven NBI classification system that can be used to predict the presence or absence of dysplasia in BE. The BING working group is composed of NBI experts from the United States, Europe and Japan, who evaluated 60 high-quality NBI images of non-dysplastic Barrett's esophagus (NDBE), high-grade dysplasia (HGD), and esophageal adenocarcinoma (EAC), to develop the BING classification [16].

Mucosal and vascular patterns in each NBI image were classified as "regular" or "irregular" based on characteristics agreed upon by the working group (◘ Table 1). Regular mucosal patterns were marked by circular, ridged/villous, or tubular patterns, and irregular mucosa was marked by absent or irregular surface patterns (◘ Fig. 1a, b). Regular vascular patterns were defined by blood vessels situated regularly along or between mucosal ridges and/or those showing normal, long branching patterns; irregular vascular patterns were marked by focally or diffusely distributed vessels not following the normal architecture of the mucosa (◘ Fig. 1c, d). Images not readily identified as regular or irregular were deemed uncertain [16].

The BING classification was able to detect cases of dysplasia with 85% accuracy, 80% sensitivity, 88% specificity, 81% positive predictive value, and 88% negative predictive value [16].

Nonetheless, the BING classification is not without its limitations. Similar to earlier classification methods, there is notable interobserver variation in agreement

Table 1 BING classification of Barrett's epithelium

Morphological characteristics	Classification
Mucosal pattern	
Circular, ridged/villous, or tubular patterns	Regular
Absent or irregular patterns	Irregular
Vascular pattern	
Blood vessels situated regularly along or between mucosal ridges and/or those showing normal, long, branching patterns	Regular
Focally or diffusely distributed vessels not following the normal architecture of the mucosa	Irregular

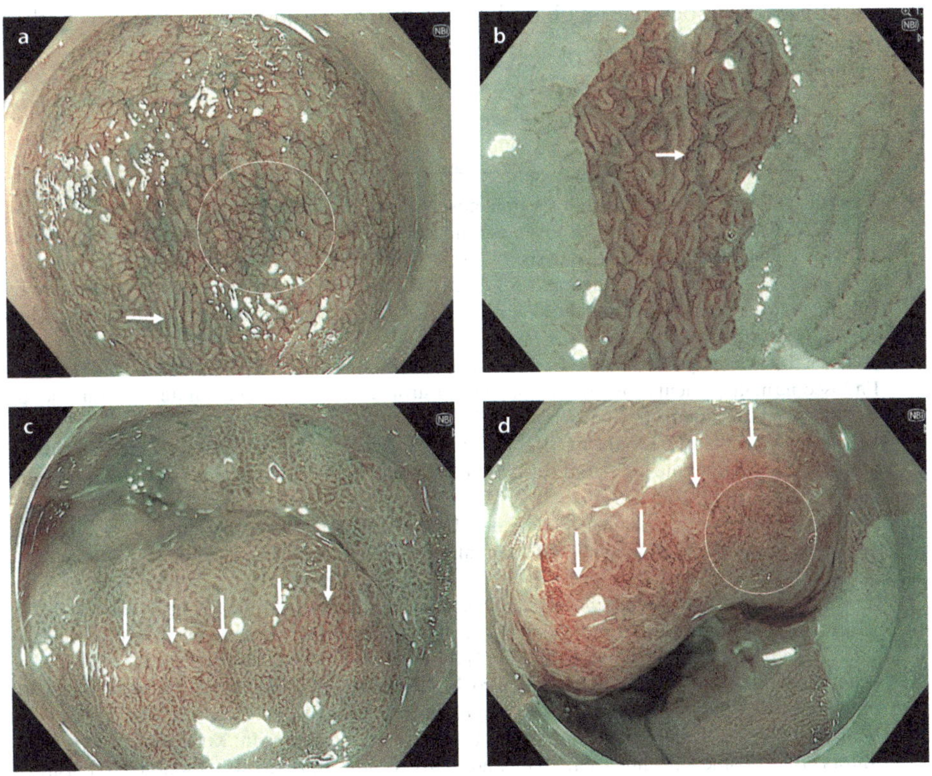

Fig. 1 **a** Image of NDBE using NBI with dual-focus magnification. Note the presence of circular (white circle) and tubular (white arrow) mucosal patterns arranged in an orderly fashion. **b** Image of NDBE using NBI with dual-focus magnification. Note the presence of blood vessels (white arrow) situated regularly along or between mucosal ridges. **c** Image of dysplastic BE using NBI with dual-focus magnification. Note the presence of irregular surface patterns below the demarcation line as marked (white arrows). **d** Image of dysplastic BE using NBI with dual-focus magnification. Note the absence of mucosal patterns below the demarcation line as marked (white arrows), as well as diffusely distributed vessels not following the normal architecture of the mucosa (white circle)

among endoscopists (with a Cohen's value of 0.681). Notably, this classification framework does not encompass low-grade dysplasia or cases of indeterminate dysplasia. Furthermore, the assessment of sub-squamous dysplasia remains elusive.

In the foreseeable future, Artificial Intelligence (AI) will have a significant role in Barrett's surveillance. A recent systematic review and meta-analysis demonstrated that AI has the capacity to enhance early Barrett's dysplasia detection, achieving an overall accuracy of 94% (95% CI: 0.92–0.96), along with a combined sensitivity of 90.3% (95% CI: 87.1–92.7%) and specificity of 84.4% (95% CI: 80.2–87.9%) [17]. Moreover, AI holds the potential to reduce interobserver variation in agreement within clinical practice.

3 Conclusion

In summary, the BING classification remains an important tool to aid endoscopists in Barrett's surveillance, assisting in identifying the presence of high-grade dysplasia and esophageal adenocarcinoma based on the mucosal and vascular patterns within the Barrett's esophagus.

References

1. Sharma P. Clinical practice. Barrett's esophagus. N Engl J Med. 2009;361(26):2548–56.
2. Njei B, McCarty TR, Birk JW. Trends in esophageal cancer survival in United States adults from 1973 to 2009: a SEER database analysis. J Gastroenterol Hepatol. 2016;31(6):1141–6.
3. Fitzgerald RC, di Pietro M, Ragunath K, Ang Y, Kang J-Y, Watson P, et al. British Society of Gastroenterology guidelines on the diagnosis and management of Barrett's oesophagus. Gut. 2014;63(1):7–42.
4. Weusten B, Bisschops R, Coron E, Dinis-Ribeiro M, Dumonceau J-M, Esteban J-M, et al. Endoscopic management of Barrett's esophagus: European Society of Gastrointestinal Endoscopy (ESGE) position statement. Endoscopy. 2017;49(2):191–8.
5. Levine DS, Haggitt RC, Blount PL, Rabinovitch PS, Rusch VW, Reid BJ. An endoscopic biopsy protocol can differentiate high-grade dysplasia from early adenocarcinoma in Barrett's esophagus. Gastroenterol. 1993;105(1):40–50.
6. Schölvinck DW, van der Meulen K, Bergman JJGHM, Weusten BLAM. Detection of lesions in dysplastic Barrett's esophagus by community and expert endoscopists. Endoscopy. 2017;49(2):113–20.
7. Nguyen TH, Thrift AP, George R, Rosen DG, El-Serag HB, Ketwaroo GA. Prevalence and predictors of missed dysplasia on index Barrett's esophagus diagnosing endoscopy in a veteran population. Clin Gastroenterol Hepatol. 2022;20(4):e876–89.
8. Waterhouse DJ, Fitzpatrick CRM, di Pietro M, Bohndiek SE. Emerging optical methods for endoscopic surveillance of Barrett's oesophagus. Lancet Gastroenterol Hepatol. 2018;3(5):349–62.
9. Mangira D, Wong S, Sharma P, Singh R. Virtual Chromoendoscopy. In: Konda VJA, Waxman I, editors. Endoscopic imaging techniques and tools. Cham: Springer International Publishing; 2016. p. 49–60.
10. Sharma P, Bansal A, Mathur S, Wani S, Cherian R, McGregor D, et al. The utility of a novel narrow band imaging endoscopy system in patients with Barrett's esophagus. Gastrointest Endosc. 2006;64(2):167–75.
11. Kara MA, Ennahachi M, Fockens P, ten Kate FJW, Bergman JJGHM. Detection and classification of the mucosal and vascular patterns (mucosal morphology) in Barrett's esophagus by using narrow band imaging. Gastrointest Endosc. 2006;64(2):155–66.

12. Singh R, Anagnostopoulos GK, Yao K, Karageorgiou H, Fortun PJ, Shonde A, et al. Narrow-band imaging with magnification in Barrett's esophagus: validation of a simplified grading system of mucosal morphology patterns against histology. Endoscopy. 2008;40(6):457–63.
13. Singh R, Jayanna M, Wong J, Lim LG, Zhang J, Lv J, et al. Narrow-band imaging and white-light endoscopy with optical magnification in the diagnosis of dysplasia in Barrett's esophagus: results of the Asia-Pacific Barrett's Consortium. Endosc Int Open. 2015;3(1):E14–8.
14. Baldaque-Silva F, Marques M, Lunet N, Themudo G, Goda K, Toth E, et al. Endoscopic assessment and grading of Barrett's esophagus using magnification endoscopy and narrow band imaging: impact of structured learning and experience on the accuracy of the Amsterdam classification system. Scand J Gastroenterol. 2013;48(2):160–7.
15. Curvers WL, Bohmer CJ, Mallant-Hent RC, Naber AH, Ponsioen CI, Ragunath K, et al. Mucosal morphology in Barrett's esophagus: interobserver agreement and role of narrow band imaging. Endoscopy. 2008;40(10):799–805.
16. Sharma P, Bergman JJGHM, Goda K, Kato M, Messmann H, Alsop BR, et al. Development and validation of a classification system to identify high-grade dysplasia and esophageal adenocarcinoma in Barrett's esophagus using narrow-band imaging. Gastroenterol. 2016;150(3):591–8.
17. Tan JL, Chinnaratha MA, Woodman R, Martin R, Chen H-T, Carneiro G, et al. Diagnostic accuracy of artificial intelligence (AI) to detect early neoplasia in Barrett's esophagus: a non-comparative systematic review and meta-analysis. Front Med (Lausanne). 2022;9:890720.

Stomach: VS Classification/ MESDA-G

Kenshi Yao

Contents

Y. Sano et al. (eds.), *Atlas of Advanced Endoscopy*, https://doi.org/10.1007/978-981-97-2732-2_9

1 Vessel Plus Surface (VS) Classification System

The VS classification system developed by the author (Kenshi Yao) [1] is the most widely accepted diagnostic system for magnifying narrowband imaging (M-NBI) of the stomach [2]. The VS classification system employs anatomical terms as endoscopic markers for analysis. Anatomical components visualized using M-NBI were divided into microvascular (MV) and microsurface (MS) patterns. As shown in ◘ Fig. 1a–c, the subepithelial capillary and collecting venule were visualized for MV pattern, while the marginal crypt epithelium (MCE), crypt opening (CO), and intervening part (IP) between the crypts were identified for the MS pattern. In addition,

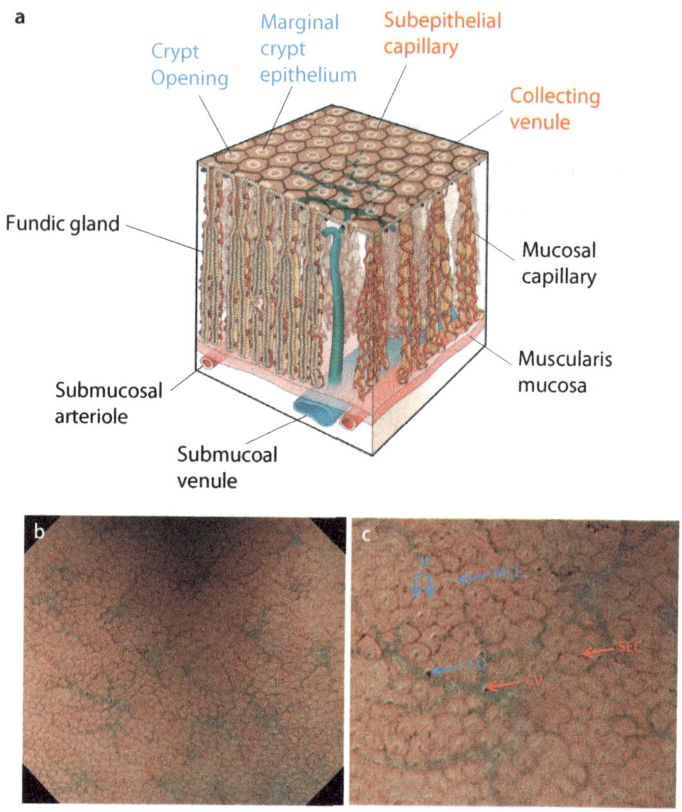

◘ **Fig. 1** Microanatomical features of the gastric mucosa and anatomical components employed in the vessel plus surface (VS) classification system. **a** Schematic diagram of the microvascular (MV) architecture and microsurface (MS) structure of the normal gastric fundic gland mucosa corresponding to surface morphology as visualized by magnifying endoscopy (ME) with narrowband imaging (NBI). Adapted from Muto et al. Dig Endosc. 2016;28:379–93, according to the Creative Commons license [2]. **b, c** The microvascular (MV) architecture is formed by subepithelial capillaries (SEC) and collecting venules (CVs), and the morphology of each SEC is that of a polygonal closed loop. These loops anastomosed repeatedly, forming a regular honeycomb-like subepithelial capillary network (SECN) pattern. The microsurface structure is composed of the marginal crypt epithelium (MCE) and the intervening part (IP) between the crypts. The epithelial morphology is visualized as a semitransparent white belt-like structure (MCE) with a circular or oval shape. In the center, the oval brownish crypt opening (CO) are visualized and surrounded by the MCE. *SEC* subepithelial capillaries, *CV* collecting venule, *SECN* subepithelial capillary network, *MCE* marginal crypt epithelium, *IP* intervening part, *CO* crypt opening in terms of the pathological status, the light blue crest [3] and white opaque substance [4] can be markers for microsurface patterns.

According to the morphology of the anatomical components, the MV and MS patterns were classified as regular, irregular, or absent (◘ Fig. 2). In principle, the MV and MS patterns must be determined separately.

DL is defined as the border between the lesion and non-lesion areas, discernible through an abrupt change in MV and/or MS patterns (◘ Fig. 2).

The criteria for high-grade dysplasia (HGD)/early cancer (EC) are as follows:

1. Presence of an irregular MV pattern with a demarcation line.
2. Presence of an irregular MS pattern with a demarcation line.

A target lesion is diagnosed as HGD/EC if it fulfills both or either of the criteria. If it does not meet either of the criteria, it is diagnosed as noncancerous.

The diagnostic performance of M-NBI, based on the VS classification system, demonstrated excellent results with high-confidence prediction; the accuracy, sensitivity, and specificity were 98% (95% confidence interval (CI) 96.9–99.6), 85.7% (95%CI: 59.8–100), and 99.4% (95%CI: 98.2–100%), respectively [5]. Common clinical applications of M-NBI according to the VS classification system are the differential diagnosis between EGC and focal gastritis during screening endoscopy [6] and determination of the horizontal extent of EGC for successful endoscopic resection [7].

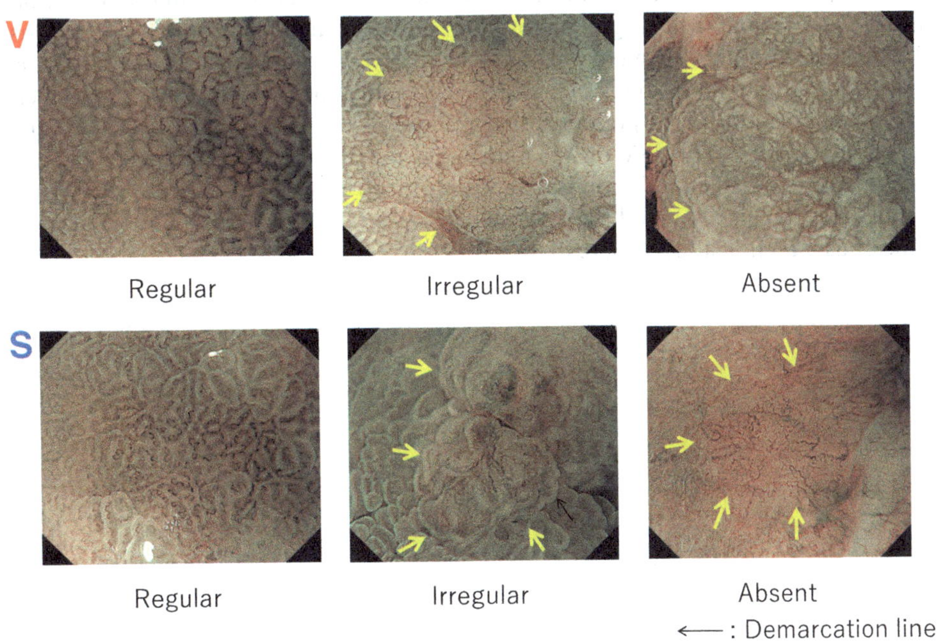

◘ **Fig. 2** Vessel plus surface (VS) classification. Both the microvascular (MV) and microsurface (MS) patterns, (V) and (S), respectively, were classified as regular, irregular, or absent, respectively. (Arrow shows the demarcation line.) Adapted from Muto et al. Dig Endosc. 2016;28:379–93, according to the Creative Commons license [2]

2 Magnifying Endoscopy Simple Diagnostic Algorithm for Gastric Cancer (MESDA-G)

Background

Since magnifying endoscopy (ME) has become common practice, numerous classifications and diagnostic systems have been proposed. To propose a unified algorithm for general use in clinical settings, the Japanese Gastroenterological Association established a working group and proposed that the Japan Gastroenterological Endoscopy Society, the Japanese Gastric Cancer Association, and the World Endoscopy Organization jointly devise a unified international algorithm for ME diagnosis of EGC based on an evidence-based approach.

Methods and Algorithm

The working group members carried out a systematic review of the English literature to determine widely accepted diagnostic systems. Consequently, the best-accepted diagnostic system was proven to be the VS classification system. Based on this system, MESDA-G was constructed using an evidence-based approach (◘ Figs. 3, 4, 5, and 6) [2]. First, lesions suspicious for EGC should be detected by careful inspection using conventional white-light endoscopy. Subsequently, magnified observations should be performed. Owing to the high sensitivity of the demarcation line [8], if it is absent, the suspicious lesion is diagnosed as noncancerous (◘ Fig. 4a, b). However, if the demarcation line is present, further assessment is required to determine the presence of an irregular MV or MS pattern. If an irregular MV and/or MS pattern is detected, the suspicious lesion is diagnosed as cancerous (◘ Fig. 6a, b) [1, 5, 6, 8]. If both the irregular MV and MS patterns are absent, the tumor is diagnosed as noncancerous (◘ Fig. 5a, b) [1, 5, 6, 8].

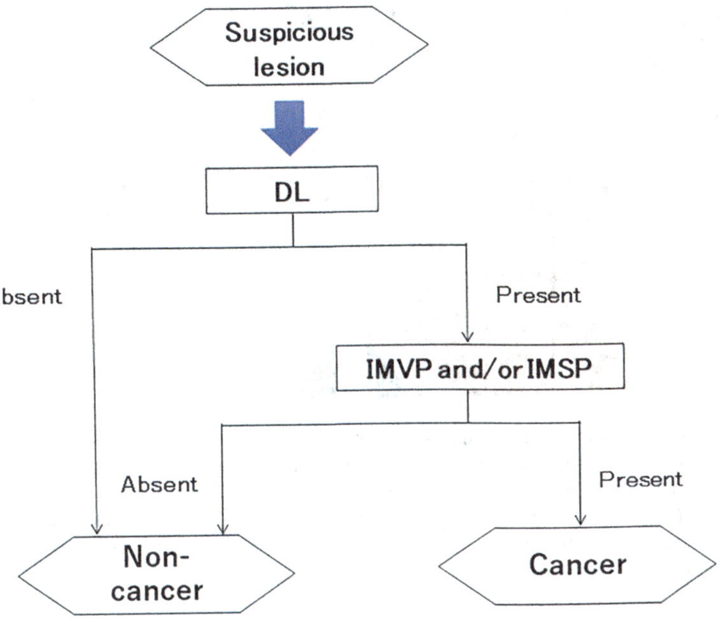

Fig. 3 Magnifying endoscopy simple diagnostic algorithm for gastric cancer (MESDA-G). Adapted from Muto et al. Dig Endosc. 2016;28:379–93, according to the Creative Commons license [2]. *DL* demarcation line, *IMVP* irregular microvascular pattern, *IMSP* irregular microsurface pattern

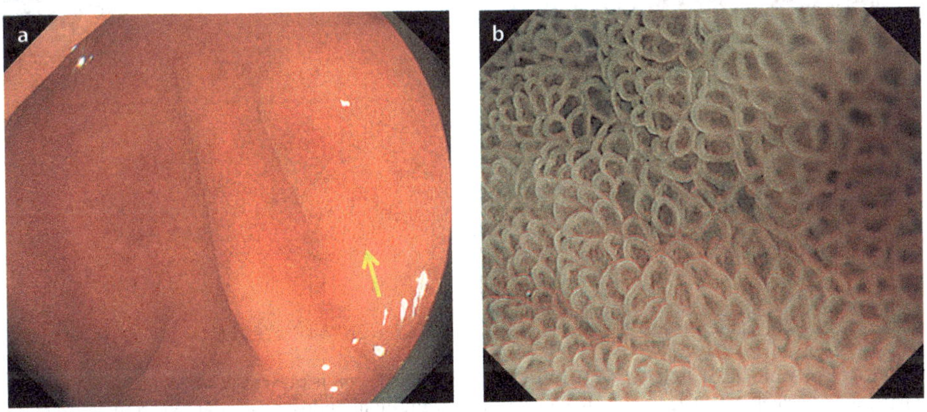

Fig. 4 An example of noncancerous lesion. **a** At the gastric antrum, reddened depressed lesion was detected using conventional white-light endoscopy (arrow). Adapted from Muto et al. Dig Endosc. 2016;28:379–93, according to the Creative Commons license [2]. **b** When the margin of the lesion was magnified with narrow-band imaging, no abrupt changes were observed in the microvascular (MV) and microsurface (MS) patterns. Accordingly, a demarcation line is absent. According to the MESDA-G, this lesion was diagnosed as noncancerous. Adapted from Muto et al. Dig Endosc. 2016;28:379–93, according to the Creative Commons license [2]

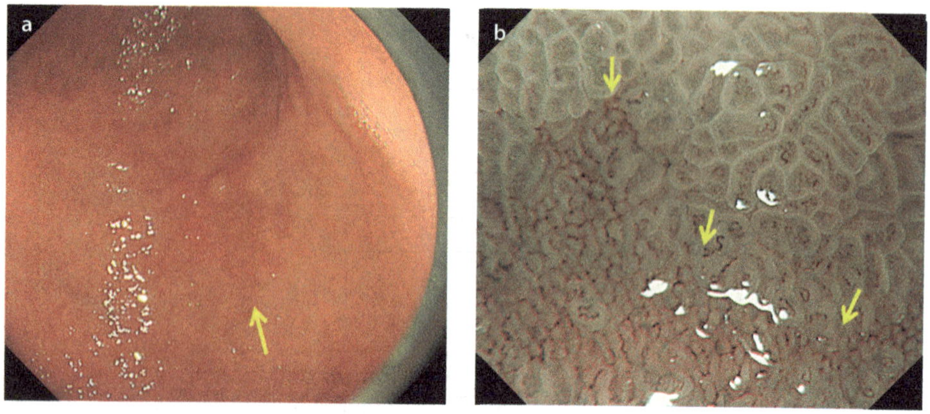

Fig. 5 An example of noncancerous lesion. **a** A slightly depressed reddened lesion is present at the gastric body as detected using conventional white-light endoscopy (arrow). Adapted from Muto et al. Dig Endosc. 2016;28:379–93, according to the Creative Commons license [2]. **b** Magnifying narrow-band imaging (M-NBI) findings of the marginal part of the lesion shows a distinct demarcation line (arrows). In addition, within the demarcation line, regular microvascular (MV) pattern and regular microsurface (MS) pattern are present. In other words, irregular MV or irregular MS patterns are absent. According to the magnifying endoscopy simple diagnostic algorithm for gastric cancer (MESDA-G), the lesion was diagnosed as noncancerous. Adapted from Muto et al. Dig Endosc. 2016;28:379–93, according to the Creative Commons license [2]

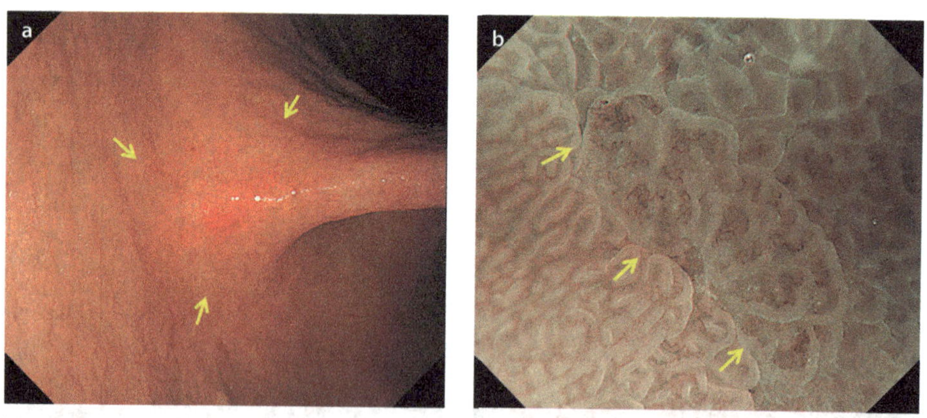

Fig. 6 An example of EGC. **a** A slightly depressed, pale lesion is present at the anterior wall of the gastric angle, which was detected using conventional white-light endoscopy. Adapted from Muto et al. Dig Endosc. 2016;28:379–93, according to the Creative Commons license [2]. **b** When we observed the margins of the lesion using magnifying endoscopy with narrow-band imaging (M-NBI), a clear demarcation line (arrows) between the background mucosa and the lesion was noted. Furthermore, there were irregular microvascular (MV) pattern and irregular microsurface (MS) pattern inside the demarcation line. According to the magnifying endoscopy simple diagnostic algorithm for gastric cancer (MESDA-G), this lesion was diagnosed as cancerous. Adapted from Muto et al. Dig Endosc. 2016;28:379–93, according to the Creative Commons license [2]

References

1. Yao K, Anagnostopoulos GK, Ragunath K. Magnifying endoscopy for diagnosing and delineating early gastric cancer. Endoscopy. 2009;41:462–7.
2. Muto M, Yao K, Kaise M, Kato M, Uedo N, Yagi K, Tajiri H. Magnifying endoscopy simple diagnostic algorithm for early gastric cancer (MESDA-G). Dig Endosc. 2016;28:379–93.
3. Uedo N, Ishihara R, Iishi H, Yamamoto S, Yamamoto S, Yamada T, Imanaka K, Takeuchi Y, Higashino K, Ishiguro S, Tatsuta M. A new method of diagnosing gastric intestinal metaplasia: narrow-band imaging with magnifying endoscopy. Endoscopy. 2006;38:819–24.
4. Yao K, Iwashita A, Tanabe H, Nishimata N, Nagahama T, Maki S, Takaki Y, Hirai F, Hisabe T, Nishimura T, Matsui T. White opaque substance within superficial elevated gastric neoplasia as visualized by magnification endoscopy with narrow-band imaging: a new optical sign for differentiating between adenoma and carcinoma. Gastrointest Endosc. 2008;68:574–80.
5. Yao K, Doyama H, Gotoda T, Ishikawa H, Nagahama T, Yokoi C, Oda I, Machida H, Uchita K, Tabuchi M. Diagnostic performance and limitations of magnifying narrow-band imaging in screening endoscopy of early gastric cancer: a prospective multicenter feasibility study. Gastric Cancer. 2014;17:669–79.
6. Ezoe Y, Muto M, Uedo N, Doyama H, Yao K, Oda I, Kaneko K, Kawahara Y, Yokoi C, Sugiura Y, Ishikawa H, Takeuchi Y, Kaneko Y, Saito Y. Magnifying narrowband imaging is more accurate than conventional white-light imaging in diagnosis of gastric mucosal cancer. Gastroenterology. 2011;141:2017–2025.e3.
7. Nagahama T, Yao K, Uedo N, Doyama H, Ueo T, Uchita K, Ishikawa H, Kanesaka T, Takeda Y, Wada K, Imamura K, Arima H, Shimokawa T. Delineation of the extent of early gastric cancer by magnifying narrow-band imaging and chromoendoscopy: a multicenter randomized controlled trial. Endoscopy. 2018;50:566–76.
8. Yamada S, Doyama H, Yao K, Uedo N, Ezoe Y, Oda I, Kaneko K, Kawahara Y, Yokoi C, Sugiura Y, Ishikawa H, Takeuchi Y, Saito Y, Muto M. An efficient diagnostic strategy for small, depressed early gastric cancer with magnifying narrow-band imaging: a post-hoc analysis of a prospective randomized controlled trial. Gastrointest Endosc. 2014;79:55–63.

Colon: The Japan NBI Expert Team (JNET) Classification

Daizen Hirata and Yutaka Saito

Contents

Y. Sano et al. (eds.), *Atlas of Advanced Endoscopy*, https://doi.org/10.1007/978-981-97-2732-2_10

1 Introduction

In 2006, Dr. Yasushi Sano developed narrowband imaging (NBI) and simultaneously proposed Sano classification, which was the world's first NBI classification [1]. NBI observation became widely recognized as an essential method for colorectal tumor diagnosis. In 2009, an international collaborative research group called the Colon Tumor NBI Interest Group (CTNIG), led by experts from Japan (Yasushi Sano, Shinji Tanaka), France (Ponchon T), the United States (Rex DK, Soetikno RM), and the United Kingdom (Saunders BP), proposed the NICE classification for non-magnified NBI observation. NICE classification is helpful but has some weaknesses, such as not having a high percentage of high confidence. Hence, a novel magnifying classification was needed to provide more accurate differential diagnoses, linking appropriate treatment strategies. Therefore, the Japan NBI Expert Team (JNET) conducted a web-based prospective study to validate the magnifying NBI findings (principal investigator: Yasushi Sano) and finally proposed the JNET classification in 2014 (Fig. 1) [2, 3].

NBI	Type 1	Type 2A	Type 2B	Type 3
Vessel pattern	· Invisible *1	· Regular caliber · Regular distribution (meshed/spiral pattern) *3	· Variable caliber · Irregular distribution	· Loose vessel areas · Interruption of thick vessels
Surface pattern	· Regular dark or white spots · Similar to surrounding normal mucosa	· Regular (tubular/branched/papillary)	· Irregular or obscure	· Amorphous areas
Most likely histology	Hyperplastic polyp/ Sessile serrated lesion *2	Low grade intramucosal neoplasia	High grade intramucosal neoplasia/ Shallow submucosal invasive cancer *4	Deep submucosal invasive cancer
Endoscopic image	Hyperplastic polyp Sessile serrated lesion	Tubular adenoma Tubulovillous adenoma	Intramucosal adenocarcinoma Shallow submucosal invasive cancer	Deep submucosal invasive cancer Deep submucosal invasive cancer

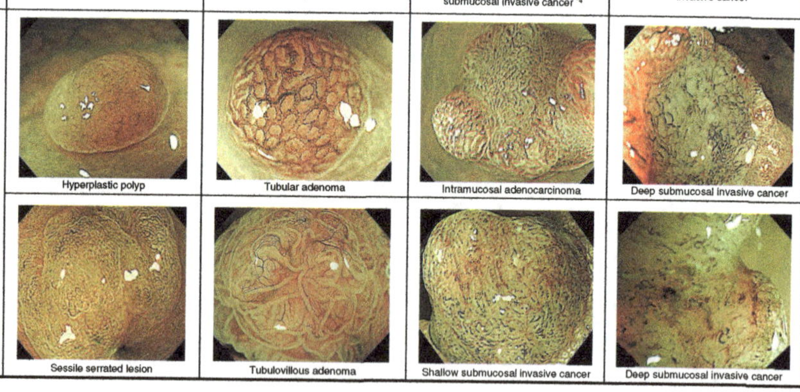

Fig 1 The JNET classification. *1. If visible, the caliber in the lesion is similar to surrounding normal mucosa. *2. The original article stated Sessile serrated polyp. We modified to Sessile serrated lesion in consideration of the 2019 revision of the WHO criteria. *3. Micro-vessels are often distributed in a punctate pattern and well-ordered reticular or spiral vessels may not be observed in depressed lesions. *4. Deep submucosal invasive cancer may be included

2 The JNET Classification

The JNET classification is a magnifying NBI classification for colorectal lesions with two diagnostic factors: vessel and surface pattern. This classification comprises four distinct categories corresponding to the histological characteristics of the polyps: JNET type 1 for hyperplastic polyp (HP) and sessile serrated lesion (SSL), type 2A for low-grade intramucosal neoplasia, type 2B for high-grade intramucosal neoplasia (Tis) and shallow submucosal invasive cancer (T1a), and type 3 for deep submucosal invasive cancer (T1b).

3 Diagnostic Performance of the JNET Classification

NBI serves dual primary purposes in differential diagnosis: qualitative and quantitative diagnosis. Qualitative diagnosis involves distinguishing between neoplastic and nonneoplastic lesions, while quantitative diagnosis accurately determines the depth of invasive cancer. The JNET classification achieves an impressive high confidence rate in qualitative/quantitative diagnosis, with rates of 93.1%/92.9%, significantly surpassing 75%/50% observed in the NICE classification [4].

Qualitative Diagnosis Using the JNET Classification

Differentiating between neoplastic and nonneoplastic lesions implies the distinction between type 1 and type 2A in the JNET classification. This distinction is relatively easy to make with careful observation of the vessel and surface pattern. The qualitative diagnostic performance of the JNET classification has achieved a sensitivity of 98.1–99.8% and an accuracy of 88.8–99.0% [4]. The diagnostic accuracy of the JNET classification was similar between ESGE and JGES and considered to be sufficient for JNET type 1 [5]. When inducing the confidence level (high or low), the rate of a low-confidence qualitative diagnosis using the JNET classification was 7%. The accuracy of low-confidence diagnosis was only 58.5%, and the addition of chromoendoscopy reportedly increased only to 66.0% [6]. Hence, the lesions diagnosed with low confidence must be resected and sent to histopathology.

Quantitative Diagnosis Using the JNET Classification

When faced with colorectal cancer, estimating its invasion depth is crucial to deciding the treatment strategy between endoscopic and surgical resections. JNET type 3 category is designed to have extremely high specificity compared with sensitivity to avoid unnecessary surgery for Tis and T1a cancer, which are curable by endoscopic resection. Reportedly, the diagnostic performance of the JNET type 3 for T1b cancer was 99.8–100% of specificity and 55.4–62.7% of sensitivity. Due to its low sensitivity, approximately 60% of T1b cancers are classified as JNET type 3 while 40% as JNET type 2B [4].

4 Treatment Strategy Based on the JNET Classification

We present the treatment strategy based on JNET classification in Fig. 2.

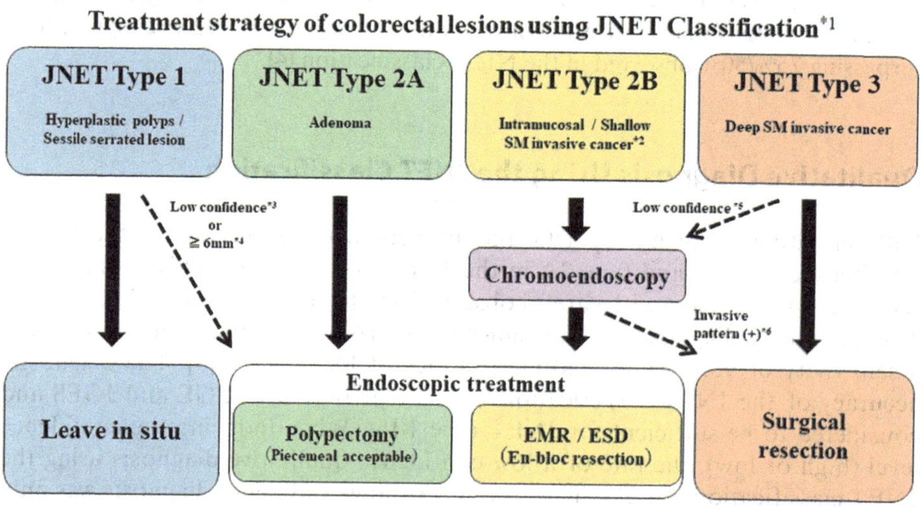

Fig 2 Treatment strategy based on the JNET classification*1. *1. Sano Y, et al. Dig. Endosc. 2016; 28: 526–33. *2. Deep submucosal invasive cancer may be included. *3. Utsumi T, et al. Dig Endosc. 2018; 30 Suppl 1:45–50. *4. Hirata D, et al. Digestion. 2023; 104(4): 262–269. *5. Iwatate M et al. Dig Endosc. 2018; 30(5): 642–651. *6. Matsuda T et al. Am J Gastroenterol 103: 2700–2706, 2008. Please refer to ANBIIG website (▶ https://anbig.org/resources-center/guide-for-nbi-diagnosis/)

JNET Type 1 Lesion

Most of the JNET type 1 lesions are HP and SSL. As JNET type 1 lesions reach a size of 6 mm or larger, the proportion of SSL and SSL with dysplasia increases significantly [7, 8]. Accurate differential diagnosis between SSL and HP for JNET type 1 lesions might enable an SSL-selective resection strategy. Actually, a multicenter prospective study by Japanese experts reveals that the diagnostic performance in differentiating SSL from HP for the JNET type 1 lesions ≥6 mm was not satisfactory (sensitivity, 79.8%; specificity, 59.1%) [9]. Hence, JNET type 1 lesions ≥6 mm are recommended to be resected because SSL-selective resection has the disadvantage of leaving approximately 20% of SSLs on site.

JNET Type 2A Lesion

The majority of the JNET type 2A lesions are low-grade adenomas, and the rate of T1b cancer in JNET type 2A lesions is extremely low. Therefore, it is generally indicated for endoscopic resection methods such as polypectomy. En bloc resection is preferred, but piecemeal resection is also acceptable. When unsure whether to diagnose JNET type 1 or 2A (low confidence), resect and send it to histopathology.

JNET Type 2B Lesion

The most frequent histopathology of the JNET type 2B lesions is Tis and T1a cancer, followed by adenoma. It is essential to recognize that 10–20% of JNET type 2B lesions involve T1b cancers with a higher risk of complication for endoscopic resection [4]. JNET type 2B lesions should be resected en bloc because of the risk of T1b cancers [5].

It would be better to diagnose T1b cancer before high-risk endoscopic resection. When invasive cancer is suspected with JNET type 2B, additional pit pattern diagnosis using chromoendoscopy is preferable for better invasion depth diagnosis. If V_N or V_I severe pit patterns are observable over a demarcated area (what is called an invasive pattern and suggestive of T1b cancer), surgical resection should also be considered [10]. Accurate histopathological assessment of invasion depth using resected specimens is also critical for curability in JNET type 2B lesions, necessitating en bloc deep resection by EMR or ESD.

JNET Type 3 Lesion

Most JNET type 3 lesions are T1b, or deeper cancer, and are generally a good indication for surgical resection. JNET type 3 lesions can be treated surgically because of its high specificity [5]. When wondering whether JNET type 3 or 2B (low confidence), additional chromoendoscopy or diagnostic endoscopic resection (EMR/ESD) is recommended. Some pedunculated JNET type 3 lesions have a low potential of lymph node metastasis unless the cancer invades into its stalk [11]. Endoscopic resection can be an option for pedunculated JNET type 3 lesions.

5 Conclusions

The JNET classification is a highly practical system designed for real-world clinical applications. Without any guidance, it may be difficult for beginners to make appropriate use of this classification. We publish an open-access video article to assist in learning this classification on *digestive endoscopy* [12]. We sincerely hope that you can understand the characteristics of the JNET classification correctly and incorporate it into your daily practice.

References

1. Sano Y, Ikematsu H, Fu KI, et al. Meshed capillary vessels by use of narrow-band imaging for differential diagnosis of small colorectal polyps. Gastrointest Endosc. 2009;69:278–83.
2. Sano Y, Tanaka S, Kudo SE, et al. NBI magnifying endoscopic classification of colorectal tumors proposed by the Japan NBI expert team (JNET). Dig Endosc. 2016;28:526–33.
3. Iwatate M, Sano Y, Tanaka S, et al. Validation study for development of the Japan NBI expert team classification of colorectal lesions. Dig Endosc. 2018;30(5):642–51.
4. Hirata D, Kashida H, Iwatate M, et al. Effective use of the Japan narrow band imaging expert team classification based on diagnostic performance and confidence level. World J Clin Cases. 2019;7(18):2658–65.
5. Saito Y, Sakamoto T, Dekker E, et al. First report from international evaluation of endoscopic classification JNET: an international multicenter web trial. Dig Endosc. 2023;36(5):591–9. https://doi.org/10.1111/den.14682.
6. Utsumi T, Iwatate M, Sunakawa H, et al. Additional chromoendoscopy for colorectal lesions initially diagnosed with low confidence by magnifying narrow-band imaging: can it improve diagnostic accuracy? Dig Endosc. 2018;30(Suppl 1):45–50.
7. Sano W, Sano Y, Iwatate M, et al. Prospective evaluation of the proportion of sessile serrated adenoma/polyps in endoscopically diagnosed colorectal polyps with hyperplastic features. Endosc Int Open. 2015;3(4):E354–8.
8. Sano W, Fujimori T, Ichikawa K, et al. Clinical and endoscopic evaluations of sessile serrated adenoma/polyps with cytological dysplasia. J Gastroenterol Hepatol. 2018;33(8):1454–60.
9. Hirata D, Kashida H, Matsumoto T, et al. A multicenter prospective validation study on selective endoscopic resection of sessile serrated lesions using magnifying colonoscopy in clinical practice. Digestion. 2023;104(4):262–9.
10. Matsuda T, Fujii T, Saito Y, et al. Efficacy of the invasive/non-invasive pattern by magnifying chromoendoscopy to estimate the depth of invasion of early colorectal neoplasms. Am J Gastroenterol. 2008;103:2700–6.
11. Matsuda T, Fukuzawa M, Uraoka T, et al. Risk of lymph node metastasis in patients with pedunculated type early invasive colorectal cancer: a retrospective multicenter study. Cancer Sci. 2011;102:1693–7.
12. Sano Y, Hirata D, Saito Y. Japan NBI expert team classification: narrow-band imaging magnifying endoscopic classification of colorectal tumors. Dig Endosc. 2018;30(4):543–5.

High-Quality Photography

Contents

How to Take High-Quality Images (H&N)

Daisuke Kikuchi

Contents

Supplementary Information The online version contains supplementary material available at https://doi.org/10.1007/978-981-97-2732-2_11. The videos can be accessed individually by clicking the DOI link in the accompanying figure caption or by scanning this link with the SN More Media App.

Y. Sano et al. (eds.), *Atlas of Advanced Endoscopy*, https://doi.org/10.1007/978-981-97-2732-2_11

1 Background

With advancements in technology such as narrowband imaging (NBI) and blue laser imaging (BLI), known as image-enhanced endoscopy (IEE), and improvements in magnified endoscopy, numerous cases of squamous cell carcinoma are now being detected through gastrointestinal endoscopy. In the detection and characterization of squamous cell carcinoma, NBI magnification (M-NBI) is significantly superior to white light endoscopy (WLE), making M-NBI an essential tool in daily clinical practice [1, 2].

However, cancers in the oral, laryngeal, and pharyngeal regions are still often discovered at an advanced stage, making early detection a challenge. There are techniques for observing these areas effectively, and this paper aims to explain the key aspects of obtaining clear endoscopic images.

Selection of High-Risk Patients

Observing the laryngeal and pharyngeal regions can be uncomfortable for patients, so it should not be performed for all cases but rather selectively for those at risk of pharyngeal cancer. Risk factors for pharyngeal cancer include a history of excessive alcohol consumption, smoking history, and a history of esophageal and pharyngeal cancer treatment [3]. The flushing of the face caused by even small amounts of alcohol intake is strongly correlated with the risk of squamous cell carcinoma and should always be inquired about. Also, since oral melanosis is associated with incidence of pharyngeal cancer, careful examination of the pharyngeal region should be conducted when oral melanosis is detected. Additionally, patients with symptoms such as pharyngeal pain and difficulty swallowing require thorough endoscopic observation.

Preparation for the Endoscopy

Strong gag reflexes can make endoscopic observation of the pharyngeal regions challenging. Thoroughly anesthetize the pharynx with Xylocaine Viscous, or spray and administer appropriate medications. During pharyngeal endoscopy, patients are required to perform actions such as vocalization, tongue movement, and Valsalva maneuver, so analgesics like pethidine hydrochloride are preferred over sedatives like midazolam or propofol. Yamasaki et al. reported in an RCT that the use of pethidine was more suitable for pharyngeal observation compared to no sedation or the use of midazolam [4].

Saliva, mucus, and sputum can also obstruct endoscopic observation. Rinsing the mouth before endoscopy can help remove these obstructions. Additionally, the use of scopolamine can reduce saliva secretion.

Endoscopic Observation of the Oral Cavity

Observing the oral cavity, such as the floor of the mouth and tongue margin, becomes impossible once a mouthpiece is inserted. Therefore, before fitting a mouthpiece on high-risk patients for squamous cell carcinoma, observe the tongue, floor of the mouth, and other areas within the oral cavity. Instruct the patient to curl their tongue and place the tip against the soft palate. Then instruct them to place the tongue tip against the left and right cheek mucosa. Practicing these tongue movements in front of a mirror before the examination can be helpful.

Endoscopic Observation of the Pharynx

The pharynx has anatomical irregularities, and capturing the entire region in a single endoscopic image is not possible. Thus, it is crucial to conduct targeted endoscopic observation using the same method each time to avoid overlooking. Traditionally, we have focused our endoscopic examinations on 12 specific areas in the pharyngeal region [5]. In recent years, we have expanded the observation area to 15 regions by including an additional 3 areas within the oral cavity (◘ Fig. 1).

After oral cavity observation, the mouthpiece is fitted. Subsequently, we observe from the hard palate to the soft palate, uvula, and both sides of the palate arches. While instructing the patient to vocalize, images are taken of the posterior oropharyngeal wall, both side walls, and the larynx. Next, images are taken of the left and right piriform sinuses, followed by the epiglottis. Finally, perform the Valsalva maneuver, and observe the posterior pharyngeal wall and the postcricoid area.

Valsalva Maneuver

Normally, the pharynx is compressed by the larynx, making it difficult to observe the posterior pharyngeal wall and the postcricoid area. As a result, many cancers in this area are discovered at an advanced stage. Performing the Valsalva maneuver can elevate the larynx and improve visibility in this region. Various methods for performing the Valsalva maneuver in gastrointestinal endoscopy have been reported in recent years. We use a dedicated mouthpiece (Valsamouth®) for observation. At the end of pharyngeal observation, instruct the patient to pucker their mouth, puff their cheeks, and exhale while narrowing their mouth. In our study, this method improved visibility in the postcricoid area in approximately 60% of patients [5]. The advantage of this method is that it allows for NBI magnification observation.

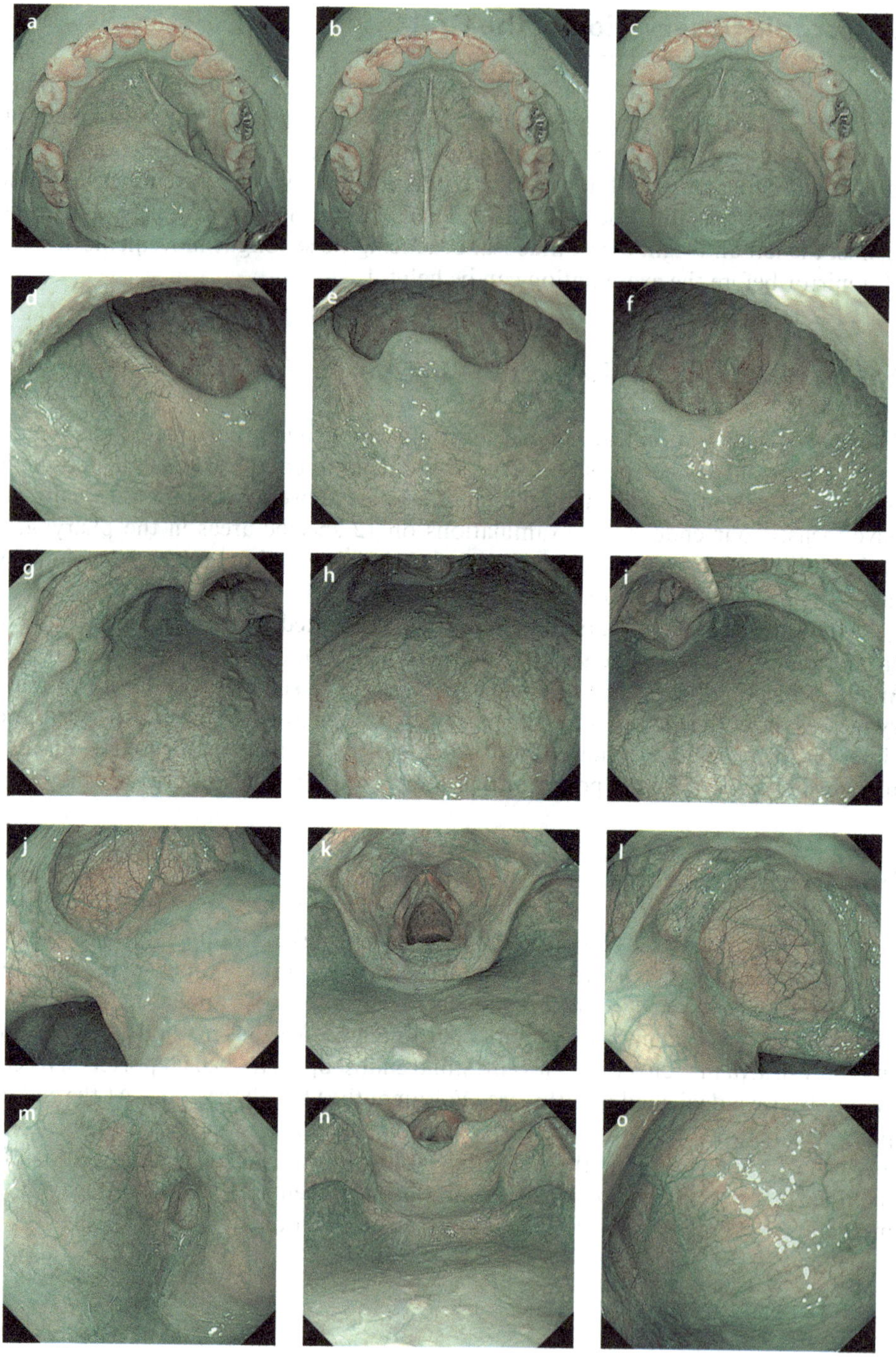

Fig. 1 Pharyngeal observation for high-risk patient. **a** Left side of the tongue. **b** Oral floor. **c** Right side of the tongue. **d** Left palate arch. **e** Hard and soft palate. **f** Right palate arch. **g** Left side of the mesopharynx. **h** Posterior wall of the mesopharynx. **i** Right side of the mesopharynx. **j** Left side of the epiglottis. **k** Larynx. **l** Right side of the epiglottis. **m** Left piriform sinus. **n** Posterior wall of the hypopharynx and postcricoid area. **o** Right piriform sinus

2 Conclusion

The prognosis for pharyngeal cancer patients undergoing endoscopic treatment is favorable [6, 7]. Maintaining the patient's quality of life and opting for curative endoscopic treatment are significantly beneficial treatment approaches (Figs. 2 and 3). However, careful endoscopic surveillance is crucial due to the high incidence of metachronous carcinogenesis after endoscopic treatment [8].

Video

A short video is available as Fig. 4

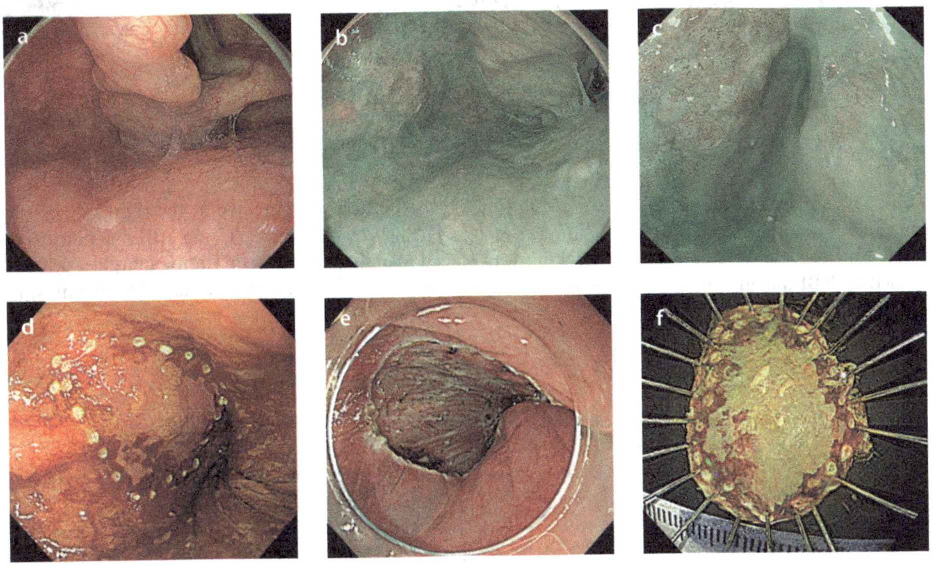

Fig. 2 Superficial hypopharyngeal cancer. **a** Slightly reddish area was detected at the left piriform sinus. **b** Using NBI, lesion was recognized as brownish area. **c** Using NBI magnification, dilated and elongated IPCL was observed. According to the JES classification, these vessels were type B1. **d** After spraying iodine staining, lesion was recognized as iodine unstained area. Marking was placed around the lesion. **e** Lesion was resected without any complication. **f** Pathological diagnosis was squamous cell carcinoma, 20 × 15 mm, subepithelial invasion (tumor thickness 300 μm), ly0, v0, HM0, VM0

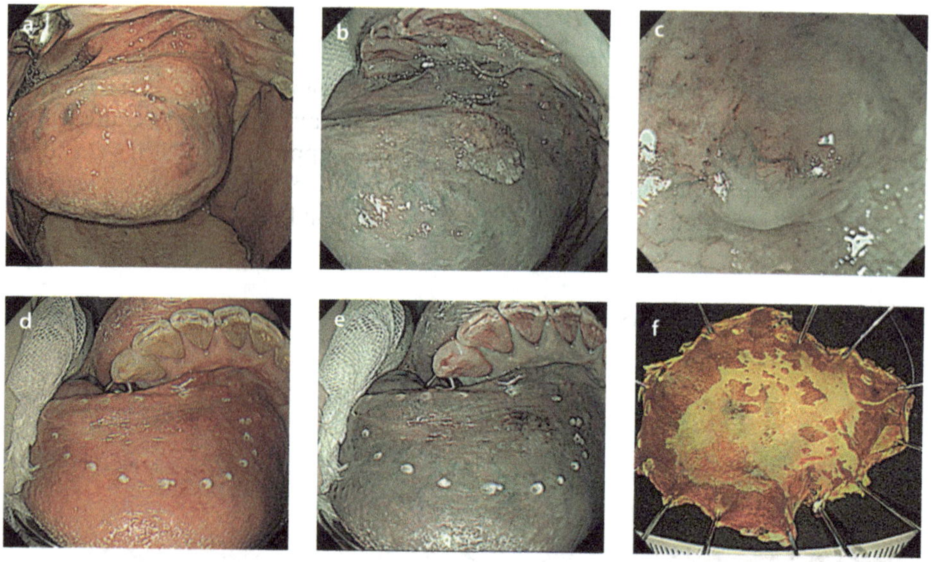

Fig. 3 Superficial cancer of the tongue. **a** Flat elevated lesion was detected at the right side of the margin of the tongue. **b** Using NBI, lesion was recognized as brownish area. **c** Using NBI magnification, dilated microvessel was detected. **d** Marking was placed around the lesion. **e** Tumor margin was confirmed by NBI magnification. **f** Pathological diagnosis was squamous cell carcinoma, 24 × 19 mm, subepithelial invasion (tumor thickness 921 μm), ly0, v0, HM0, VM0

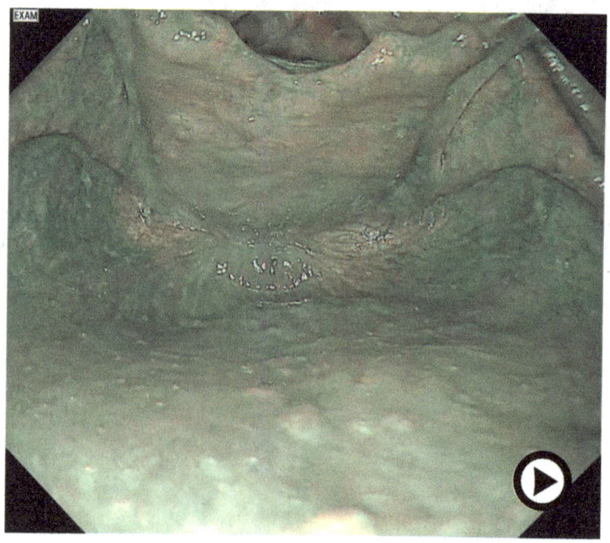

Fig. 4 Surveillance endoscopic examination for the patient after pharyngeal ESD. Superficial pharyngeal cancer at the left palate arch was resected by ESD 2 years ago (▶ https://doi.org/10.1007/000-c7p)

References

1. Muto M, Minashi K, Yano T, et al. Early detection of superficial squamous cell carcinoma in the head and neck region and esophagus by narrow band imaging: a multicenter randomized controlled trial. J Clin Oncol. 2010;28:1566–72.
2. Kikuchi D, Iizuka T, Yamada A, et al. Utility of magnifying endoscopy with narrow band imaging in determining the invasion depth of superficial pharyngeal cancer. Head Neck. 2015;37:846–50.
3. Katada C, Yokoyama T, Yano T, et al. Alcohol consumption and multiple dysplastic lesions increase risk of squamous cell carcinoma in the esophagus, head, and neck. Gastroenterology. 2016;151:860–9.
4. Yamasaki Y, Ishihara R, Hanaoka N, et al. Pethidine hydrochloride is a better sedation method for pharyngeal observation by transoral endoscopy compared with no sedation and midazolam. Dig Endosc. 2017;29:39–48.
5. Kikuchi D, Tanaka M, Suzuki Y, et al. Utility of Valsalva maneuver in the endoscopic pharyngeal observation. Esophagus. 2020;17:323–9.
6. Iizuka T, Kikuchi D, Suzuki Y, et al. Clinical relevance of endoscopic treatment for superficial pharyngeal cancer: feasibility of techniques corresponding to each location and long-term outcomes. Gastrointest Endosc. 2021;93:818–27.
7. Ogasawara N, Kikuchi D, Tanaka M, et al. Long-term outcome of cervical lymph node metastasis in superficial pharyngeal squamous cell carcinoma after endoscopic submucosal dissection. Gastrointest Endosc. 2023;98:524–533.e2 (online ahead of print).
8. Ogasawara N, Kikuchi D, Tanaka M, et al. Comprehensive risk evaluation for metachronous carcinogenesis after endoscopic submucosal dissection of superficial pharyngeal squamous cell carcinoma. Esophagus. 2022;19:460–8.

How to Take High-Quality Images (UGI)

Noriya Uedo

Contents

Supplementary Information The online version contains supplementary material available at https://doi.org/10.1007/978-981-97-2732-2_12. The videos can be accessed individually by clicking the DOI link in the accompanying figure caption or by scanning this link with the SN More Media App.

Y. Sano et al. (eds.), *Atlas of Advanced Endoscopy*, https://doi.org/10.1007/978-981-97-2732-2_12

1 Introduction

Why is it important to record high-quality endoscopic images? Compared to text description, a good endoscopic image contains more objective and precise information on mucosal/lesion characteristics. If such valuable medical information on mucosal/lesion characteristics is properly recorded, it can be effectively shared (1) with medical colleagues inside/outside the hospital to provide learning opportunities, (2) with your own future to review and analyze the information for research, and (3) with readers worldwide if published in a medical journal. High-quality endoscopic images are better beautiful, but they must be informative and narrative. In this sense, endoscopic photodocumentation is similar to photojournalism.

Historically, in Japan, upper gastrointestinal endoscope was developed as a gastro-camera in the 1950s [1]. In this system, a 4-mm film was loaded into a small camera at the endoscope tip, the endoscope was inserted through the mouth into the stomach, and photographs were taken without actually observing the mucosa. The film is then developed, and the endoscopic photographs are used to make a diagnosis. Since there was no biopsy channel at that time, the endoscopists had to diagnose whether surgery was indicated based solely on the recorded images. This is why Japanese endoscopists highly value the quality of endoscopic photographs. On the other hand, the development of upper gastrointestinal endoscopy in Western countries begun with Adolf Kussmaul's rigid scope in "a sword swallower" in 1868 [2]. In this situation, gastrointestinal mucosa was only observed and was not recorded as images. Even after the development of flexible fiber-optic endoscopes in the 1950s or early videoendoscopes in the 1980s, endoscopic findings in Western countries were essentially documented as text in endoscopy reports. With the current availability of mass storage of video images, medical information on the gastrointestinal mucosa should be recorded and stored as high-quality endoscopic images.

2 Sedation

The method and type of sedation vary in circumstances, including local guidelines and the number and services of anesthesiologists. Sedation is often recommended when detailed examination is necessary and stable high-quality image need to be captured over a certain amount of time. Benzodiazepines such as midazolam are the most widely used, but care needs to be taken for occasional restlessness by disinhibition. Analgesics such as pethidine can be used in combination with sedatives to reduce pharyngeal reflexes. A randomized controlled trial comparing the completion rate of pharyngeal observation between three groups (no sedation, midazolam, and pethidine) showed that the completion rate of pharyngeal observation was the same in the midazolam group as in the no sedation group but significantly higher in the pethidine group [3].

3 Preparation

It has been recommended for the patients to refrain from drinking or eating after dinner 1 day before the esophagogastroduodenoscopy (EGD). However, this instruction makes saliva or mucous sticky and strongly adhered to the mucosa, making it difficult to remove from the mucosal surface. Patients are better to be allowed to drink transparent liquids (water or tea without sugar, etc.) up to 1 h before EGD.

It is important to remove mucous or froth on the surface of the mucosa. If the mucosa/lesions are covered with such stuffs, detailed characteristics cannot be appreciated. If a mucolytic agent such as pronase or *N*-acetylcysteine is available, the mixed solution is administered 30 min before the procedure. The pronase mixture consists of 20,000 U pronase (Pronase MS, Kaken Pharmaceutical, Tokyo, Japan), 1 g sodium bicarbonate, and 0.5 mL of dimethylpolysiloxane (2% Gascon Drop Oral Liquid, Horii Pharmaceutical Inc., Osaka Japan), and the *N*-acetylcysteine mixture comprises of 2 mL of acetylcysteine (200 mg/mL, Parvolex, Celltech, UK, or Mucomyst, Bristol-Myers Squibb, USA) and 0.5-mL activated dimethicone (Infacol, Forest Laboratories, Slough, Berkshire, UK). The mixture should be diluted in enough amount (100–200 mL) of water. Recent meta-analysis suggested that only simethicone solution provides comparable mucosal visibility with simethicone + *N*-acetylcysteine mixture solution [4]. Therefore, only 100–200 mL of simethicone solution works sufficiently.

When active peristalsis interferes endoscopic observation, an antispasmodic agent, such as 20-mg scopolamine butylbromide (Buscopan, Nippon Boehringer Ingelheim, Tokyo, Japan) or 1-mg glucagon (Glucagon G Novo; Eisai, Tokyo), is administered. A topical application of peppermint oil or 0.8% l-menthol (Mincrea, Nihon Pharmaceutical Co. Ltd., Tokyo, Japan) is reported to be effective as antispasmodics [5].

4 Equipment and Settings

There are some commercial lens cleaners (SL cleaner, Sugimura Laboratory Inc., Tokyo, Japan, and Cleash, Fujifilm Co., Tokyo, Japan and Nagase Medicals Co., Ltd., Hyogo, Japan). These avoid cloudiness or formation of water droplets on the lens during endoscopic observation [6].

To capture high-resolution image, a high-definition videoendoscope with the latest image processor is preferred. As for image resolution, the newer the better, but even with standard definition videoendoscopes, it is important to strive to achieve the highest image quality for the videoendoscope being used. High definition and high resolution come from technology, but high quality is dependent on the endoscopist's skill.

Black soft hood (MAJ-1989, Olympus, Co, Ltd., Tokyo, Japan) is essential for magnifying observation. Before insertion of the magnifying scope, distance between the lens and subject need to be adjusted by the protruding length of the cap by measuring the long dimension of the endoscopic image, e.g., 4.5 mm for EVIS-GIF H290Z under maximum magnification (◘ Fig. 1). Even with near-focus endoscopes (EVIS-GIF HQ190 or EVIS-GIF HQ290), a black soft hood can be used to stabilize the scope.

Recommended setting for structural enhancement is B4 or B6 in white light image and B8 for NBI.

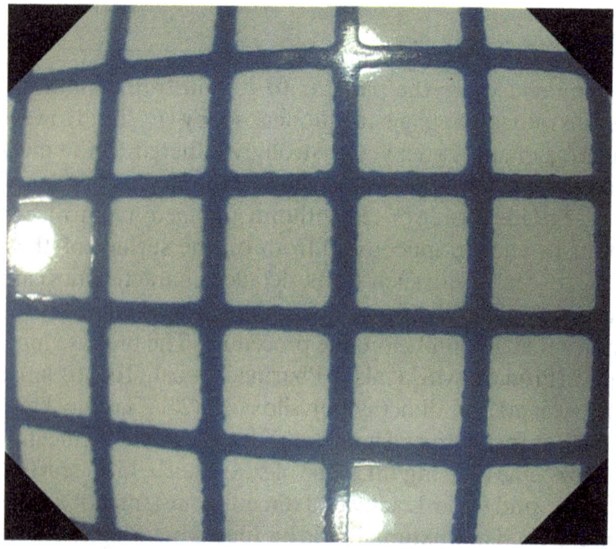

Fig. 1 Calibration of the distance between the lens and the subject using a scale chart

5 Technique

Videos are sometimes mistakenly thought to be superior to still images. While video is good for sharing technical information about endoscopy, still images are often better than videos for sharing cognitive information. Detailed features of the lesion can be well observed, shared, and discussed on high-quality still images.

Any remaining mucus or froth on the surface of the mucosa/lesions is cleaned by a scope water jet or by syringe flushing with simethicone solution through the working channel. Because the cancerous mucosa is fragile and strong water flushing by water jet may cause oozing hemorrhage, delicate water flushing by syringe is preferred.

Observations should be made with an optimal amount of air. A low air volume in the stomach may result in overlooking of lesions between the gastric folds of the greater curvature (Fig. 2). However, superficial lesions in the esophagus [7] or lesser curvature of the stomach may be better visualized with less air, as they become flattened with more air (Fig. 3). Less air condition is also important during magnifying endoscopic observation. By deflating air, mucosal injuries and contact bleeding can be avoided (Fig. 4). To avoid contact bleeding, it is important to contact the surrounding noncancerous mucosa with the tip of the black soft hood.

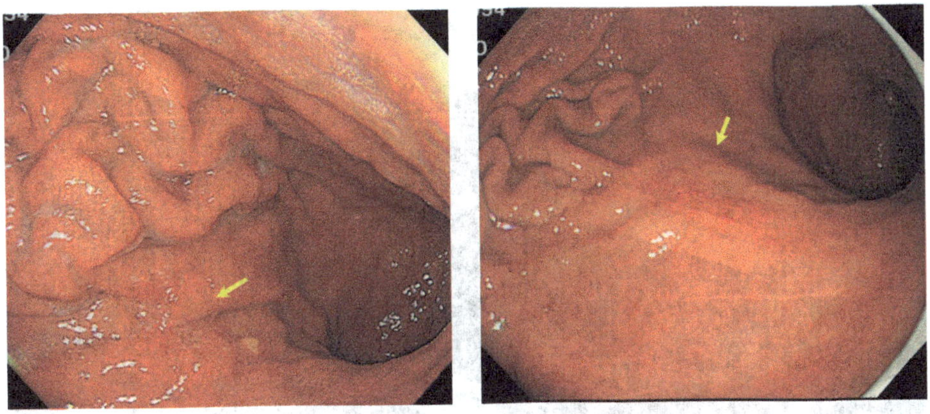

Fig. 2 Overlooking of the lesion at the greater curvature in less air condition

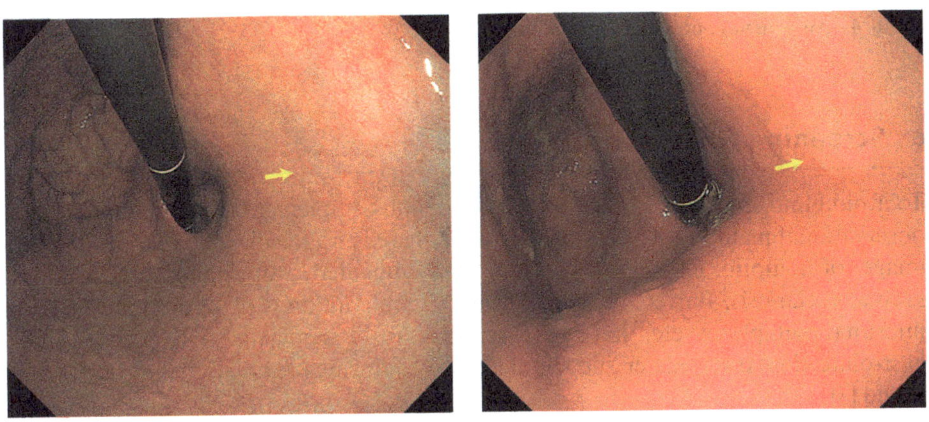

Fig. 3 Overlooking of the lesion at the lesser curvature in condition of over-insufflation. (Image courtesy of Dr. Abe S)

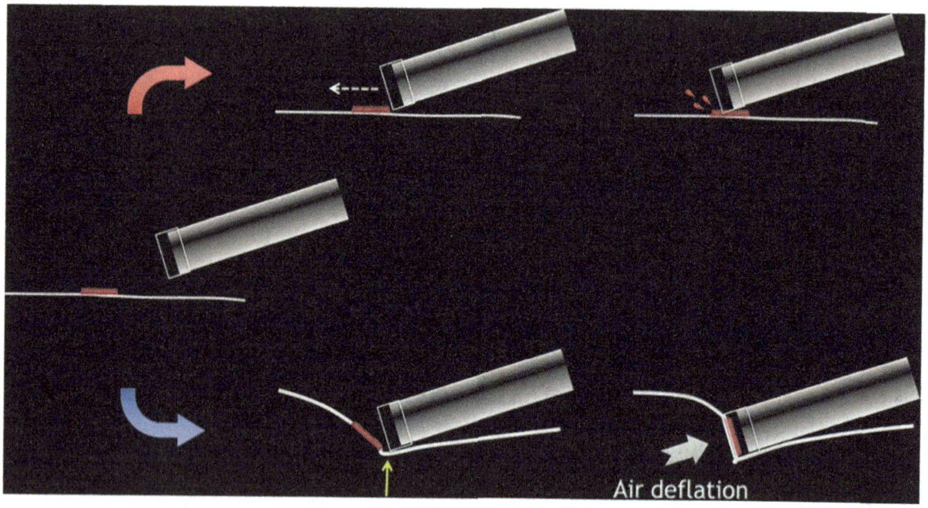

Fig. 4 Air volume control during magnification

6 Screening Observation

To avoid blind spots and missed suspicious lesions, implementation of standardized observational methods is important. Even if there is no localized lesion, endoscopic images of anatomic checkpoints should be documented. Some endoscopic findings, such as melanosis, Barrett's esophagus, endoscopic findings of atrophy and gastric intestinal metaplasia, etc., have been reported to be associated with risk of upper gastrointestinal tract cancer and should be noted during observation and recorded if found [8].

In high-risk patients for squamous cell carcinoma, systematic observation of the oral cavity, the oropharynx, the larynx, and the hypopharynx is desired [9]. In the esophagus, photodocumentation of anatomical landmarks, i.e., the left main bronchus and the left atrium, facilitates identifying radial and longitudinal orientation of the esophagus [10]. Yao has proposed the systematic screening protocol for the stomach (Fig. 5) [11]. In this method, regardless of the order of observation, inspection and photodocumentation of at least 4-3 quadrant directions (the anterior wall, posterior wall, greater curvature, and lesser curvature) at the middle upper corpus, the lower corpus, and the antrum are proposed.

All recorded images should be contiguous, and certain part of image is better to overlap with one previous image.

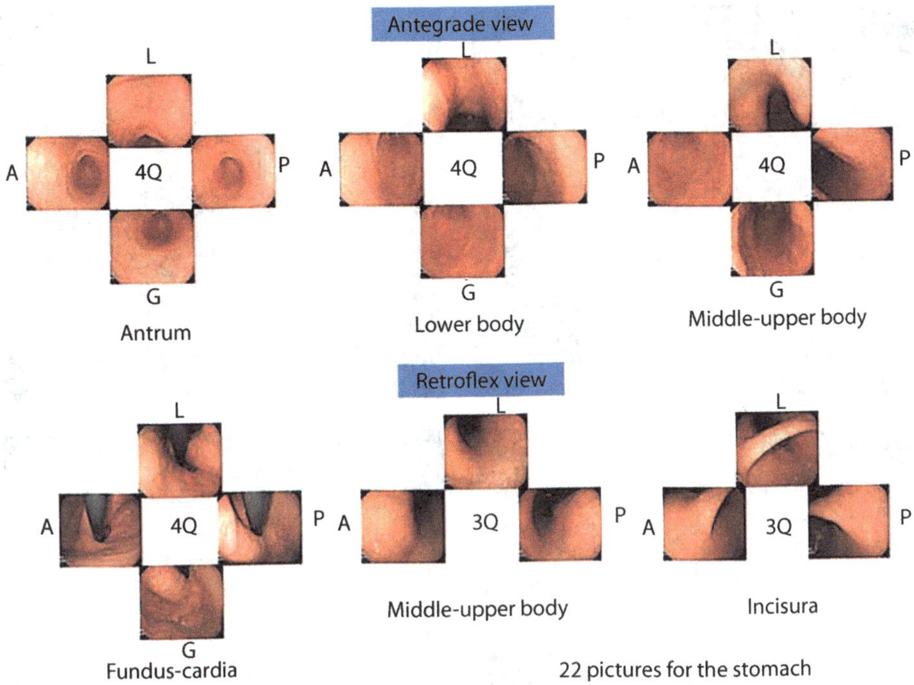

Fig. 5 The systematic screening protocol for the stomach. (Reprinted from Yao K. Ann Gastroenterol. 2013;26:11–22)

7 Inspection for Suspicious Lesions

For each suspicious lesion detected, a detailed characterization of the lesion is performed at different distances and angles (Figs. 6 and 7). The region of interest should be placed in the center of the image. For small lesions, the entire circumference of the lesion should be included in a single image; large lesions that cannot be depicted in a single image should be divided into multiple images, each overlapping in part. Observations are first made on a long-distance view that includes anatomical landmarks to represent the lesion location (Fig. 6a–d) and then on a short-distance view to characterize the lesion (Fig. 6e–j). The surface topography is better seen when observed from an oblique angle, and the differences in color are better seen when observed from a vertical angle. The entire margin of the lesion is represented in a single image for small lesions, but it is represented in multiple consecutive images for large lesions. The characteristics of the lesion necessary for treatment decisions, i.e., location, size, histology, depth of tumor invasion, presence or absence of ulcer findings, etc., should be recorded on the endoscopic image.

Further characterization is performed with dye- or equipment-based image-enhanced endoscopy (IEE) with or without magnification. IEE observation should be performed after enough morphological assessment by white light images. Magnifying equipment-based IEE such as narrowband imaging, blue light imaging, etc. enables us to observe microsurface structure and microvessel architecture of the superficial mucosa which corresponds to histological findings.

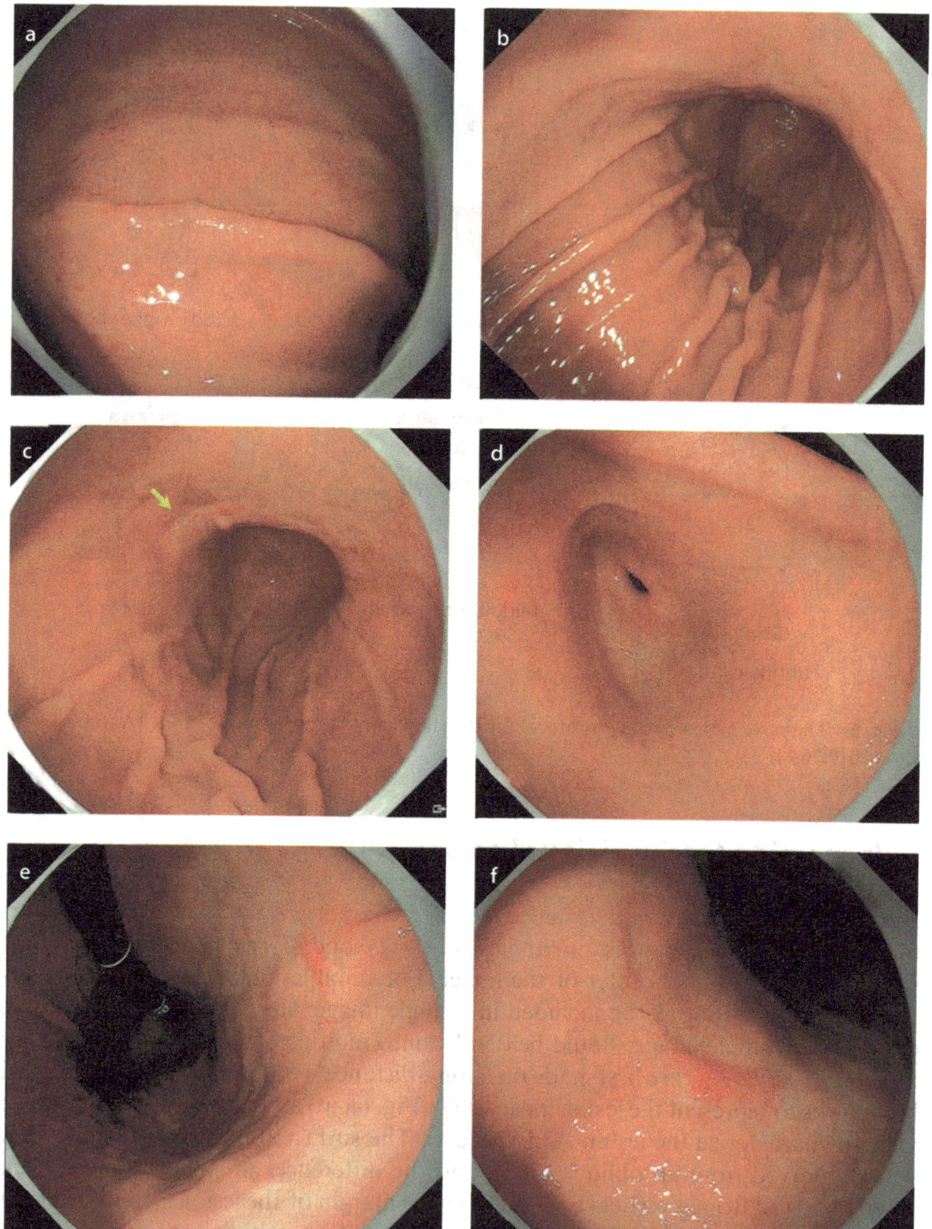

Fig. 6 Inspecting the undifferentiated type 0-IIc (superficial depressed) lesion at the lower gastric body. Background mucosa has active atrophic gastritis represented by spotty redness **a**. Endoscopic atrophy is seen in the corpus lesser curvature **b**. Asymmetric distribution of pale mucosal color indicates the presence of abnormality **c**. No endoscopic finding of gastric intestinal metaplasia in the antrum **d**. Retroflex view of the lesion from the anterior wall side **e**. Retroflex of the lesion from the lesser curvature side by turning a small knob to the right **f**. During observation, flat (type 0-IIb) extension of the lesion toward the anterior side is suspected as slightly pale mucosal color (yellow arrow, **g**). The flat extension was also noticed in narrowband imaging (NBI) (yellow arrow, **h**). The flat extension was confirmed by magnifying observation (**i**, white box in **h**). Demarcation line is shown in water immersion technique **j**. After examining the lesion characteristics by endoscopy, a biopsy was performed from the mucosa outside the lesion on the oral side to confirm negative lesion extension **k**, **l**

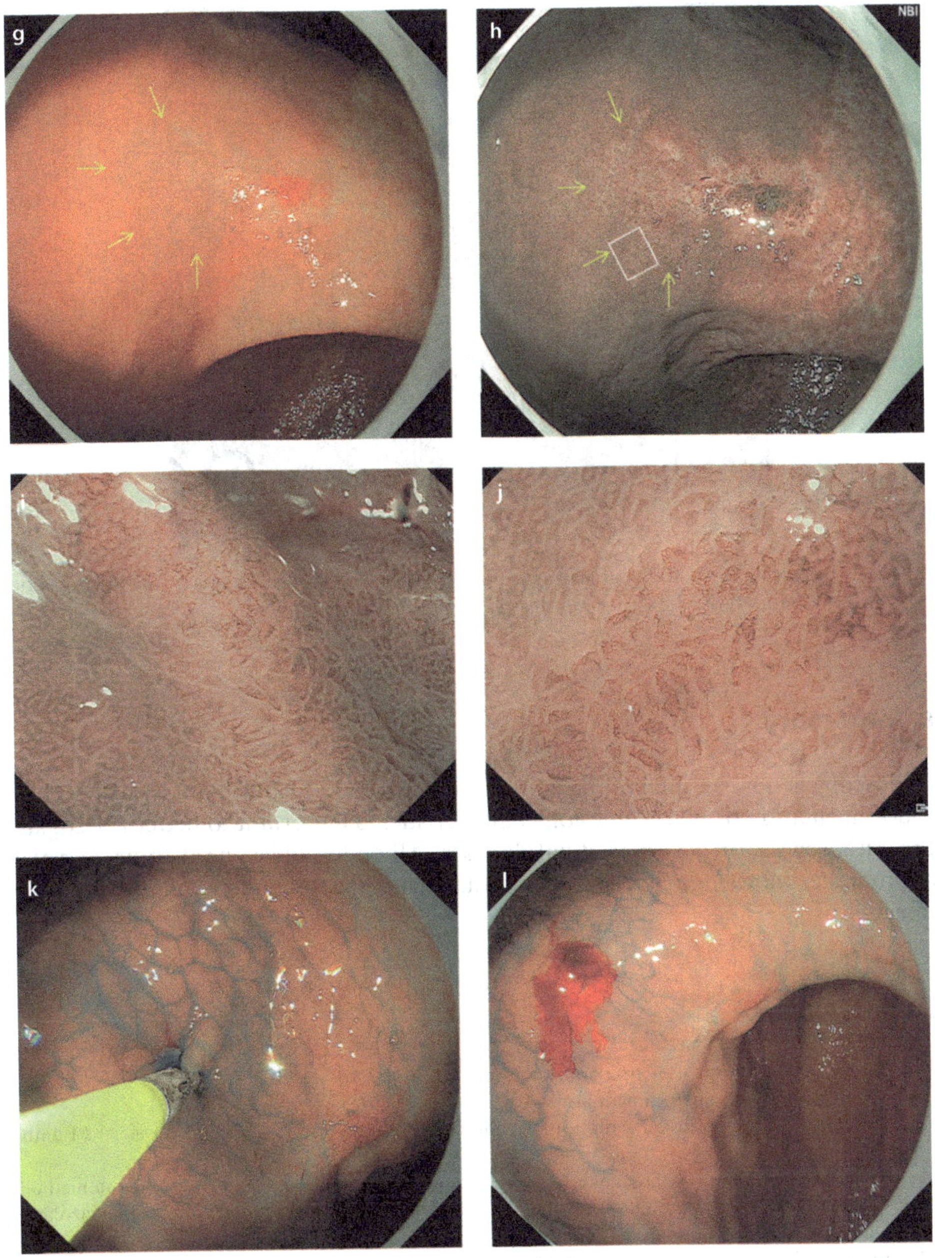

Fig. 6 (continued)

After characterization of the lesion/mucosa by endoscopy, a targeted biopsy may be performed to evaluate histological findings. Photographs taken before and after the forceps biopsy will allow for correlation of the endoscopic findings with the histological findings of the biopsy specimen (Fig. 6k, l).

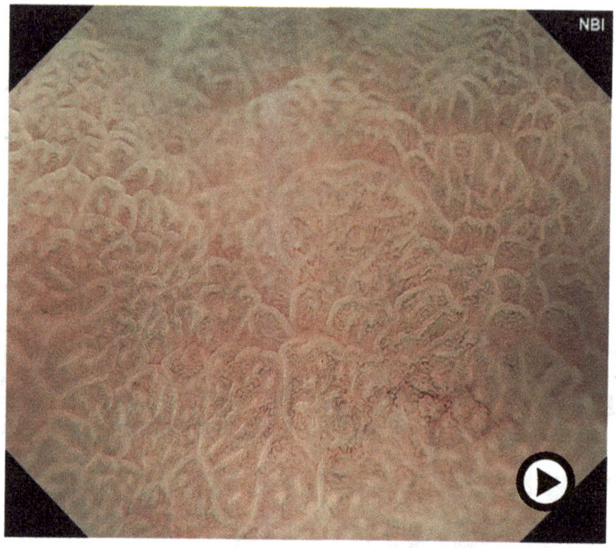

Fig. 7 Inspecting the undifferentiated type 0-IIc (superficial depressed) lesion at the lower gastric body (▶ https://doi.org/10.1007/000-c7q)

8 Summary

Endoscopy is now an integral part of gastroenterology. Detailed analysis and photodocumentation of mucosal/lesion characteristics is important to better investigate the pathogenesis of gastrointestinal diseases. All endoscopists need to acquire the appropriate knowledge and skills to acquire and record high-quality endoscopic images.

References

1. https://www.olympus-europa.com/company/en/news/stories/2019-08-19t13-00-37/in-the-beginning-was-the-gastrocamera.html. Available on Oct 2023.
2. Sircus W. Milestones in the evolution of endoscopy: a short history. J R Coll Physicians Edinb. 2003;33(2):124–34.
3. Yamasaki Y, Ishihara R, Hanaoka N, et al. Pethidine hydrochloride is a better sedation method for pharyngeal observation by transoral endoscopy compared with no sedation and midazolam. Dig Endosc. 2017;29:39–48.
4. Burke E, Harkins P, Moriarty F, et al. Does premedication with mucolytic agents improve mucosal visualization during oesophagogastroduodenoscopy: a systematic review and meta-analysis. Surg Res Pract. 2021;2021:1570121.
5. Hiki N, Kaminishi M, Yasuda K, et al. Antiperistaltic effect and safety of L-menthol sprayed on the gastric mucosa for upper GI endoscopy: a phase III, multicenter, randomized, double-blind, placebo-controlled study. Gastrointest Endosc. 2011;73:932–41.
6. Yoshida N, Naito Y, Yasuda R, et al. A novel lens cleaner to prevent water drop adhesions during colonoscopy and esophagogastroduodenoscopy. Endosc Int Open. 2017;5:E1235–41.
7. Iwatsubo T, Ishihara R, Yamasaki Y, et al. Narrow band imaging under less-air condition improves the visibility of superficial esophageal squamous cell carcinoma. BMC Gastroenterol. 2020;20:389.

8. Chiu PWY, Uedo N, Singh R, et al. An Asian consensus on standards of diagnostic upper endoscopy for neoplasia. Gut. 2019;68:186–97.
9. Hamada K, Ishihara R, Yamasaki Y, et al. Transoral endoscopic examination of head and neck region. Dig Endosc. 2018;30:516–21.
10. Emura F, Gomez-Esquivel R, Rodriguez-Reyes C, et al. Endoscopic identification of endoluminal esophageal landmarks for radial and longitudinal orientation and lesion location. World J Gastroenterol. 2019;25:498–508.
11. Yao K. The endoscopic diagnosis of early gastric cancer. Ann Gastroenterol. 2013;26:11–22.

How to Take High-Quality Images (LGI)

Yasushi Sano

Contents

Supplementary Information The online version contains supplementary material available at https://doi.org/10.1007/978-981-97-2732-2_13. The videos can be accessed individually by clicking the DOI link in the accompanying figure caption or by scanning this link with the SN More Media App.

Y. Sano et al. (eds.), *Atlas of Advanced Endoscopy*, https://doi.org/10.1007/978-981-97-2732-2_13

1 Why Do We Need High-Quality Image Documentation?

Affordable digital imaging systems have enabled endoscopists to easily document and report their images. Image documentation should be accurate, comprehensive, and clear to reflect endoscopic findings and interventions faithfully [1]. Moreover, it is the best method to verify the completeness of an endoscopic procedure. Therefore, photo documentation is an essential aspect of endoscopic quality control. In addition, taking high-quality endoscopic images daily will improve both right- and left-hand endoscope control and fine endoscope movement during therapeutic procedures. This kind of control training is very useful for recent magnifying endoscopy and endocytoscopy observation. In fact, we often find that endoscopists who are good at magnifying colonoscopy observation are also skilled in techniques such as EMR/ESD. In the future, AI diagnosis may become widespread, but endoscopists need to capture appropriate and high-quality images to use AI assistance effectively. For the novice endoscopist, image documentation is the first step toward becoming an expert endoscopist.

2 Suggested Systematic Image Documentation in Colonoscopy

It is recommended that the endoscopist accurately document the endoscopic landmarks of the cecum, appendiceal orifice, ileocecum, ileocecal valve, end of the ileum if end ileal intubation is possible, ascending colon, transverse colon, descending colon, sigmoid colon, and rectum both in the anterior and in the posterior view (◘ Fig. 1) [2–5].

The following are the locations of the nine landmarks (◘ Fig. 2) [6]:

1. Terminal ileum (if possible)
2. Cecum and appendiceal orifice
3. Ascending colon just proximal to the hepatic flexure
4. Transverse colon just distal to the hepatic flexure
5. Transverse colon—just below splenic flexure
6. Descending colon just distal to splenic flexure
7. Middle sigmoid colon
8. Lower part of the rectum
9. Lower rectum (retroflexed view)

Recommendations	Terminal ileum	Colon	Rectum	Anal canal perianal area
ESGE [1] (2012)	Optional (if intubated)	Cecum and appendiceal orifice (taken from a distance of 2–4 cm to encompass the cecal strap fold) Cecum and ileocecal valve Ascending colon under the hepatic flexure Transverse colon just distal to the hepatic flexure Transverse colon just proximal to the splenic flexure Descending colon below the splenic flexure Middle part of the sigmoid*	Lower part of the rectum in both forward (taken 2 cm above the anal line) and retroflexed views	–
ASGE/ACG/ AGA [2,3] (2015)	Optional (in alternative to cecal landmarks)	Cecum, appendicular orifice, and cecal strap fold Cecum and ileocecal valve	Rectum in retroflexed view	–
Tang et al. [4] (2015)	Optional (if intubated)	Cecum, appendiceal orifice, and ileocecal valve Ascending colon Transverse colon Descending colon Sigmoid colon	Rectum in both forward and retroflexed views	Optional (if they are relevant to the presenting symptoms)

Fig. 1 Recommended images of landmarks in colonoscopy

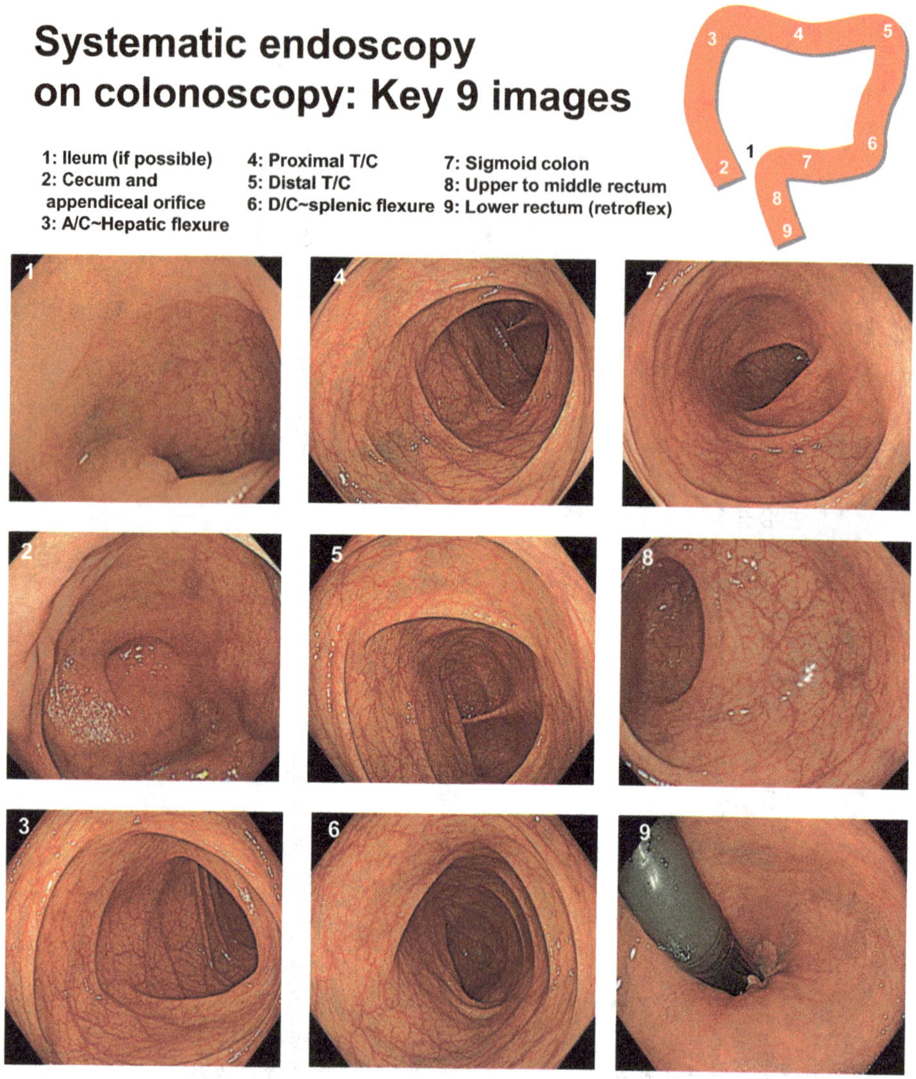

Fig. 2 Systematic endoscopy on colonoscopy: key nine images

3 How to Capture Meaningful and Beautiful Endoscopic Photography

In general, professional photographers have a few tips for photographing models and landscapes. Similarly, there are some tips for endoscopic photography. The following points are important when taking pictures:

1. Clean the lesion with a water jet or scope tip to remove mucus and feces (see below).
2. Position the lesion at 6 o'clock (use forceps if necessary).

3. Focus on the lesion.
4. Center the lesion and capture the entire image.
5. Take multiple images at gradually increasing magnification.
6. Take a magnified view of the area of interest.
7. Perform IEE (NBI, dye endoscopy) if necessary.

Figures 3, 4 and 5 shows a typical poor example of novice endoscopists, which they should try to improve.

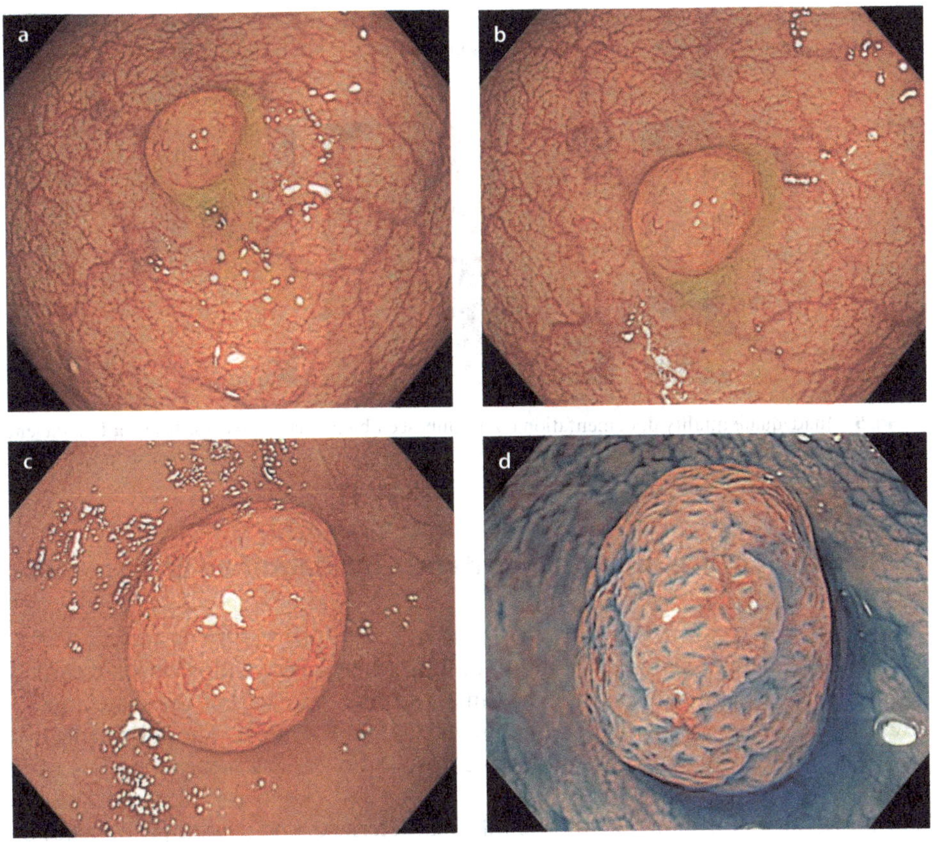

Fig. 3 Inadequate quality documentation commonly seen by novice endoscopists (1). **a**, **b** Mucus or feces is present on the surface of the lesion. **c** Clean off the mucus with a water jet. **d** Pit pattern can be observed cleanly using chromoendoscopy

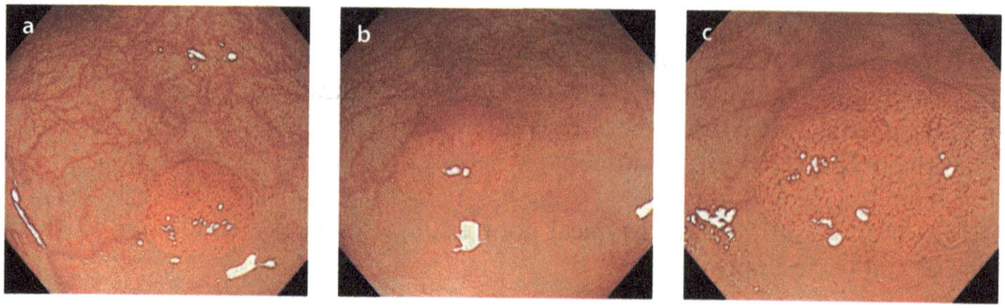

Fig. 4 Inadequate quality documentation commonly seen by novice endoscopists (2). **a** Lesion is not centered. **b** Out of focus. **c** Part of the lesion is not captured in the picture (right side)

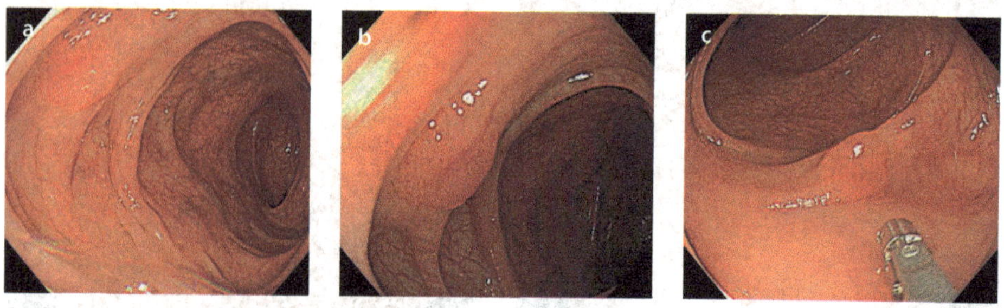

Fig. 5 Inadequate quality documentation commonly seen by novice endoscopists (3). **a** Insufficient bowel extension, lesion not centered. **b** Out of focus. **c** Water droplets on the endoscope lens

4 How to Capture the Area of Interest by Magnified Endoscopic Observation?

At clinical conferences, it is often disappointing when novice endoscopists recorded endoscopic images of rare cases. Many times, dozens of images of the same area of interest are taken and recorded without any meaningful context. The key to effective recording through magnifying endoscopic observation is to capture the diagnostic process in the pictures.

Firstly, the lesion should be thoroughly characterized, and its size, prominence or depression, erythema, and color changes should be carefully observed and recorded as endoscopic images. Subsequently, dye and magnification observations should be applied to the most significant areas (areas of interest). Endoscopic images should be taken while gradually increasing the magnification from weak to medium and strong magnification.

In addition, the process of clinical judgment, such as "whether it is a neoplasia or non-neoplasia" and "whether resection is necessary," should be recorded as a series of photographs so that anyone can understand them. Taking dozens of meaningless images serves no purpose. Novices need to understand that good images taken by experts are the result of thoughtful and deliberate consideration (Fig. 6).

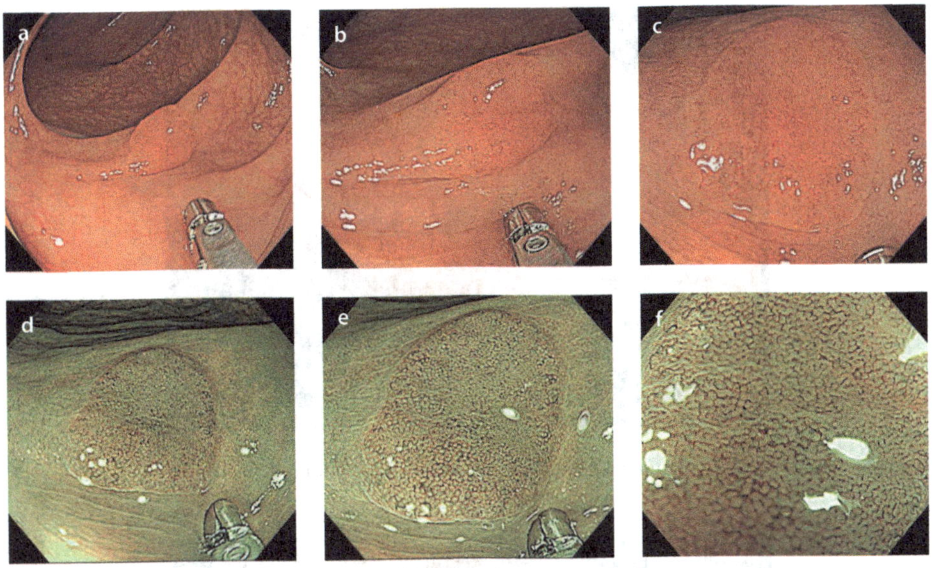

Fig. 6 High-quality documentation. Generally, the endoscopist will record the images in the following order: weak magnification, medium magnification, and strong magnification. The endoscopist should take pictures so that other physicians can later understand which part of the lesion was recorded. **a** The lesion should preferably be located at 6 o'clock using a catheter or similar device. **b** Photograph with slightly increased magnification. **c** Medium magnification, so that all of the lesion is visible. **d** Observe the surface and vessel using NBI. **e** Observation of the central depression. **f** Record a strongly magnified photograph if necessary

5 Tips for Improving Quality

Mucus Cleaning Method Using Endoscope Tip (Fig. 7)

This is a cleaning method using the endoscope tip when there is mucus on the lesion, such as a flat lesion or SSL. The endoscope tip is contacted softly with the lesion, and the water supply button on the left-hand side of the endoscope control unit is pressed. The endoscope lens cleaning nozzle sends water to remove the mucus from the lesion.

Lesion Cleaning with Water Jet (Fig. 8)

When there is mucus on the surface of the lesion, it is necessary to clean it with a water jet. However, in the case of early-stage cancer, if the jet is applied directly to the lesion, the water pressure will cause bleeding. To avoid this, water is pumped into the normal mucosa at the margins of the lesion, and the water jet is used to clean the lesion using the flowing water that spreads on the mucosal surface. This method avoids bleeding and cleans the lesion.

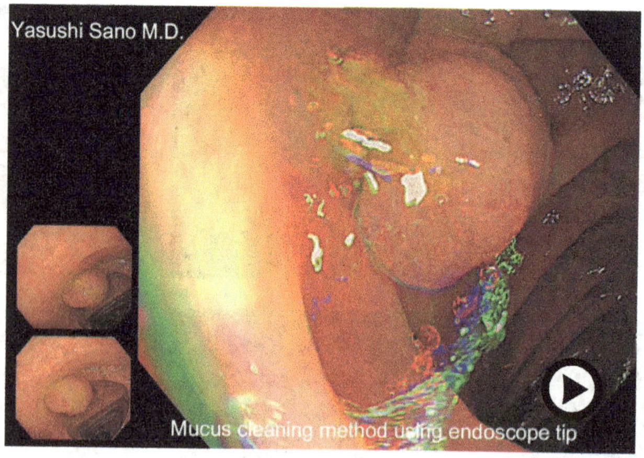

Fig. 7 Mucus cleaning method using endoscope tip (▶ https://doi.org/10.1007/000-c7s)

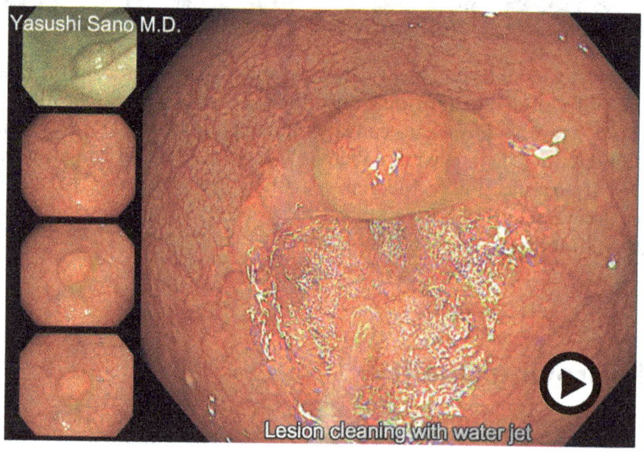

Fig. 8 Lesion cleaning with water jet (▶ https://doi.org/10.1007/000-c7r)

Magnifying Observation Using the NT Catheter [7] (Fig. 9)

Nontraumatic, globular tip catheter: This catheter is used to remove mucus and to drop dye solution onto the lesion. Better positioning for magnifying observation can be obtained by pushing and holding the surrounding mucosa.

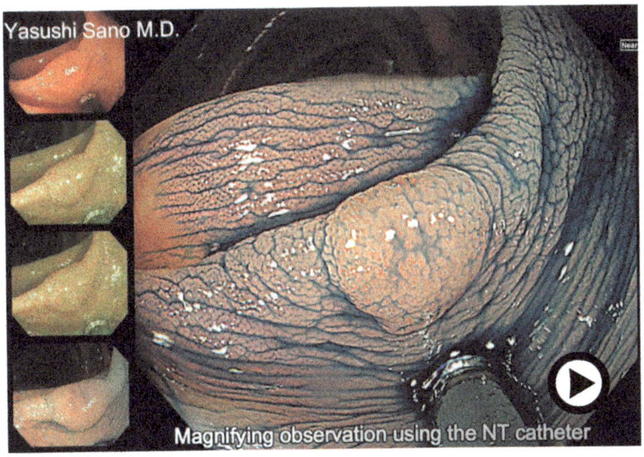

Fig. 9 Magnifying observation using the NT catheter (▶ https://doi.org/10.1007/000-c7t)

6 Summary

Accurate and adequate image documentation is an essential part of GI endoscopic reporting. At the present time, systematic photo documentation is recommended by major endoscopic societies and has become an integral aspect of quality control. Therefore, systematic imaging should be routinely performed in GI endoscopy, and information systems should incorporate this relevant data.

References

1. Sano Y, Chiu HM, Li XB, Khomvilai S, Pisespongsa P, Co JT, Kawamura T, Kobayashi N, Tanaka S, Hewett DG, Takeuchi Y, Imai K, Utsumi T, Teramoto A, Hirata D, Iwatate M, Singh R, Ng SC, Ho SH, Chiu P, Tajiri H. Standards of diagnostic colonoscopy for early-stage neoplasia: recommendations by an Asian private group. Dig Endosc. 2019;31(3):227–44.
2. Rembacken B, Hassan C, Riemann JF, Chil-ton A, Rutter M, Dumonceau JM, et al. Quality in screening colonoscopy: position statement of the ESGE. Endoscopy. 2012;44:957–68.
3. ASGE/ACG Task Force on Quality in Endoscopy. Quality indicators for GI endoscopic procedures. Gastrointest Endosc. 2015;81:1–80.
4. Rex DK, Schoenfeld PS, Cohen J, Pike IM, Adler DG, Fennerty MB. Quality indicators for colonoscopy. Gastrointest Endosc. 2015;81:31–53.
5. Tang SJ, Raju G. Endoscopic photography and image documentation. Gastrointest Endosc. 2015;82:925–31.
6. Marques S, Bispo M, Pimentel-Nunes P, Chagas C, Dinis-Ribeiro M. Image documentation in gastrointestinal endoscopy: review of recommendations. GE Port J Gastroenterol. 2017;24(6):269–74.
7. Sano Y, Saito Y, Fu KI, Matsuda T, Uraoka T, Kobayashi N, Ito H, Machida H, Iwasaki J, Emura F, Hanafusa M, Yoshino T, Kato S, Fujii T. Efficacy of magnifying chromoendoscopy for the differential diagnosis of colorectal lesions. Dig Endosc. 2005;17:105–16.

Head and Neck

Contents

Pharyngeal Lymphoid Follicle

Chikatoshi Katada and Hirokazu Higuchi

Contents

Supplementary Information The online version contains supplementary material available at https://doi.org/10.1007/978-981-97-2732-2_14. The videos can be accessed individually by clicking the DOI link in the accompanying figure caption or by scanning this link with the SN More Media App.

Y. Sano et al. (eds.), *Atlas of Advanced Endoscopy*, https://doi.org/10.1007/978-981-97-2732-2_14

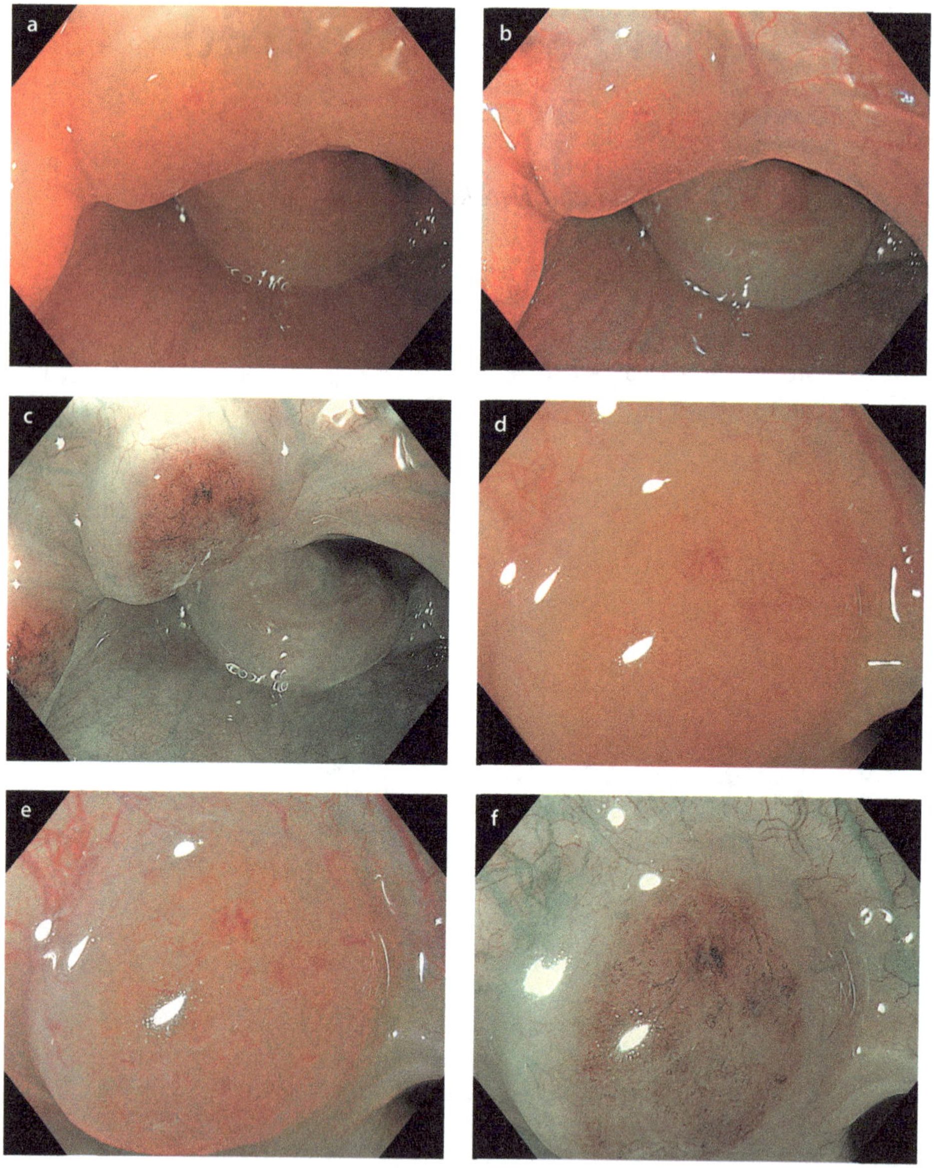
a
b
c
d
e
f

▪ Pathological Diagnosis

Lymphoid follicle.

g

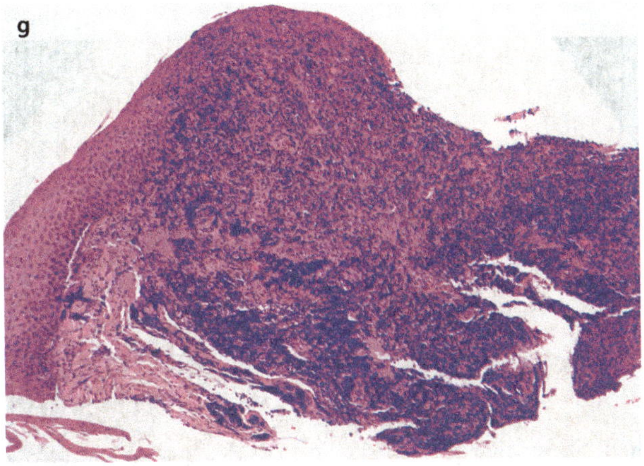

▪ Explanation

A flat elevated lesion measuring approximately 8 mm was detected in the anterior wall of the oropharynx. When viewed by non-magnifying WLI, the color of the lesion was reddish (Fig. a). Mode 1 in non-magnifying TXI enhanced the lesion's color more effectively than WLI (Fig. b). Non-magnifying NBI depicted the lesion as a brownish area with microvascular irregularity (Fig. c). When combined with full magnification, WLI could not evaluate the presence or absence of a clear demarcation line in the lesion or the degree of microvascular irregularity (Fig. d). When combined with full magnification, TXI enhanced the demarcation line and microvessels more effectively than WLI (Fig. e). When combined with full magnification, NBI visualized a brownish area without a clear demarcation line and microvessels without severe irregularity due to inflammation (Fig. f). The histopathological findings were consistent with the lymphoid follicle (Fig. g).

Tips

Lymphoid follicles are commonly located in the posterior wall of the oropharynx, valleculae, and pyriform sinus. Lymphoid follicles are usually multiple lesions, covered with squamous epithelium, and often associated with inflammation. Typically, NBI with magnifying endoscopy in superficial head and neck cancer reveals a well-demarcated brownish area and scattered foci of microvascular proliferation [1]. In the lymphoid follicle, IEE with magnifying endoscopy shows an indistinct demarcation line and microvessels without severe irregularity.

▪ Video

A short video is available as ◘ Fig. 1.

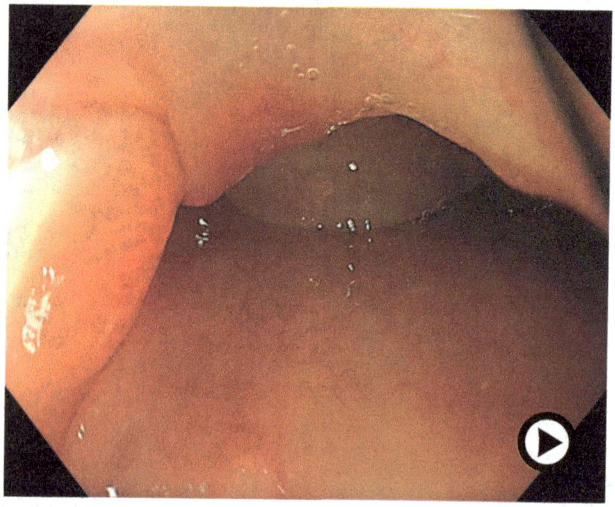

◘ **Fig. 1** Practical observation and diagnosis of lymphoid follicle with magnifying TXI and NBI (▶ https://doi.org/10.1007/000-c7v)

Reference

1. Muto M, Nakane M, Katada C, et al. Squamous cell carcinoma in situ at oropharyngeal and hypopharyngeal mucosal sites. Cancer. 2004;101:1375–581.

Pharyngeal Papilloma

Yasuaki Furue and Koichi Kano

Contents

Supplementary Information The online version contains supplementary material available at https://doi.org/10.1007/978-981-97-2732-2_15. The videos can be accessed individually by clicking the DOI link in the accompanying figure caption or by scanning this link with the SN More Media App.

Y. Sano et al. (eds.), *Atlas of Advanced Endoscopy*, https://doi.org/10.1007/978-981-97-2732-2_15

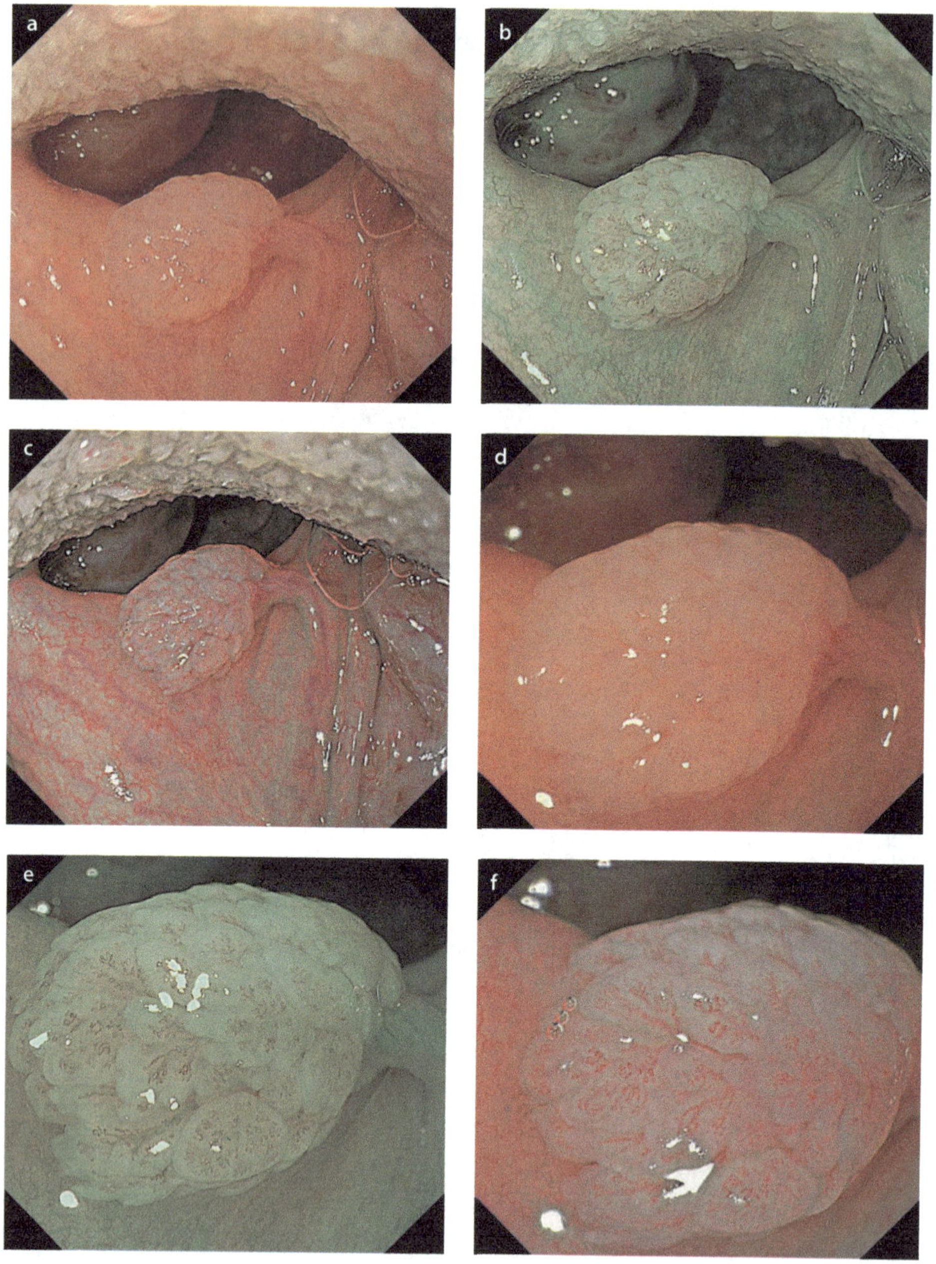
a
b
c
d
e
f

Pathological Diagnosis

Squamous papilloma, pharynx.

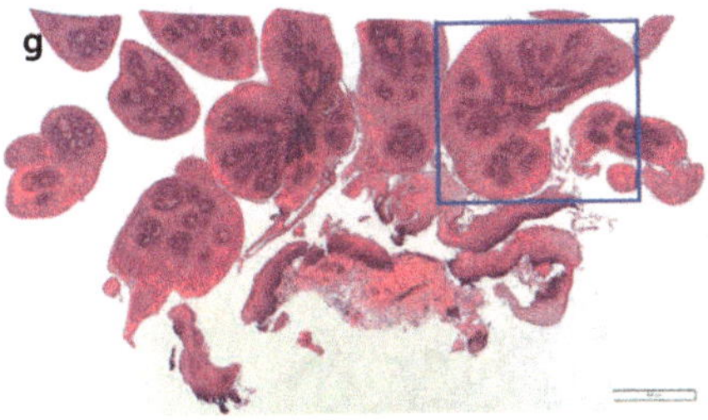

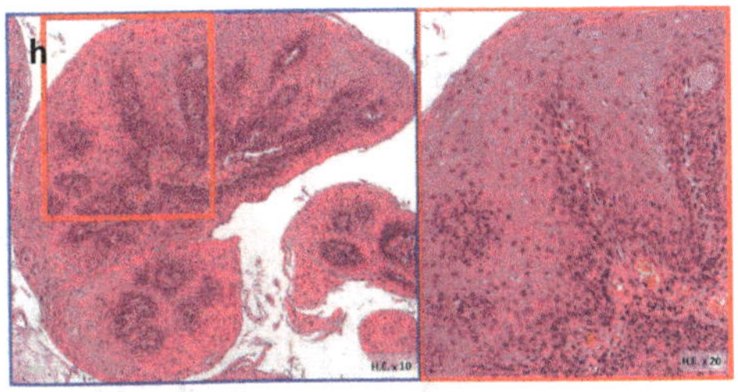

Explanation

A polypoid lesion measuring approximately 5 mm was detected in the right palatoglossal arch. WLI non-magnifying endoscopy revealed that the lesion displayed a uniform whitish coloration (Fig. a), along with the presence of papillary surface structures (Fig. b). In contrast to WLI, TXI (mode 1) resulted in enhanced visualization of the surface structure and color of the lesion (Fig. c). TXI magnifying endoscopy revealed the presence of multiple papillary structures on the lesion's surface, with relatively smooth and uniformly distributed internal vessels extending toward the lesion surface (Fig. d). Using NBI non-magnifying endoscopy, the lesion displayed a uniform whitish coloration (Fig. e), highlighting the internal vessels more sharply than that observed using TXI magnifying endoscopy (Fig. f). Based on these findings, the lesion was diagnosed as a papilloma. Subsequent histopathological examination confirmed the diagnosis as squamous papilloma with papillary growth of stratified squamous epithelium (Figs. g and h).

Tips

Papilloma is the most common benign neoplasm of the pharynx, with an equal distribution between male and female sex. It is associated with HPV infection, particularly with types 6 and 11. Squamous papilloma presents as lobular or sessile growths. Lobular lesions are characterized by clusters of fingerlike projections that appear white or pale pink to red in color, depending on the degree of keratinization, whereas sessile lesions are dome-shaped exhibiting a nodular, papillary, or verrucous surface. These lesions are exophytic, composed of papillary proliferations of hyperplastic stratified epithelium, varying in thickness, covered with a layer of parakeratin or orthokeratin, or may be nonkeratinized. The fingerlike epithelial projections extend from a narrow base and are supported by a fibrovascular core containing enlarged capillaries [1].

▪ Video

A short video is available as ◘ Fig. 1.

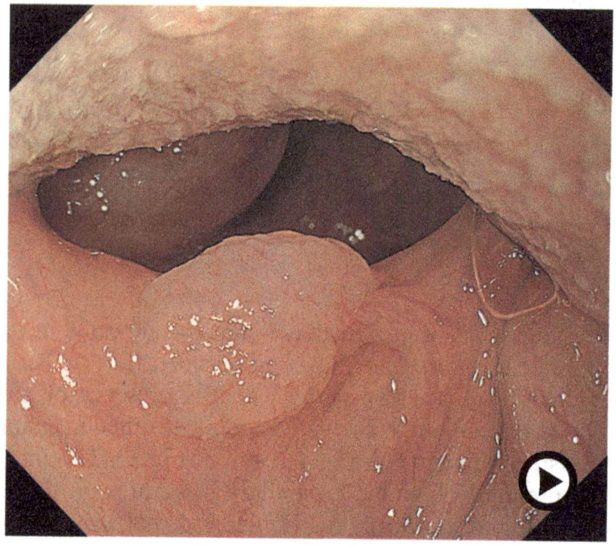

◘ **Fig. 1** Practical observation and diagnosis of papilloma with magnifying TXI and NBI (▶ https://doi.org/10.1007/000-c7w)

Reference

1. El-Naggar AK, Chan JKC, Grandis JR, Takata T, Slootweg PJ. WHO classification of head and neck tumours. 4th ed. Lyon: IARC; 2017.

Pharyngeal Superficial Cancer (0-I)

Keiichiro Nakajo and Tomonori Yano

Contents

Supplementary Information The online version contains supplementary material available at https://doi.org/10.1007/978-981-97-2732-2_16. The videos can be accessed individually by clicking the DOI link in the accompanying figure caption or by scanning this link with the SN More Media App.

Y. Sano et al. (eds.), *Atlas of Advanced Endoscopy*, https://doi.org/10.1007/978-981-97-2732-2_16

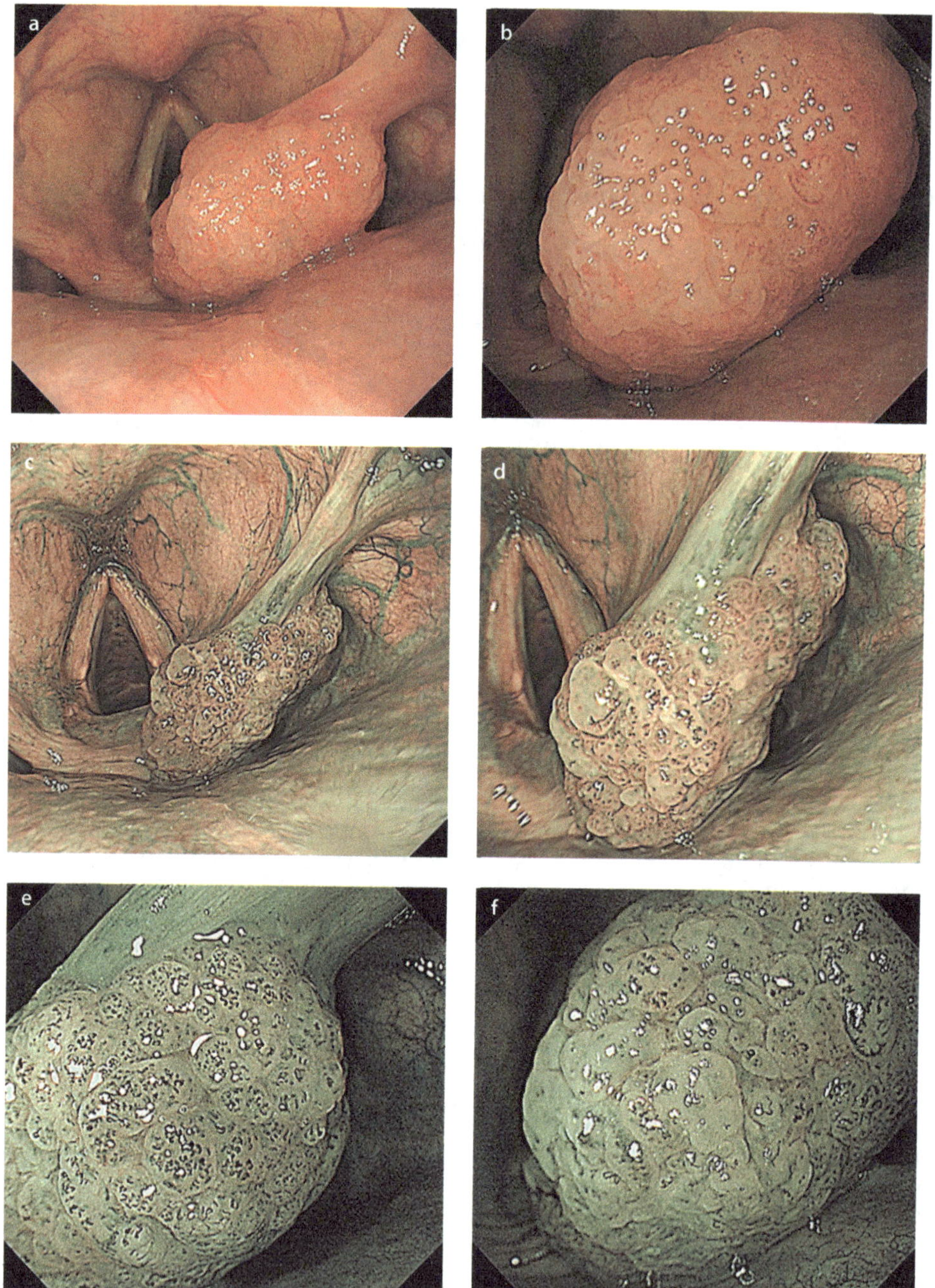
a
b
c
d
e
f

Pathological Diagnosis

Squamous cell carcinoma: type 0-I, 25 × 24 mm, ly1, v1, pHM0, pVM0; tumor thickness, 6280 μm.

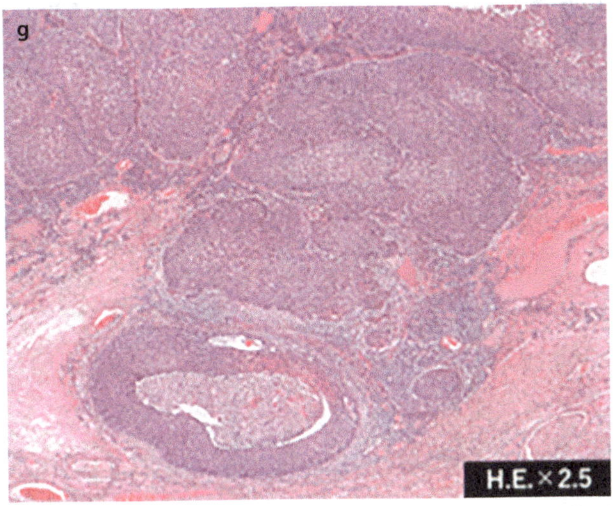

Explanation

A 25-mm protruding lesion was detected in the right piriform sinus of the hypopharynx, extending to the aryepiglottic fold and arytenoid region with clear margins. Under magnifying WLI, the lesion appeared uniformly pale reddish in color (Figs. a and b), displaying granular surface texture without ulcers and depressions. NBI revealed the entire lesion as a brownish area (Figs. c and d), with magnifying NBI exposing dot-like vessels of type B1 microvascular structures according to the Japan Esophageal Society's classification, visible in each granule (Figs. e and f). Although WLI suggested a protruded lesion, which is possibly subepithelial invasive carcinoma, the vessel findings in NBI raised the possibility of intraepithelial carcinoma. Consequently, an endoscopic laryngopharyngeal surgery was performed, resulting in en bloc resection of the lesion. Histopathological diagnosis confirmed moderately differentiated squamous cell carcinoma with subepithelial invasion (Fig. g).

Tips

The depth of type 0-I lesions using WLI is evaluated based on findings, such as lesion size, surface characteristics, shape of the elevated portion, and tumor hardness. Katada et al. reported that 92.3% of type 0-I lesions in pharyngeal cancer were subepithelial invasive cancer [1]. Conversely, Sunakawa et al. noted that 47.2% of elevated lesions with only B1 vessels had subepithelial invasive cancer, revealing limitations in depth assessment based on vascular morphology [2].

Video

A short video is available as Fig. 1.

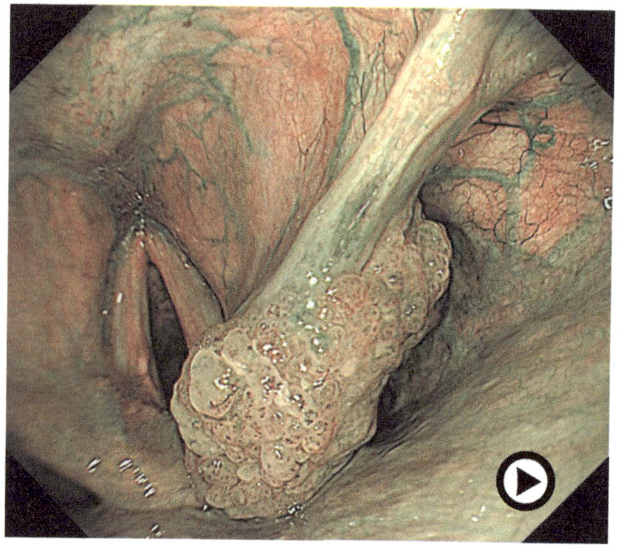

Fig. 1 Practical observation and diagnosis of superficial cancer (0-I) with NBI (▶ https://doi.org/10.1007/000-c7x)

References

1. Katada C, Okamoto T, Ichinoe M, et al. Prediction of lymph node metastasis and lymphatic invasion of superficial pharyngeal cancer on narrow band imaging with magnifying endoscopy. Auris Nasus Larynx. 2020;47:128–34.
2. Sunakawa H, Hori K, Kadota T, et al. Relationship between the microvascular patterns observed by magnifying endoscopy with narrow-band imaging and the depth of invasion in superficial pharyngeal squamous cell carcinoma. Esophagus. 2021;18:111–7.

Pharyngeal Superficial Cancer (0-IIa)

Yasuaki Furue and Koichi Kano

Contents

Supplementary Information The online version contains supplementary material available at https://doi.org/10.1007/978-981-97-2732-2_17. The videos can be accessed individually by clicking the DOI link in the accompanying figure caption or by scanning this link with the SN More Media App.

Y. Sano et al. (eds.), *Atlas of Advanced Endoscopy*, https://doi.org/10.1007/978-981-97-2732-2_17

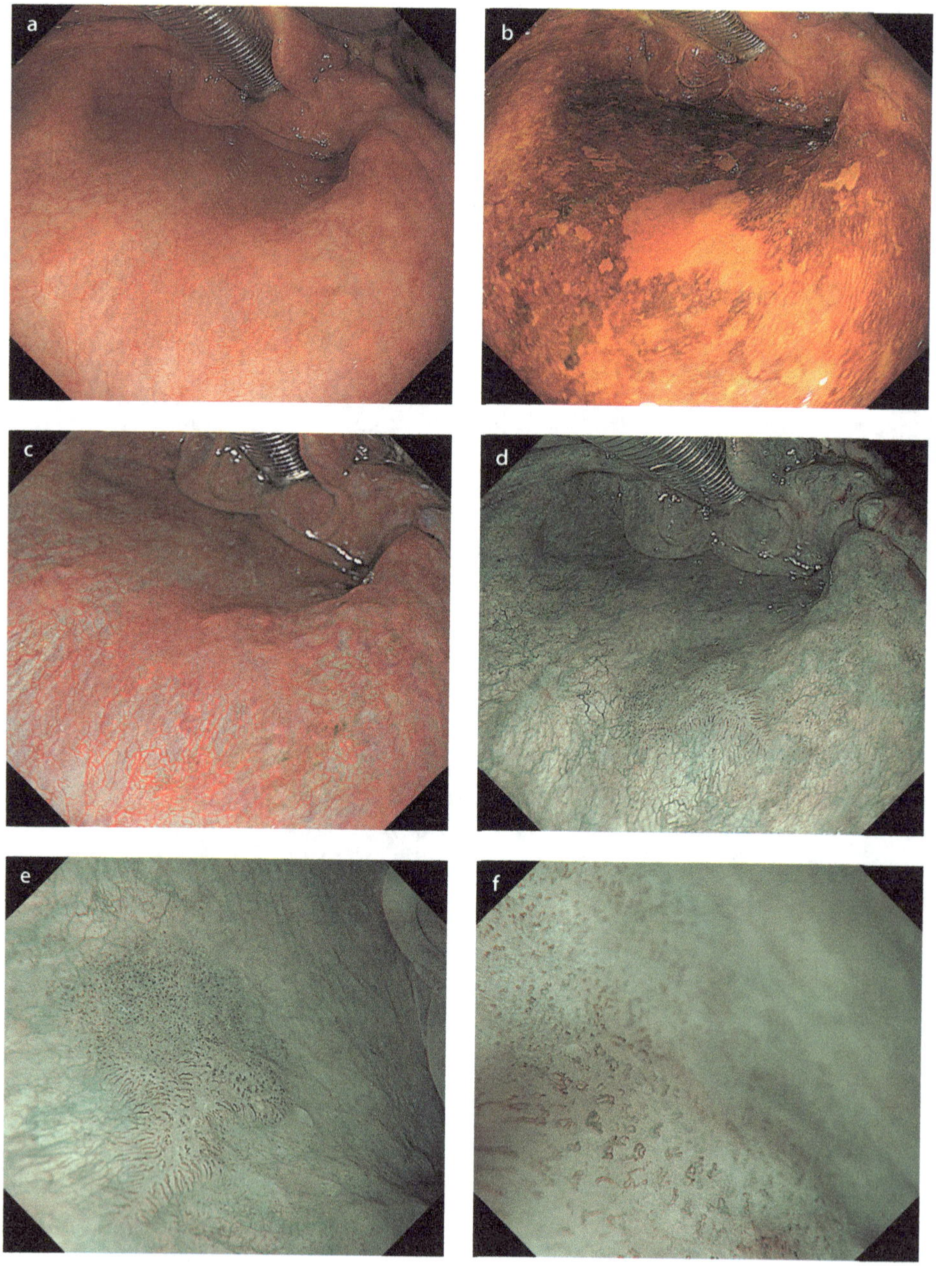
a
b
c
d
e
f

Pathological Diagnosis

Squamous cell carcinoma: type 0-IIa, 10 × 8 mm, ly0, v0, pn0, pHM0, pVM0; tumor thickness, 400 μm.

g

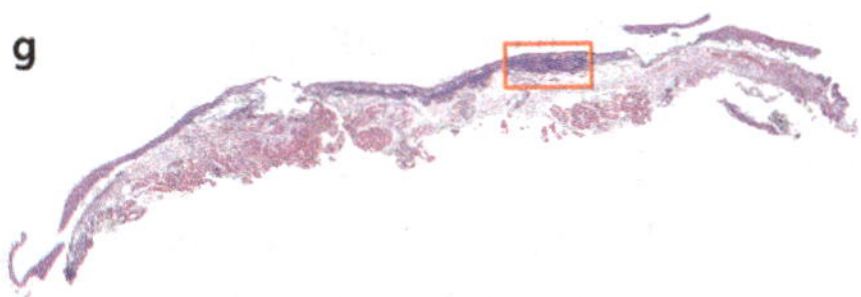

h

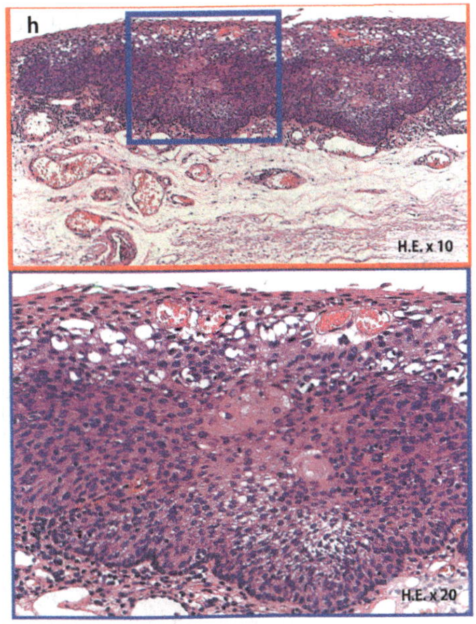

Explanation

A flat lesion measuring approximately 15 mm was detected in the hypopharynx. The lesion exhibited a reddish coloration when observed using non-magnifying WLI (Fig. a). After being sprayed with 0.75% iodine solution, the lesion became apparent as a Lugol-voiding lesion (Fig. b). In contrast to WLI, TXI (mode 1) resulted in enhanced visualization of the lesion's margin and color (Fig. c). Non-magnifying NBI endoscopy revealed the lesion as a brownish area (Fig. d). Saline immersion magnifying endoscopy under general anesthesia with intubation revealed the presence of irregular microvascular structures within the brownish area (Fig. e). NBI combined with full magnification provided a clear visualization of type B1 microvascular structures, according to the classification of magnifying endoscopy established by the Japan Esophageal Society (Fig. f) [1]. Based on these findings, the lesion was diagnosed as a superficial hypopharyngeal cancer and was resected en bloc using endoscopic laryngopharyngeal surgery. Subsequent histopathological examination confirmed the diagnosis as moderately differentiated squamous cell carcinoma with subepithelial invasion (Figs. g and h).

Tips

Since muscularis mucosa is absent in the pharynx, the depth of superficial cancer of the hypopharynx can be evaluated by assessing tumor thickness. The frequencies of types 0-Is, 0-Ip, 0-IIa, 0-IIb, and 0-IIc are 14.9%, 4.5%, 68.6%, 9.0%, and 3.0%, respectively. The tumor thicknesses of types 0-Is, 0-Ip, 0-IIa, 0-IIb, and 0-IIc are 3575 ± 2660 μm, 7267 ± 3288 μm, 848 ± 798 μm, 325 ± 220 μm, and 2148 ± 2169 μm, respectively [2]. Superficial cancers with a tumor height of up to 1 mm are classified as type 0-IIa and are the most common macroscopic type of hypopharyngeal cancer.

▪ Video

A short video is available as ◘ Fig. 1.

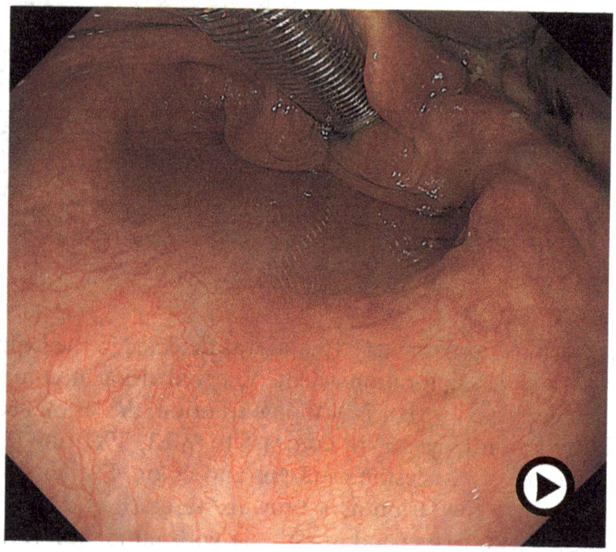

◘ **Fig. 1** Practical observation and diagnosis of superficial cancer (0-IIa) with magnifying TXI and NBI (▶ https://doi.org/10.1007/000-c7y)

References

1. Oyama T, Inoue H, Arima M, et al. Prediction of the invasion depth of superficial squamous cell carcinoma based on microvessel morphology: magnifying endoscopic classification of the Japan Esophageal Society. Esophagus. 2017;14:105–12.
2. Katada C, Okamoto T, Ichinoe M, et al. Prediction of lymph-node metastasis and lymphatic invasion of superficial pharyngeal cancer on narrow band imaging with magnifying endoscopy. Auris Nasus Larynx. 2020;47:128–34.

Pharyngeal Superficial Cancer (0-IIb)

Keiichiro Nakajo and Tomonori Yano

Supplementary Information The online version contains supplementary material available at https://doi.org/10.1007/978-981-97-2732-2_18. The videos can be accessed individually by clicking the DOI link in the accompanying figure caption or by scanning this link with the SN More Media App.

Y. Sano et al. (eds.), *Atlas of Advanced Endoscopy*, https://doi.org/10.1007/978-981-97-2732-2_18

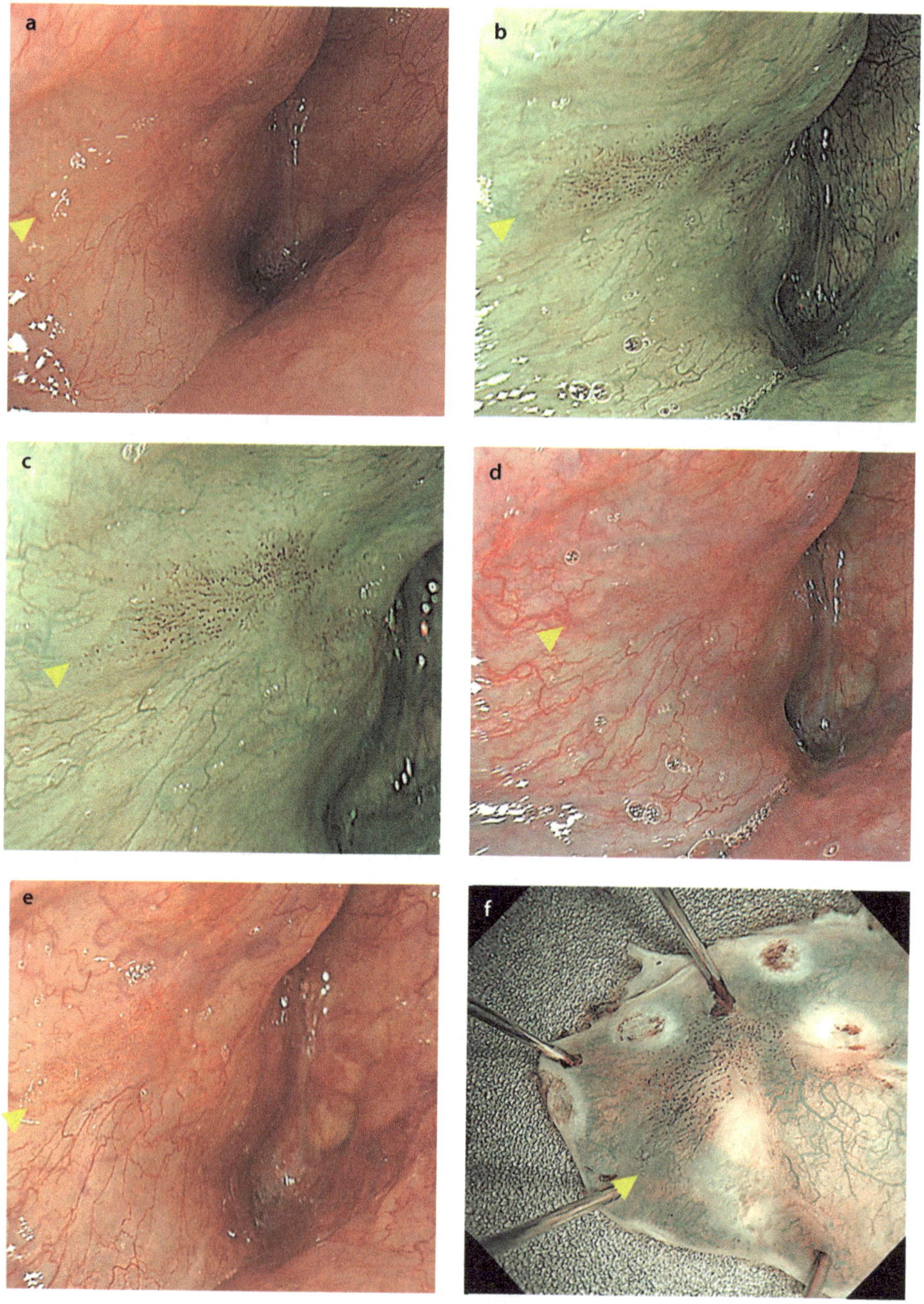
a
b
c
d
e
f

Pathological Diagnosis

Squamous cell carcinoma in situ: type 0-IIb, 7 × 4 mm, ly0, v0, pHM0, pVM0.

g

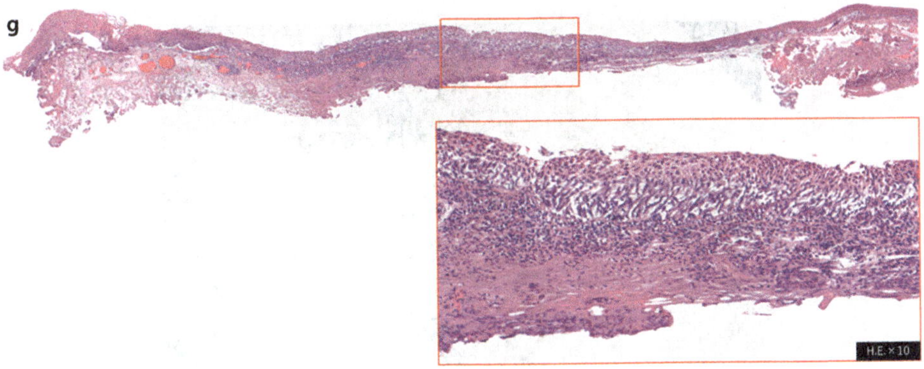

Explanation

A 7-mm flat lesion was detected in the right piriform sinus of the hypopharynx. By WLI, the lesion was recognized as a faint reddish-toned area, thus, confirming the entire lesion was challenging (Fig. a). Using NBI observation, it was depicted as a well-defined brownish area. Magnifying NBI revealed dot-like vessels of type B1 microvascular structures according to the Japan Esophageal Society's classification (Figs. b and c). Observation using TXI modes 1 (Fig. d) and 2 (Fig. e) emphasized the background color tones within the lesion and surrounding structures, making it easier to recognize the lesion compared to WLI. However, when compared to NBI, the recognition of blood vessels and structures was unclear. When iodine was sprayed during the treatment, the lesion exhibited clear nonstaining, and the lesion border was clear. An endoscopic laryngopharyngeal surgery was performed, resulting in en bloc resection of the lesion. Observing the resected specimen using NBI, type B1 vessels within the lesion were clearly identified (Fig. f). The histopathological diagnosis confirmed squamous cell carcinoma in situ (Fig. g).

Tips

Primarily, the detection of 0-IIb lesions using WLI relies on the observation of interrupted vascular patterns, loss of glossiness, and changes in color tone, which are important findings. However, detecting small 0-IIb lesions with endoscopy is extremely challenging. Conversely, diagnosis for 0-IIb using NBI observation on well-defined brownish areas within the lesion, known as the "brownish area," is the key diagnostic feature. In this case, it was easy to identify its presence using NBI. In routine screening endoscopic examinations, using NBI is recommended to identify small and flat lesions, such as 0-IIb.

▪ Video

A short video is available as ◘ Fig. 1.

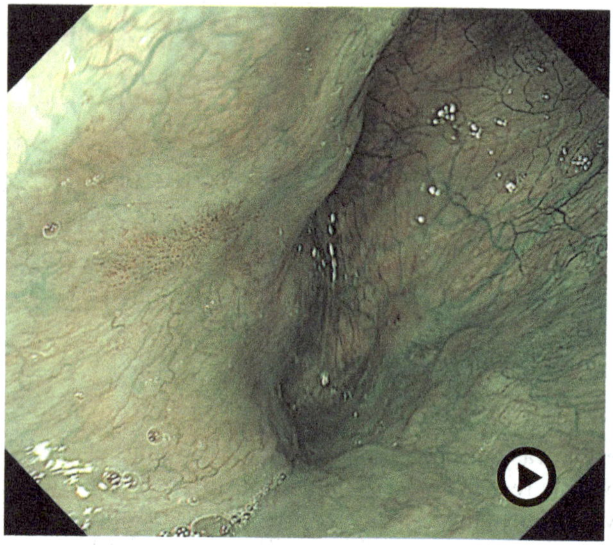

◘ **Fig. 1** Practical observation and diagnosis of superficial cancer (0-IIb) with magnifying TXI and NBI (▶ https://doi.org/10.1007/000-c7z)

Esophagus (Non-Barrett)

Contents

Reflux Esophagitis

Kasenee Tiankanon and Rapat Pittayanon

Y. Sano et al. (eds.), *Atlas of Advanced Endoscopy*, https://doi.org/10.1007/978-981-97-2732-2_19

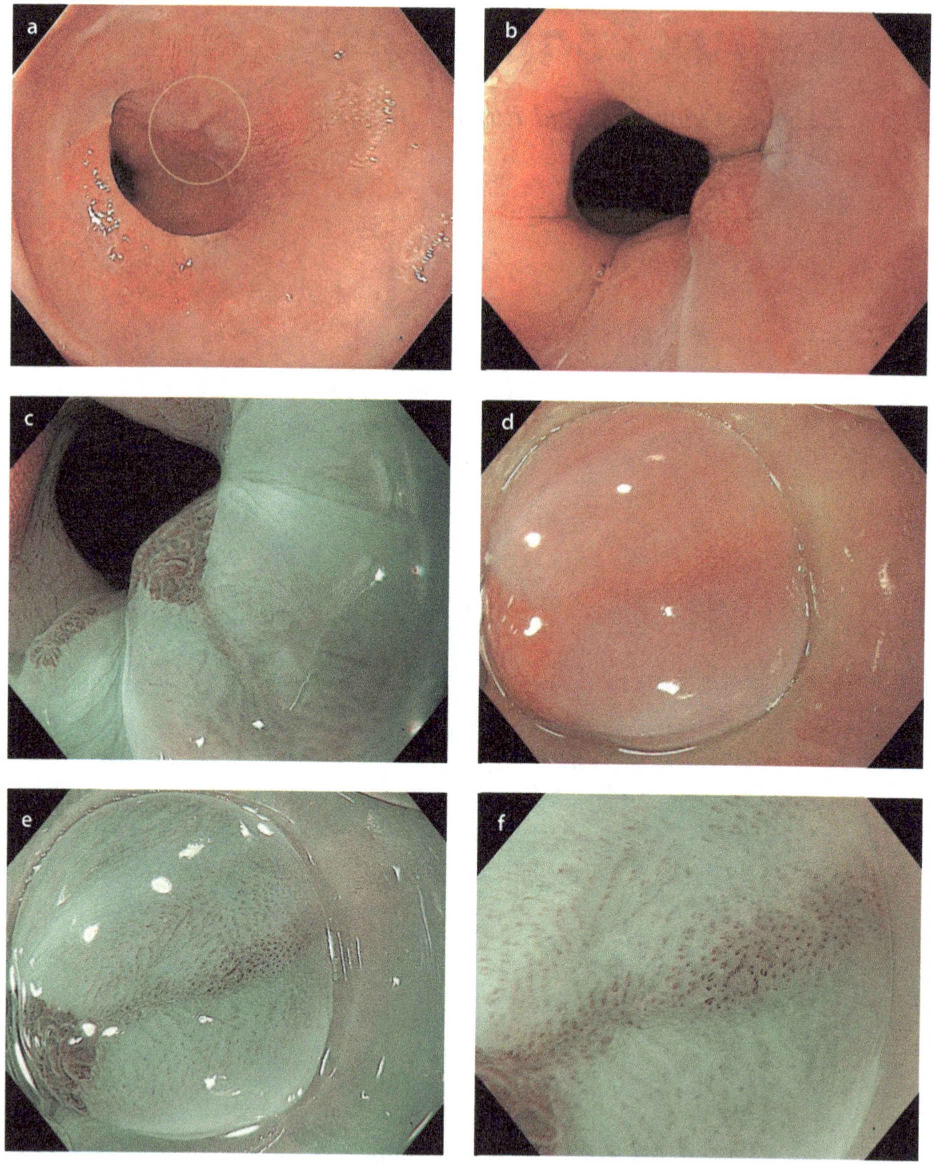

Pathological Diagnosis

Elongation of the lamina propria papillae and basal cell hyperplasia (Fig. g). Inflammatory cells in the squamous epithelium (Fig. h) composed of eosinophils (red arrow heads) and lymphocytes.

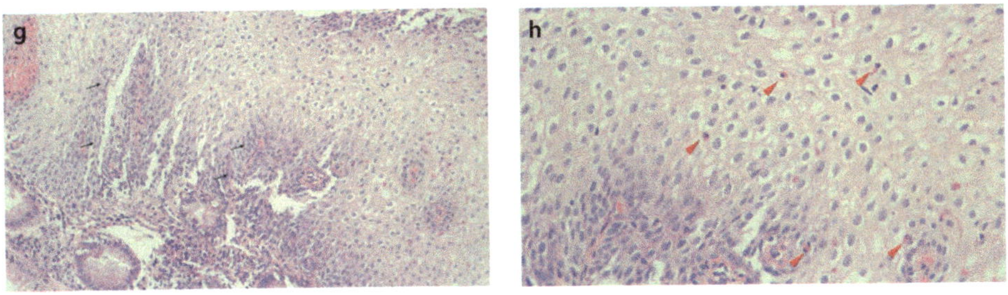

Explanation

A linear mucosal breakage at the squamocolumnar junction (SCJ) less than 5 mm in length that does not extend between the tops of two mucosal folds (Fig. a in yellow circle). Non-magnifying white light endoscopy showed punctate erythema proximal to the SCJ and ridge or villous-appearing mucosa (Fig. b). Non-magnifying NBI showed increased vascular markings proximal and distal to the SCJ, with regular microvasculature distal to the SCJ (Fig. c). When compared to nearby normal mucosa by using transparent cap, standard white light EGD showed increased vascular marking (Fig. d). When applied NBI with magnification, it revealed an increase in the number of intrapapillary capillary loops (IPCLs) (Fig. e). When switching to high magnification, white light EGD showed only linear erythema, but NBI could reveal a group of regularly dilated IPCL in a linear pattern extending from the mucosal breaking (Inoue's classification IPCL type III) and surrounded by smooth-running elongated tortuous IPCL without dilatation (Inoue's classification IPCL types I and II) indicative of inflammatory without dysplastic changes (Fig. f). Based on these findings, the lesion was compatible with the Los Angeles (LA) classification grade A reflux esophagitis and was treated with a proton-pump inhibitor. Histopathological examination indicated papillae elongation of the lamina propria and basal cell hyperplasia (Fig. g in arrows). Multiple inflammatory cells infiltrated the squamous epithelium (Fig. h), including eosinophils (red arrow heads) and lymphocytes.

Tips

Correctly classifying dilated and/or tortuous IPCLs necessitates familiarity and a learning curve. The non-magnified standard scope cannot be utilized since the zoom mode is required for optimal inspection of IPCLs. When utilizing magnification, the scope tip should be focused close to the mucosa, but not too close or too distant, in order to examine the lesion clearly. A transparent plastic cap attachment may help to maintain the correct distance between the scope tip and the mucosa.

The advancement of NBI with magnification endoscopic assessment of reflux esophagitis has a high correlation with histology results. The physical biopsy is unnecessary when optical diagnosis confidently establishes non-dysplastic regions.

Acknowledgments The authors would like to thank Dr. Anapat Sanpavat, Department of Pathology, Faculty of Medicine, Chulalongkorn University and King Chulalongkorn Memorial Hospital, The Thai Red Cross Society, Bangkok, Thailand, for providing the pictures and description of pathology.

Esophageal Papilloma

Shiaw-Hooi Ho and Seiichiro Abe

Contents

Supplementary Information The online version contains supplementary material available at https://doi.org/10.1007/978-981-97-2732-2_20. The videos can be accessed individually by clicking the DOI link in the accompanying figure caption or by scanning this link with the SN More Media App.

Y. Sano et al. (eds.), *Atlas of Advanced Endoscopy*, https://doi.org/10.1007/978-981-97-2732-2_20

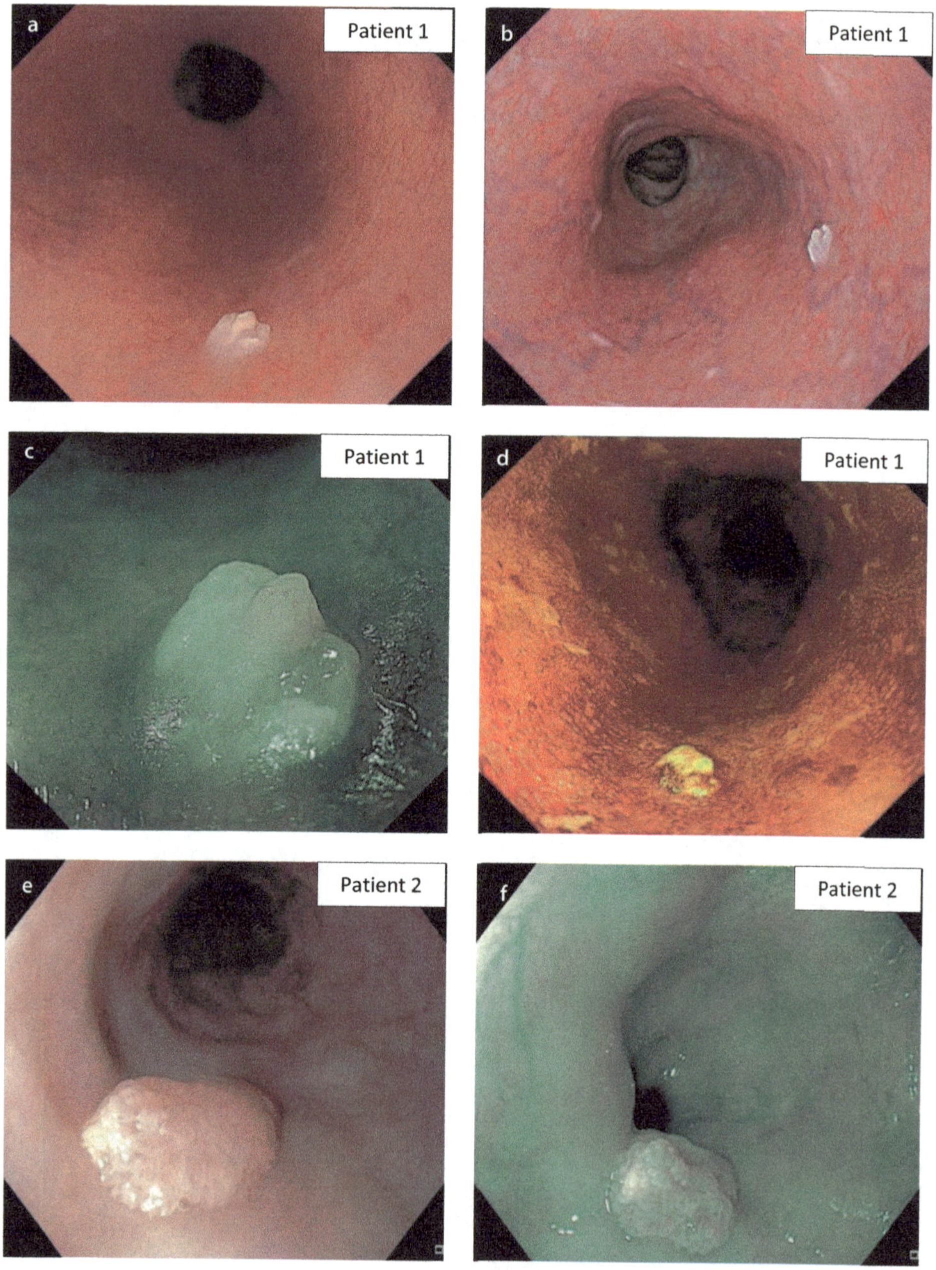
a
Patient 1
b
Patient 1
c
Patient 1
d
Patient 1
e
Patient 2
f
Patient 2

■ Pathological Diagnosis

Benign esophageal squamous papilloma.

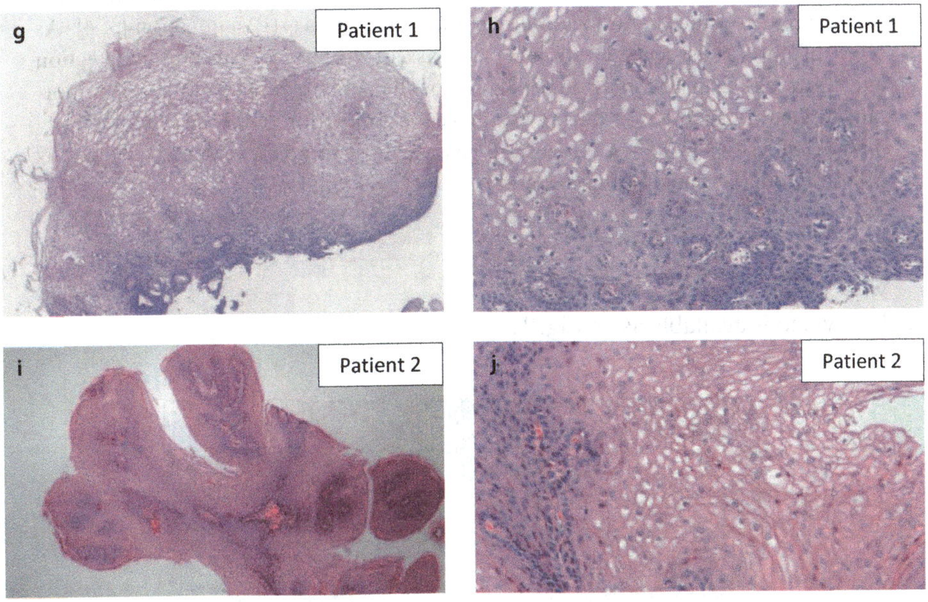

■ Explanation

Shown here were two esophageal squamous papillomas (ESP) of varying sizes detected at the level of the mid and lower esophagus in two different patients. The first lesion (Figs. a–d, g, and h) was about 2 mm, while the second lesion (Figs. e, f, i, and j) was about 8 mm in size. Both lesions exhibited the typical feature of whitish exophytic growth with wartlike projections under WLI (Figs. a and e). Enhancement of color and texture was observed in TXI mode 1 (Fig. b). Magnifying NBI revealed more clearly the dilated and elongated microvessels in the projections (Fig. f). However, such observation under magnifying NBI was not seen in the smaller ESP (Fig. c). This lesion remains unstained under Lugol's chromoendoscopy (Fig. d). These findings were highly suggestive of a diagnosis of ESP. Due to paucity in the literature evidence, there is no standard recommendation for the resection of this lesion. In patients with concurrent esophageal symptoms or in patients with concern of malignancy, ESP can be effectively removed with cold polypectomy technique or other advanced techniques as well [1, 2]. Histologically, the wartlike projections consisted of matured acanthotic squamous epithelium overlying the branched-like fibrovascular stalks (Figs. g and i). Under high-level magnification, sections showed parakeratosis and thickened stratum corneum. The upper-level epithelium showed koilocytic cells with clear cytoplasm (Figs. h and j).

Tips

ESP is a rare benign growth with uncertain etiologies. Two proposed causative factors are reflux-induced and local human papilloma virus (HPV) infection [3, 4]. As HPV infection is associated with increased risk of esophageal cancer, the detection and resection of an ESP may be beneficial for the patient [5]. Therefore, it is important to identify such lesion during esophageal examination. Other morphologically mimicking lesions in the esophagus are glycogenic acanthosis, verrucous squamous cell carcinoma, granulation tissue, and papillary leukoplakia.

▪ Video

A short video is available as Fig. 1.

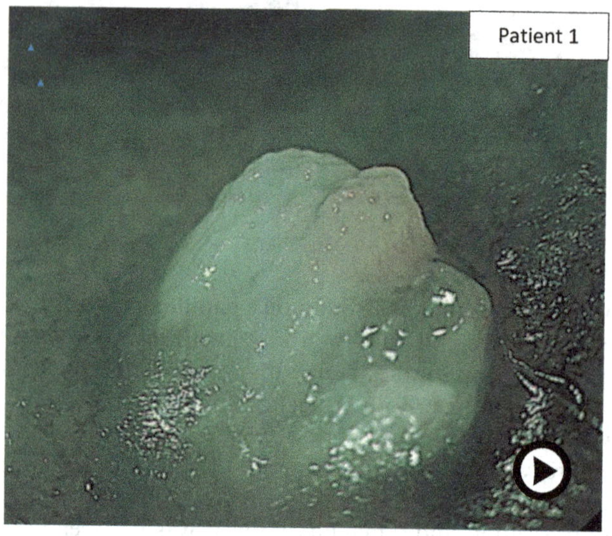

Fig. 1 Observation and diagnosis of an esophageal squamous papilloma using various image-enhanced endoscopy modalities, namely, TXI modes, NBI mode, magnifying NBI mode, and Lugol's chromoendoscopy followed by endoscopic resection using the cold biopsy forceps polypectomy technique in patient 1 (▶ https://doi.org/10.1007/000-c80)

Acknowledgments Dr. Shigeki Sekine, Department of Diagnostic Pathology, National Cancer Center Hospital, Tokyo, Japan.

Dr. Alex Hwong Ruey Leow, Pantai Hospital KL, Kuala Lumpur, Malaysia.

Dr. Bang Rom Lee, Pantai Hospital KL, Kuala Lumpur, Malaysia.

References

1. Attila T, Fu A, Gopinath N, Streutker CJ, Marcon NE. Esophageal papillomatosis complicated by squamous cell carcinoma. Can J Gastroenterol. 2009;23(6):415–9.
2. Benisch BM, Mantell C. Multiple squamous papillomas of the esophagus. Arch Otolaryngol. 1974;100(5):379.
3. Carr NJ, Monihan JM, Sobin LH. Squamous cell papilloma of the esophagus: a clinicopathologic and follow-up study of 25 cases. Am J Gastroenterol. 1994;89(2):245–8.
4. Szanto I, Szentirmay Z, Banai J, Nagy P, Gonda G, Voros A, et al. [Squamous papilloma of the esophagus. Clinical and pathological observations based on 172 papillomas in 155 patients]. Orv Hetil. 2005;146(12):547–52.
5. Syrjanen KJ. HPV infections and oesophageal cancer. J Clin Pathol. 2002;55(10):721–8.

Multiple Lugol-Voiding Lesions

Daisuke Kikuchi

Y. Sano et al. (eds.), *Atlas of Advanced Endoscopy*, https://doi.org/10.1007/978-981-97-2732-2_21

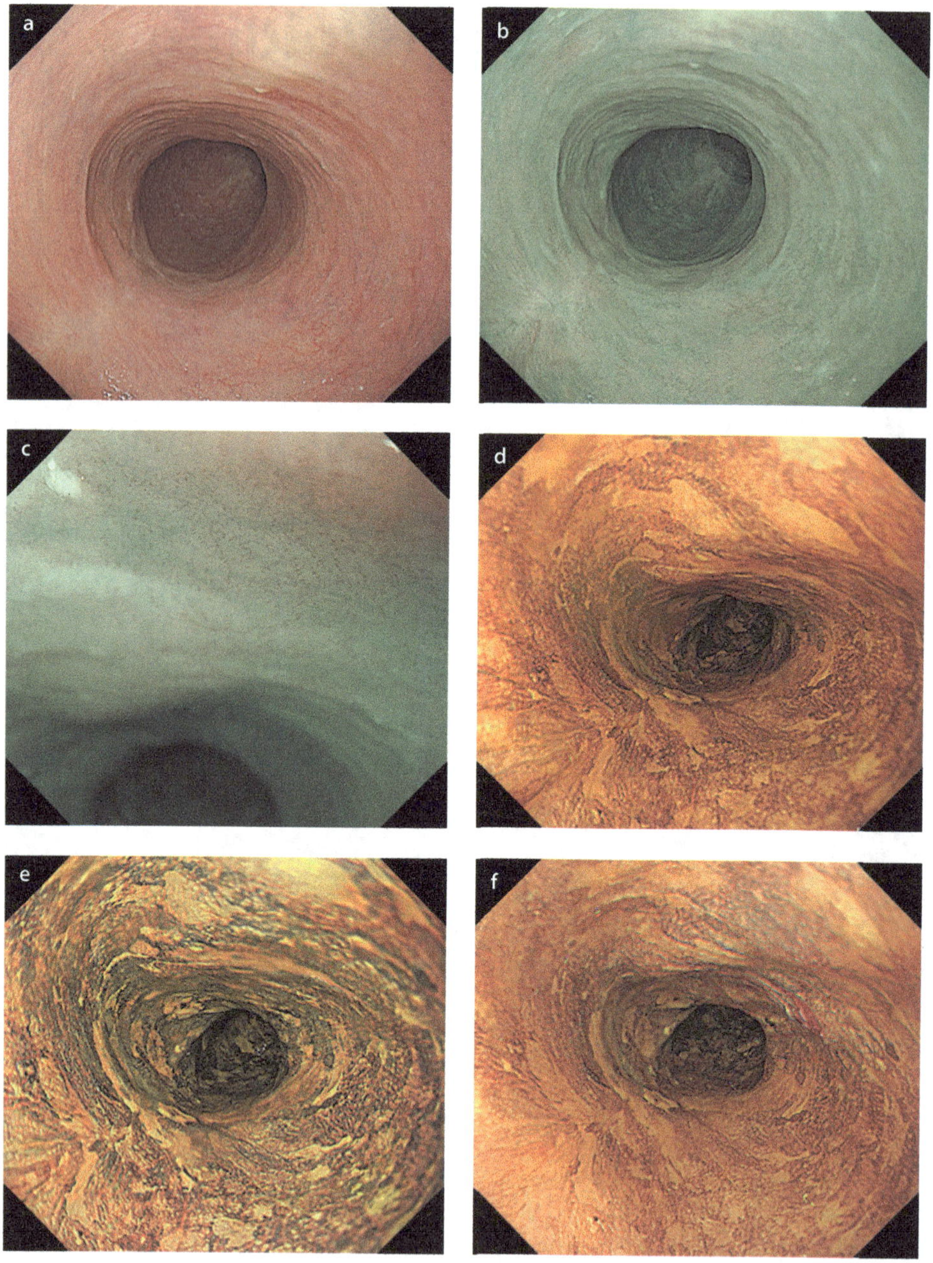

Pathological Diagnosis

Mild inflammatory lymphocytic infiltration was observed, and no neoplastic changes were observed. The final pathological diagnosis was esophagitis (Fig. g).

g

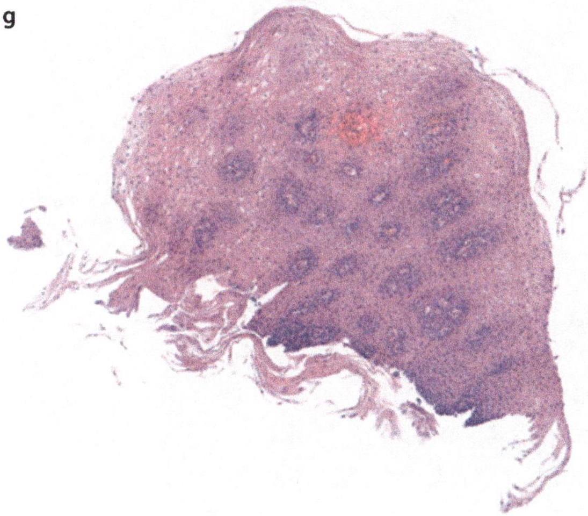

Explanation

In this series of endoscopic images, we examine a lesion in the esophagus during the surveillance endoscopy after ESD. A non-magnifying WLI image revealed an ESD scar on the posterior wall of the middle esophagus, along with a slightly reddish and depressed area on the anterior wall (Fig. a). A non-magnifying NBI image highlighted the reddish and depressed lesion as a brownish area (Fig. b). Switching to a magnifying NBI image (Fig. c), slightly dilated IPCL with relatively low vascular density led to a diagnosis of inflammation or low-grade intraepithelial neoplasia. A chromoendoscopy image, with 0.8% iodine staining, revealed multiple iodine-unstained areas, all of which were negative for the pink color sign (Fig. d). Many times, iodine-unstained lesions can only be detected using chromoendoscopy, and they may not be visible with WLI or NBI. A chromoendoscopy image with TXI offered enhanced visibility of iodine-unstained areas compared to WLI (Fig. e). A biopsy was performed for the largest iodine-unstained area on the anterior wall, and the pathological result showed esophagitis (Fig. f).

Tips

Squamous cell carcinomas, such as pharyngeal and esophageal cancer, have a high incidence of metachronous carcinogenesis. Therefore, surveillance endoscopy should be conducted with care. Lesions that cannot be detected by WLI or NBI are sometimes identified through iodine staining. Lesions larger than 10 mm in size or irregularly shaped are indicators of a potential cancerous lesion. The pink color sign is also useful in distinguishing between cancerous and noncancerous conditions.

High-Grade Squamous Dysplasia of the Esophagus (1)

Wen-Lung Wang and Ching-Tai Lee

Supplementary Information The online version contains supplementary material available at https://doi.org/10.1007/978-981-97-2732-2_22. The videos can be accessed individually by clicking the DOI link in the accompanying figure caption or by scanning this link with the SN More Media App.

Y. Sano et al. (eds.), *Atlas of Advanced Endoscopy*, https://doi.org/10.1007/978-981-97-2732-2_22

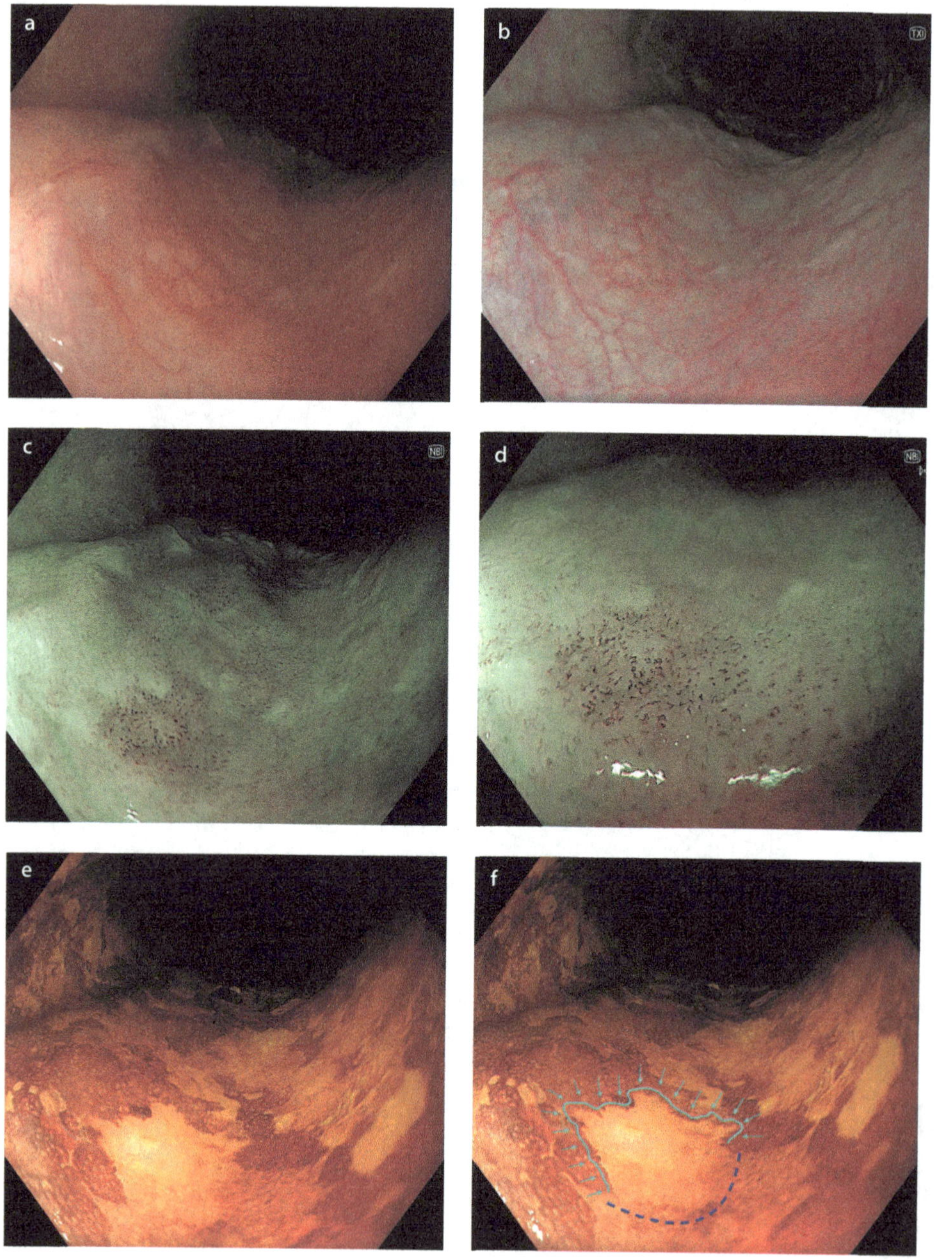
a
b
TXI
c
NBI
d
NBI
e
f

Pathological Diagnosis

High-grade dysplasia of squamous epithelial cells. No stromal invasion. Resection margins are free of dysplastic cell involvement.

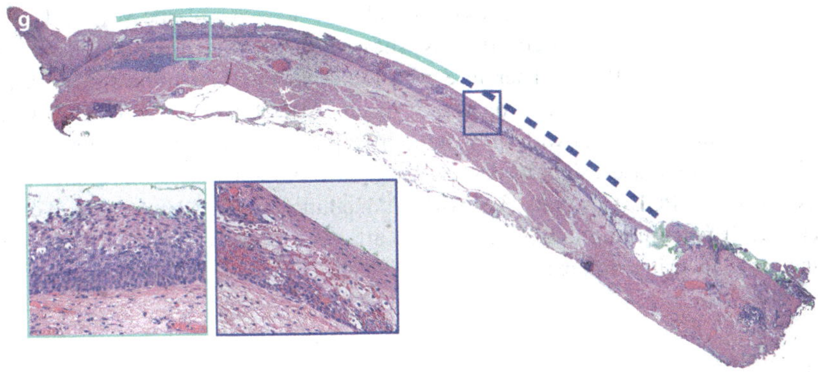

Explanation

Olympus X1 system with GIF-EZ1500 endoscope was used for examination. In WLI observation (Fig. a), a 5-mm IIb lesion with ill-defined margin and minute discoloration of the esophageal squamous mucosa over the middle esophagus could be seen. When switching to TXI mode (Fig. b), some abnormal dilated vascular structures (e.g., IPCL) within the discoloration area were further enhanced. However, the margin of the lesion was still hard to be identified. In non-magnifying NBI observation (Fig. c), the lesion could be recognized as a well-defined brownish area. The presence of abnormal dilated neoplastic microvessels (e.g., IPCL) was also evidenced. When using the NBI mode with dual-focus (DF) magnification (Fig. d), assessment of the intra-papillary capillary loops (IPCLs) was possible. Here, the type B1 IPCLs with loop structure and small avascular area (AVA) were recognized. Chromoendoscopy for the suspicious lesion with 1% Lugol's iodine spray in WLI mode (Figs. e and f) reveals an unstained area. The lesion had clear demarcation over left side (shown as light blue solid line with arrow) and relative unclear margin with partial staining over right side (shown as dark blue dotted line) after Lugol's iodine spray. Equivocal pink color change of the mucosa was recognized over left side of the lesion. High-grade squamous dysplasia was suspected according the all these features. ESD was performed for en bloc resection and detailed pathological analysis. Histopathological findings of the specimen after ESD (Fig. g) show high-grade dysplasia of squamous epithelial cell over left side of resected specimen (shown as light blue solid line) and low-grade dysplasia of squamous epithelial cell over right side (shown as dark blue dotted line). There was no stromal invasion. Resection margins are free of dysplastic cell involvement.

Tips

Small-size LGIN/HGIN of esophageal squamous epithelium is difficult to be detected under WLI due to completed flat shape with only minute (but easily ignored) color change of background mucosa and subtle dilated neoplastic vessels. Under NBI with magnification, LGIN/HGIN shows slightly increased thickness and dilated, loop-shaped IPCLs. Therefore, we should pay more attention to any suspicious brownish area with small dark spot inside. After Lugol spray, LGIN/HGIN may reveal unstained area with only diminutive pink color change. Margin may be ill-defined. In this case, using the new NBI and TXI (mode 1), the subtle dilated neoplastic vessels (IPCLs) are seen, even without magnification. After switching to dual-focus (DF) magnification, type B1 IPCLs are easy to be identified without any difficulty in focusing with extended depth of field (EDOF) technology.

Video

A short video is available as Fig. 1.

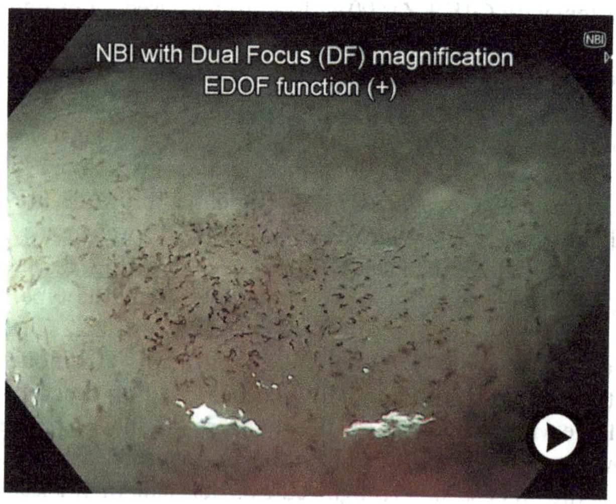

Fig. 1 Practical observation and diagnosis of small esophageal squamous high-grade dysplasia with WLI, TXI, magnifying NBI, and chromoendoscopy (▶ https://doi.org/10.1007/000-c81)

High-Grade Squamous Dysplasia of the Esophagus (2)

Hon Chi Yip and Ni Yunbi

Contents

Supplementary Information The online version contains supplementary material available at https://doi.org/10.1007/978-981-97-2732-2_23. The videos can be accessed individually by clicking the DOI link in the accompanying figure caption or by scanning this link with the SN More Media App.

Y. Sano et al. (eds.), *Atlas of Advanced Endoscopy*, https://doi.org/10.1007/978-981-97-2732-2_23

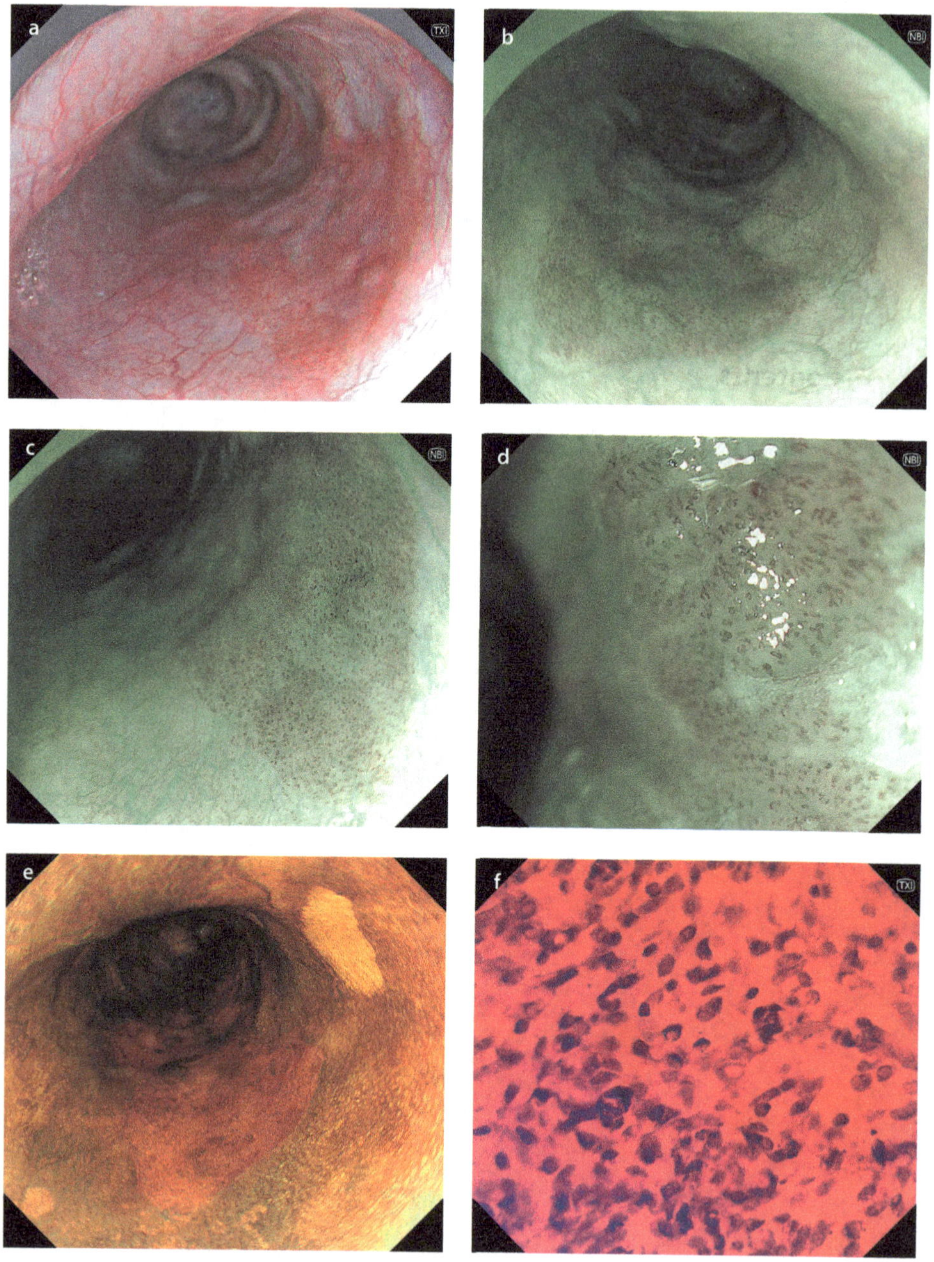
a
TXI
b
NBI
c
NBI
d
NBI
e
f
TXI

▪ Pathological Diagnosis

High-grade squamous dysplasia

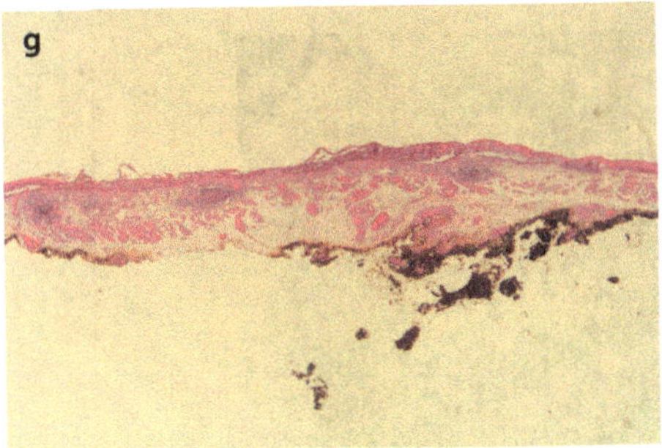

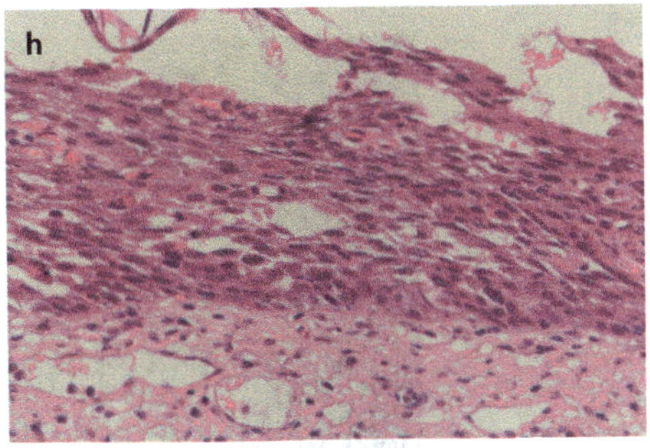

▪ Explanation

A demarcated flat lesion of approximately 2.5 cm, occupying one-third circumference of the middle esophagus, was detected. On TXI (mode 1), the lesion was observed to be erythematous with prominent capillaries on the surface (Fig. a). The demarcation line was even more obvious on non-magnified NBI image, with the lesion appearing as brownish discolored area with increase in density of dilated brownish "dots" representing intra-papillary capillary loop (IPCL) (Figs. b and c). On high-magnification NBI, the morphology of IPCL was observed to be dilated, tortuous with various calibers and shapes (Fig. d). The irregular IPCL remained its loop-like structures and was diagnosed as JES B1 vessels. Chromoendoscopy with 0.5% Lugol's iodine demonstrated well-demarcated voiding area corresponding to the demarcation seen on NBI image (Fig. e). No definite pink color sign was observed. In vivo endocytoscopic examination was performed with ultrahigh-magnification endoscope (GIF-H290EC, Olympus Medical Corporations, Tokyo, Japan) after CM double staining (Crystal violet and Methylene blue) plus TXI (mode 1). Increased cellularity with irregular shaped nuclei and high NC ratio could be observed, suggestive of high-grade neoplastic lesion, corresponding to ECA3 based on classification system proposed by Shimamura et al. [1] (Fig. f). The overall endoscopic diagnosis was an early esophageal squamous cell carcinoma, predicted depth of invasion at T1a-EP/LPM (video). The lesion was resected en bloc using ESD technique with clip-line-assisted traction method. Histopathological examination showed high-grade squamous dysplasia with clear horizontal and vertical resection margins (Figs. g and h).

Tips

Superficial esophageal squamous cell carcinoma frequently only has subtle color and surface changes on conventional WLI. Detection of suspicious lesion could be enhanced with the systematic use of NBI. It has been reported that reducing lumen distension by changing the amount of gas within the esophagus could improve visibility of early flat esophageal squamous neoplastic lesions. The IPCL pattern would require examining the esophageal mucosa under magnification. Anchoring the tip of the endoscope with soft black hood to the adjacent normal mucosa while adjusting the magnification and scope angulation would help to obtain a clear image of the IPCL pattern.

Video

A short video is available as ◻ Fig. 1.

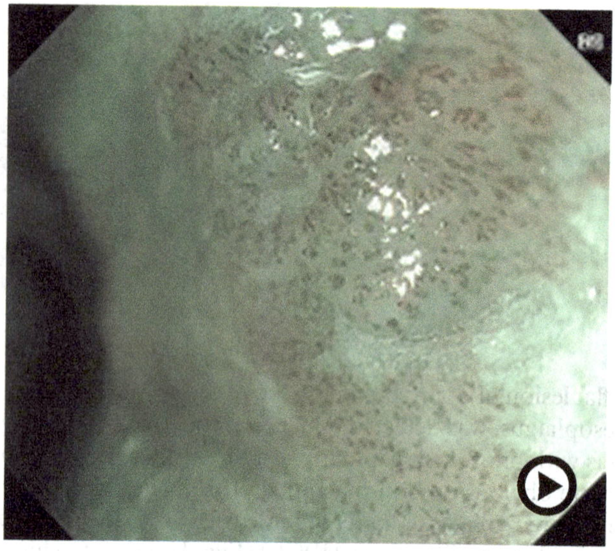

◻ **Fig. 1** Practical observation and diagnosis of high-grade squamous dysplasia of the esophagus with NBI, TXI, and ultrahigh magnification (▶ https://doi.org/10.1007/000-c82)

Reference

1. Shimamura Y, Inoue H, Rodriguez de Santiago E, Abad MRA, Fujiyoshi Y, Toshimori A, et al. Diagnostic yield of fourth-generation endocytoscopy for esophageal squamous lesions using a modified endocytoscopic classification. Dig Endosc. 2021;33(7):1093–100. https://doi.org/10.1111/den.13914.

Esophageal Squamous Cell Carcinoma (0-I)

Noriko Matsuura and Noriya Uedo

Supplementary Information The online version contains supplementary material available at https://doi.org/10.1007/978-981-97-2732-2_24. The videos can be accessed individually by clicking the DOI link in the accompanying figure caption or by scanning this link with the SN More Media App.

Y. Sano et al. (eds.), *Atlas of Advanced Endoscopy*, https://doi.org/10.1007/978-981-97-2732-2_24

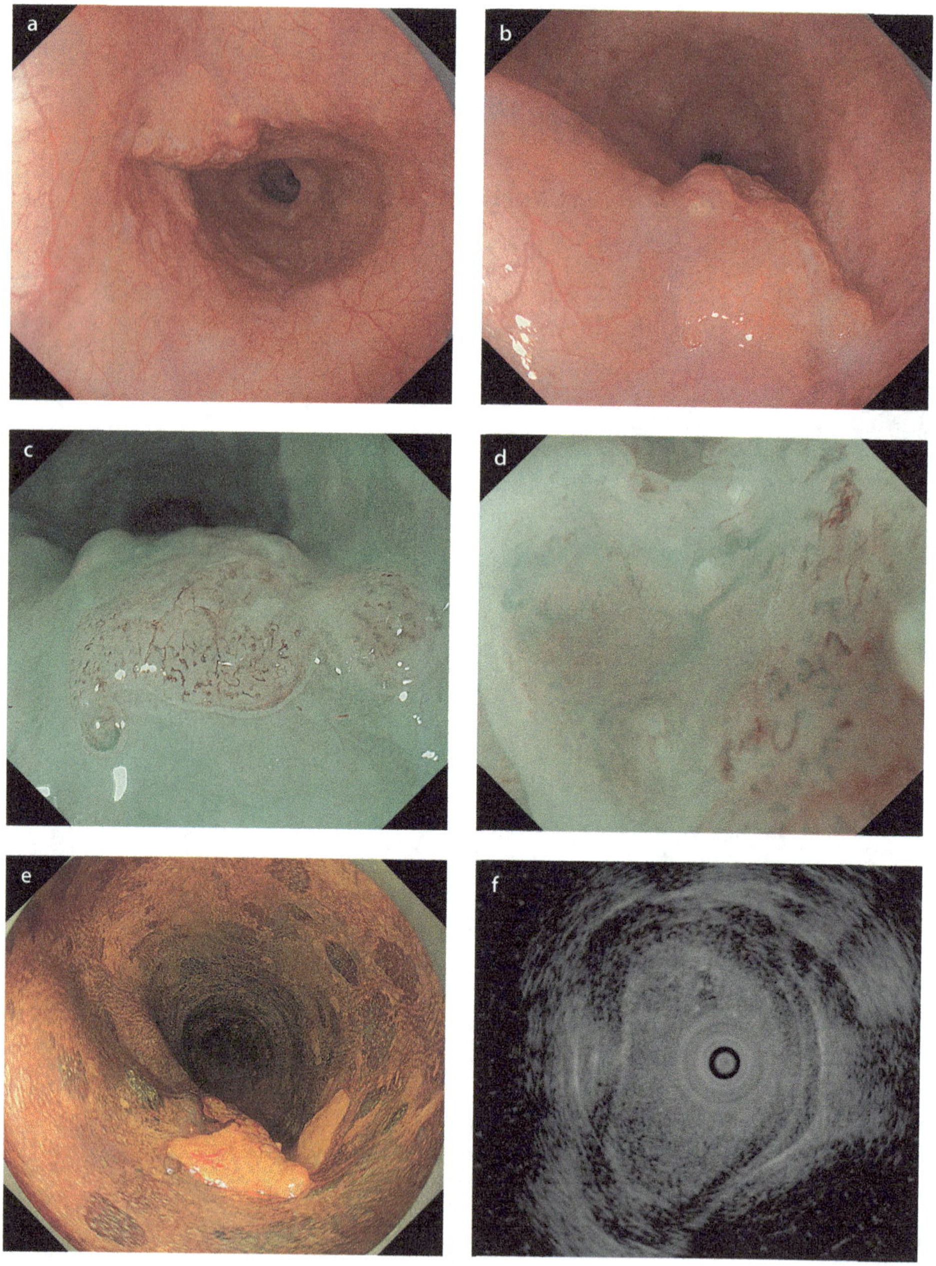
a
b
c
d
e
f

▪ Pathological Diagnosis

Squamous cell carcinoma, pT1b-SM3

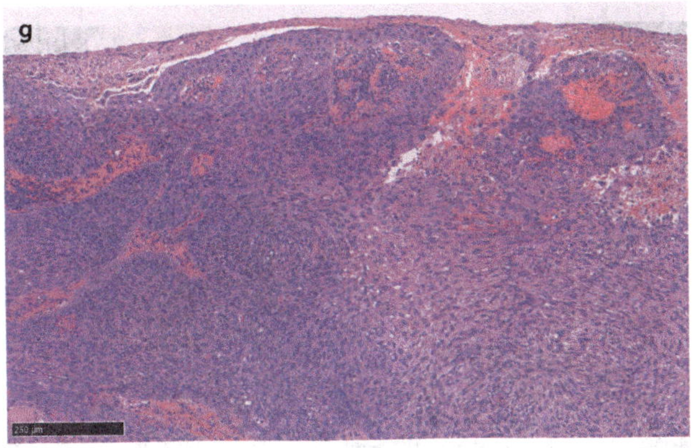

▪ Explanation

A 20-mm erythematous protrusion was detected in the middle thoracic esophagus (Fig. a). The high protrusion exhibited hardness. Submucosal tumor (SMT)-like protrusion was observed on the anterior wall side on white-light imaging (WLI) (Fig. b). Magnifying narrowband imaging (ME-NBI) revealed Japan Esophageal Society (JES) classification type B1 and B2 vessels on the oral side of the lesion (Fig. c), as well as type B3 vessels in the anterior wall (Fig. d). B3 vessels often appear green in color. Iodine staining showed an unstained area with a pink color sign (Fig. e). Endoscopic ultrasonography (EUS) revealed a low-echoic area invading the submucosal layer (Fig. f). The biopsy showed squamous cell carcinoma; based on these findings, the lesion was diagnosed as cT1b-SM2 squamous cell carcinoma. The patient underwent a subtotal esophagectomy with lymph node dissection. Pathological examination of the resected specimen revealed a squamous cell carcinoma invading the submucosal layer (SM3) and lymphatic vessels, with lymph node metastasis. The resected lesion measured 17 mm in diameter (Fig. g).

Tips

Diagnosis of superficial esophageal cancer invasion depth is based on a combination of WLE and ME-NBI. On WLE, SMT-like elevations and deep depressions predict clinical SM2 or deeper invasion. On ME-NBI, B3 vessels predict clinical SM2 or deeper invasion. Although ME-NBI is useful, care should be exercised when the superficial cancer is shallow and the deep submucosal invasion cancer is not continuous with superficial carcinoma. Evaluating the invasion depth only with ME-NBI is impossible; therefore, a combination of WLE and ME-NBI is required to diagnose the cancer invasion depth. Chromoendoscopy with Iodine staining is useful for evaluating the lesion range but cannot accurately diagnose cancer invasion depth. EUS allows tomographic evaluation of the gastrointestinal wall. However, caution is required as EUS often leads to overdiagnosis for cancer invasion depth. Ultimately, endoscopic resection is considered for cancers suspected of clinical EP to clinical SM1, while surgery and chemoradiation therapy are considered for cancers suspected deeper than clinical SM2.

▪ Video

A short video is available as ◘ Fig. 1.

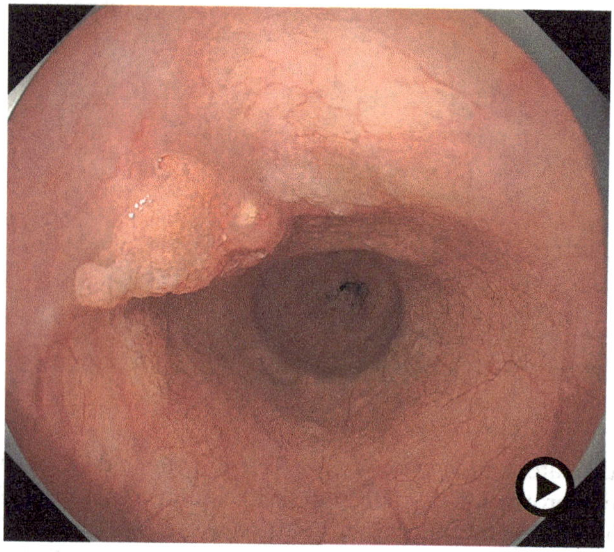

◘ **Fig. 1** Practical observation and diagnosis of superficial esophageal cancer with WLI NBI, iodine staining, and EUS (▶ https://doi.org/10.1007/000-c83)

Esophageal Squamous Cell Carcinoma (0-IIa)

Yuto Shimamura and Haruhiro Inoue

Supplementary Information The online version contains supplementary material available at https://doi.org/10.1007/978-981-97-2732-2_25. The videos can be accessed individually by clicking the DOI link in the accompanying figure caption or by scanning this link with the SN More Media App.

Y. Sano et al. (eds.), *Atlas of Advanced Endoscopy*, https://doi.org/10.1007/978-981-97-2732-2_25

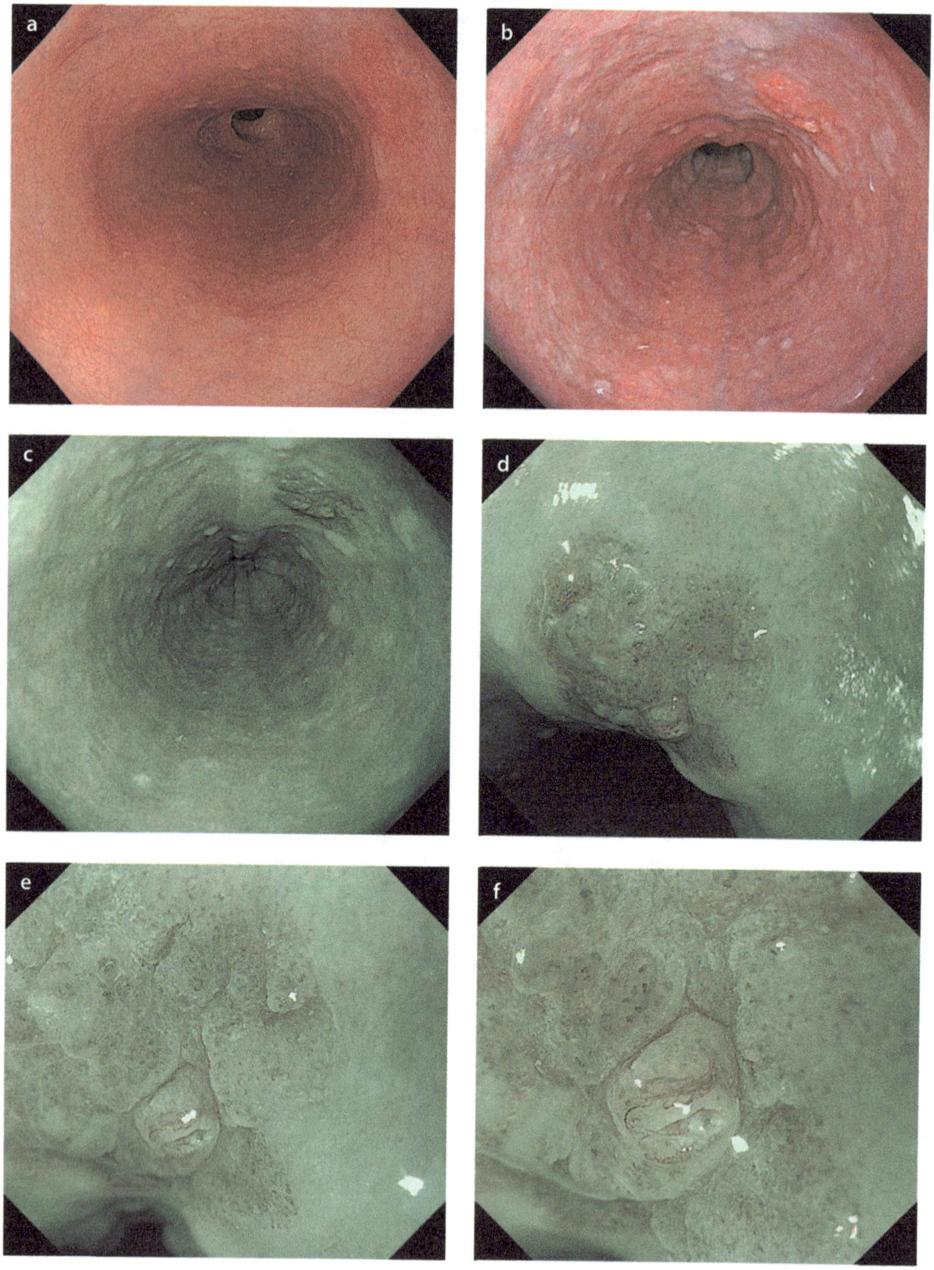
a
b
c
d
e
f

■ Pathological Diagnosis

Esophageal squamous cell carcinoma, staged as pT1a (invasion limited to muscularis mucosae), demonstrating clear negative margins both horizontally and vertically, and an absence of lympho-vascular invasion.

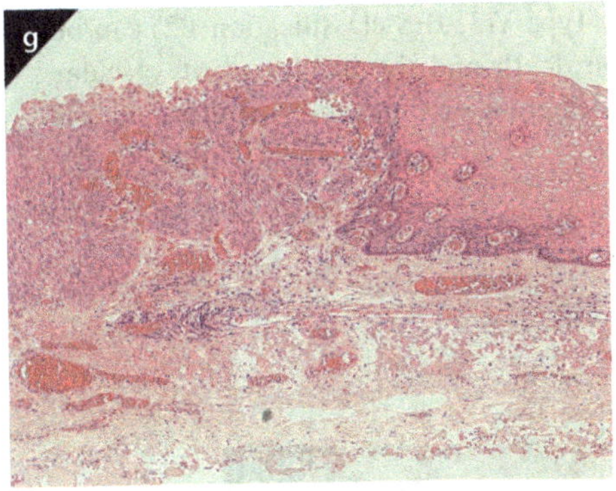

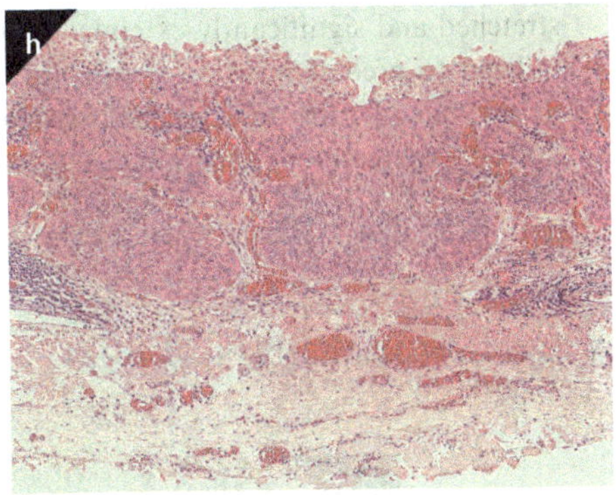

■ Explanation

A slightly raised lesion measuring 15 mm in size exhibits a reddish hue in contrast to the color tone of the surrounding normal esophageal mucosa when observed under WLI (Fig. a). Within the lesion, observable interruptions can be noted in the vasculature. When utilizing TXI mode, images are optimized to accentuate texture and color variations (Fig. b). With non-magnifying NBI observation, the lesion presents itself as a well-defined brownish area (Fig. c). Under low-magnification NBI, the intrapapillary capillary loops (IPCL) manifest as dot-like vessels (Fig. d). High-magnification NBI enables a more intricate assessment of IPCL, such as dilation, tortuosity, variations in vessel caliber, and nonuniformity (IPCL Classification Type V-1, Japanese esophageal society [JES] classification B1) (Fig. e). Additionally, certain areas show an avascular area (AVA) along with elongated microvessels that appear stretched and markedly extended (IPCL Classification Type V-3, JES classification B2) (Fig. f). These collective observations strongly suggest the presence of superficial squamous cell carcinoma (SCC) limited to invasion within the muscularis mucosa. The lesion was removed en bloc using the ESD technique. Subsequent histopathological evaluation confirmed the diagnosis of SCC (Figs. g and h).

Tips

Improving the quality of observation is vital for detecting lesions and assisting in the qualitative diagnosis of esophageal SCC. Detection commonly relies on NBI, where lesions manifest as distinctive brownish areas. Furthermore, WLI and TXI facilitate the differentiation of clear color contrasts between the lesion and the surrounding normal mucosa. It also aids in identifying vascular interruptions within the lesion. While most superficial SCC are categorized as type 0-II, submucosal carcinomas exhibit various types, including 0-I, 0-III, or predominantly mixed type 0-IIc. Tumors with other atypical histological characteristics demonstrate diverse growth patterns.

In the context of magnifying NBI, a detailed assessment of IPCL irregularities is valuable. This assessment enables the recognition of four IPCL vascular irregularities: dilation, tortuosity, caliber changes, and nonuniformity (IPCL Type V-1, JES classification B1). Furthermore, identifying AVA and elongated microvessels that are stretched and significantly extended (IPCL Type V-3, JES classification B2) can be utilized to predict invasion depth. Invasion depth and the likelihood of vascular invasion and lymph node metastasis are intricately interconnected, forming the basis for making treatment decisions.

▪ Video

A short video is available as Fig. 1.

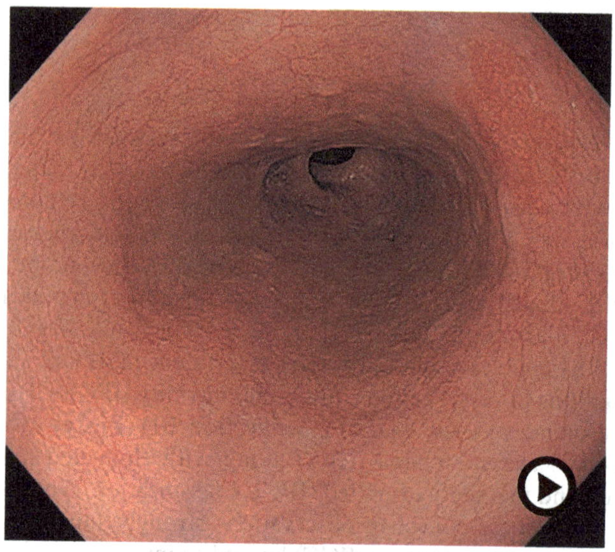

Fig. 1 Practical observation and diagnosis of superficial esophageal cancer (0-IIa, M) with WLI, TXI, and magnifying NBI (▶ https://doi.org/10.1007/000-c84)

Esophageal Squamous Cell Carcinoma (0-IIc) (1)

Daisuke Kikuchi and Philip Chiu

Y. Sano et al. (eds.), *Atlas of Advanced Endoscopy*, https://doi.org/10.1007/978-981-97-2732-2_26

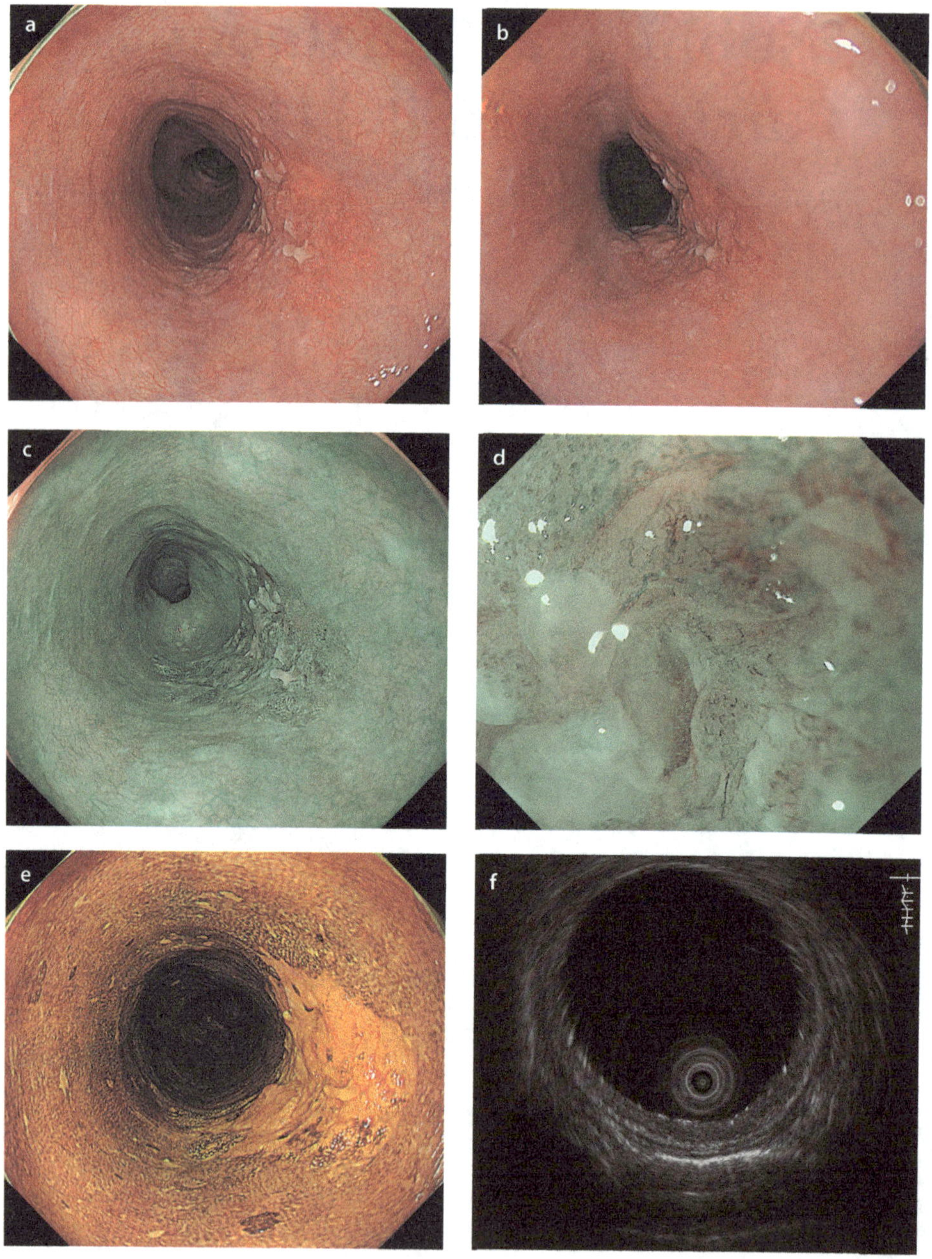
a
b
c
d
e
f

▪ Pathological Diagnosis

The patient was treated by esophagectomy. The final pathological diagnosis was moderately differentiated squamous cell carcinoma, pT1b (SM3), ly1, v1, pPM0, pDM0, and pRM0 (Figs. g and h). The lymph node metastases (3/56) were pathologically detected in the mediastinum.

g

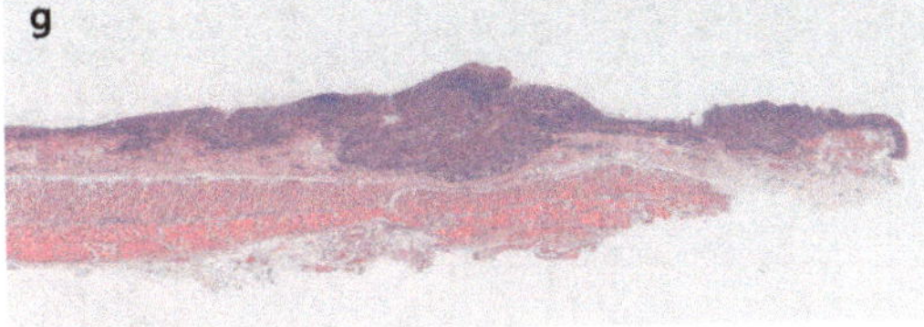

h

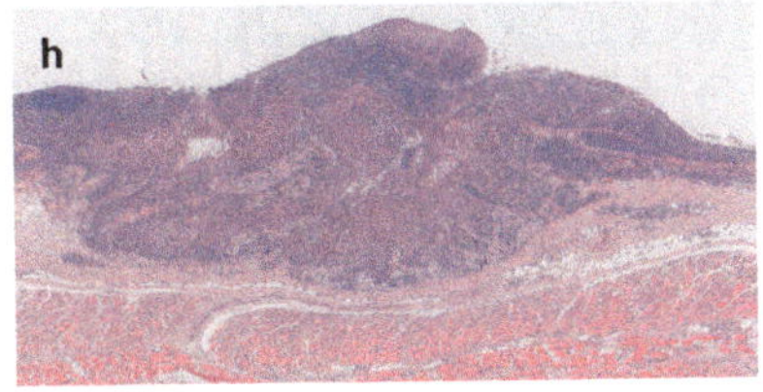

▪ Explanation

A reddish and uneven depressed lesion was detected on the right side of the middle esophagus by non-magnifying WLI (Fig. a). After deflation of the air, uneven component at the center of the lesion became more distinct (Fig. b). Based on the WLI image, this lesion was diagnosed as SM2. This lesion appeared as a brownish area in non-magnifying NBI (Fig. c). Utilizing magnifying NBI image, dilated irregular microvessels were identified at the center of the lesion. According to the Japan Esophageal Society's classification, these microvessels were categorized as type B3 vessel, confirming a diagnosis of SM2 (Fig. d). After applying 0.8% iodine staining, the lesion was identified as an iodine-unstained area, and the pink color sign was positive for this lesion (Fig. e). Using a miniature probe 20 MHz, lesion was recognized as a hypo-echoic area. The third layer, corresponding to the muscularis mucosae, was disrupted, and the fifth layer, corresponding to the deeper layer of the submucosa, appeared thin. Based on the EUS finding, this lesion was diagnosed as SM2 (Fig. f).

Tips

The invasion depth of superficial esophageal cancer should be comprehensively diagnosed using three modalities: WLI, NBI, and EUS. The diagnosis begins with WLI. During this phase, it is crucial to consider whether the lesion is affected by changes in air volume or peristalsis. Next, NBI magnification is performed. Irregular microvessels exhibiting dilatation, tortuosity, irregular caliber, and various shapes are categorized as type B vessels. Type B1 vessels display a loop-like formation, while type B2 vessels do not exhibit a loop-like structure. Type B3 vessels are characterized by highly dilated abnormal vessels with a caliber more than three times that of typical B2 vessels. The presence of type B3 vessels within the lesion predicts a histological invasion depth of SM2. Finally, EUS is also a valuable tool. Although a multicenter trial (JCOG1604) did not conclusively establish the routine use of EUS for diagnosing the depth of invasion in superficial esophageal cancer, it is considered beneficial for detecting SM2 cancers and providing an objective diagnosis of SM2 lesions.

Esophageal Squamous Cell Carcinoma (0-IIc) (2)

Stefan Seewald and Tiing Leong Ang

Supplementary Information The online version contains supplementary material available at https://doi.org/10.1007/978-981-97-2732-2_27. The videos can be accessed individually by clicking the DOI link in the accompanying figure caption or by scanning this link with the SN More Media App.

Y. Sano et al. (eds.), *Atlas of Advanced Endoscopy*, https://doi.org/10.1007/978-981-97-2732-2_27

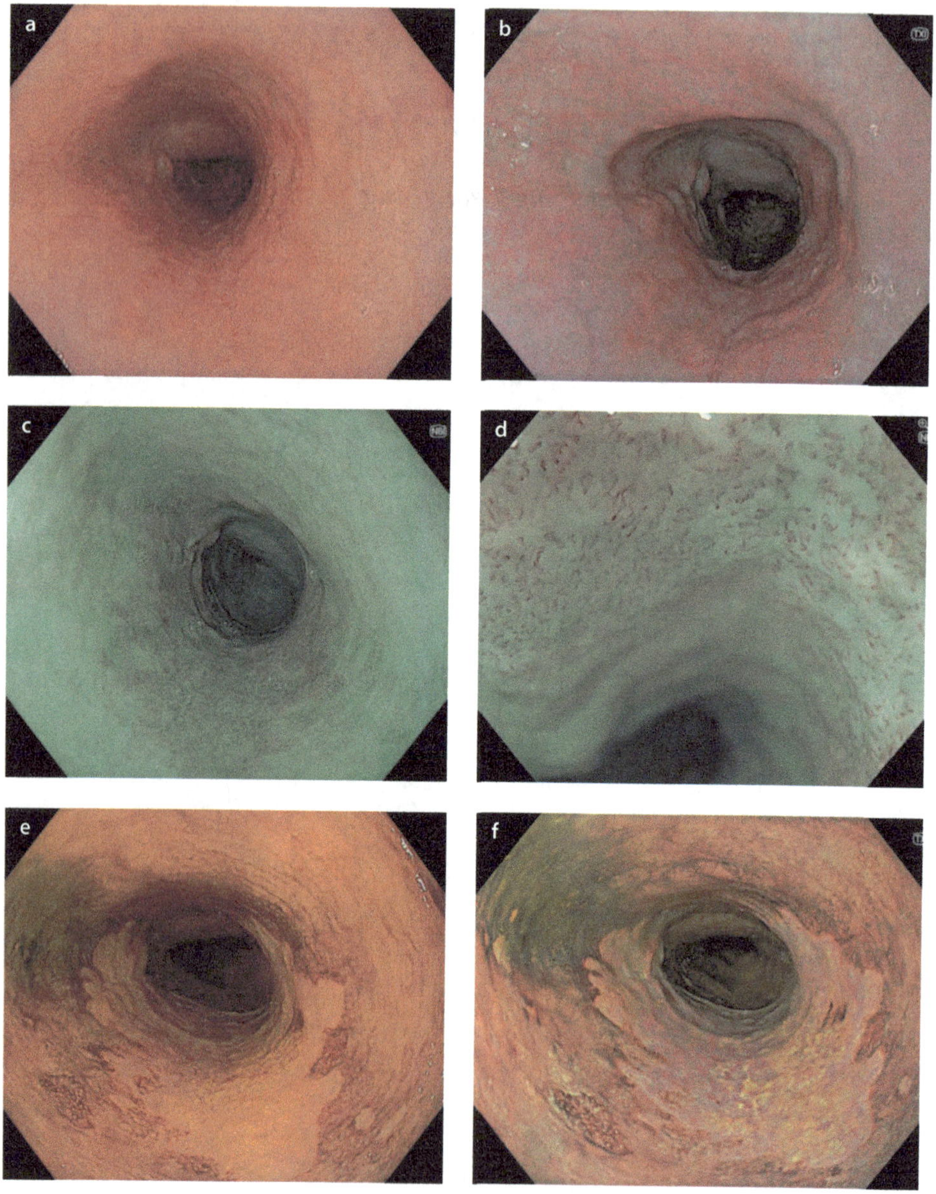
a
b
c
d
e
f

▪ Pathological Diagnosis

Moderately differentiated squamous cell carcinoma with superficial submucosa invasion (pT1b (sm1, 195 mm) G2-3 L0 V0 R0).

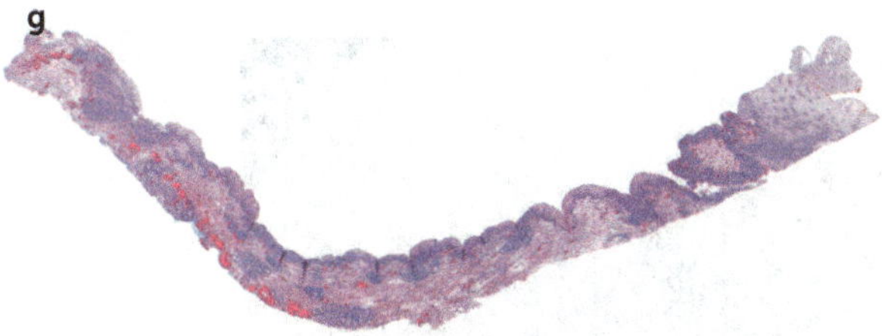

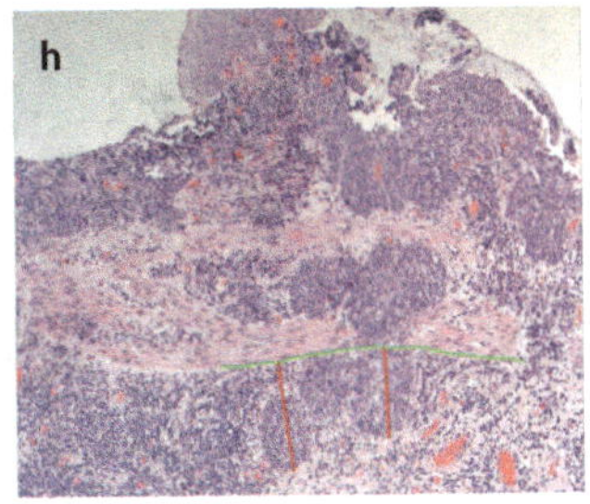

▪ Explanation

While using non-magnifying WLI observation, a vague area of abnormal discoloration of the esophageal squamous mucosa with 0_IIa morphology was detected (Fig. a). After switching to TXI mode, the abnormal reddish mucosa was further enhanced, and more than two-third of the circumference appeared involved (Fig. b). The lesion was then further characterized by switching to non-magnifying NBI mode. The abnormal area appeared darker, due to the presence of neoplastic microvessels (Fig. c). With the use of the magnifying NBI mode, the abnormal intrapapillary capillary loops (IPCL) were clearly demonstrated (Fig. d). The lesion was also examined by chromoendoscopy using Lugol's iodine in WLI mode, and this revealed an extensive unstained area due to the underlying esophageal squamous cell carcinoma (Fig. e). Chromoendoscopy with Lugol's iodine in the TXI mode showed further enhancement of the unstained neoplastic lesion, when compared to WLI (Fig. f). The patient was diagnosed to have extensive superficial esophageal squamous cell carcinoma and underwent treatment by endoscopic submucosa dissection. The histopathological findings revealed moderately differentiated squamous cell carcinoma with superficial submucosa invasion (pT1b (sm1, 195 mm) G2-3 L0 V0 R0) (Figs. g and h).

Tips

TXI mode is useful for initial screening. Next, use non-magnifying NBI and check the change in vessel/surface pattern. Focal areas of change are often suggestive of cancer. Finally, switch to magnifying NBI to evaluate the individual vessels. Chromoendoscopy with Lugol's iodine is very useful for establishing the extent of the lesion.

▪ Video

A short video of a case of extensive squamous cell carcinoma of the esophagus is available as ◘ Fig. 1.

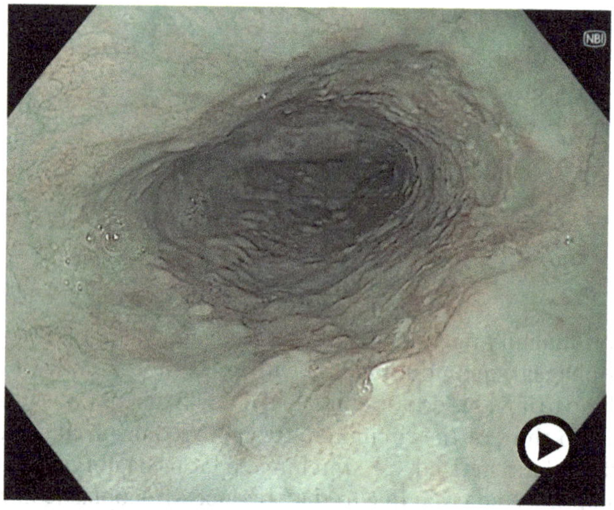

◘ **Fig. 1** Non-magnifying and magnifying NBI views of extensive squamous cell carcinoma of the esophagus (▶ https://doi.org/10.1007/000-c85)

Barrett's Esophagus

Contents

Long- and Short-Segment Barrett's Esophagus

Chin Kiming Tan and Lai Mun Wang

Content

Supplementary Information The online version contains supplementary material available at https://doi.org/10.1007/978-981-97-2732-2_28. The videos can be accessed individually by clicking the DOI link in the accompanying figure caption or by scanning this link with the SN More Media App.

Y. Sano et al. (eds.), *Atlas of Advanced Endoscopy*, https://doi.org/10.1007/978-981-97-2732-2_28

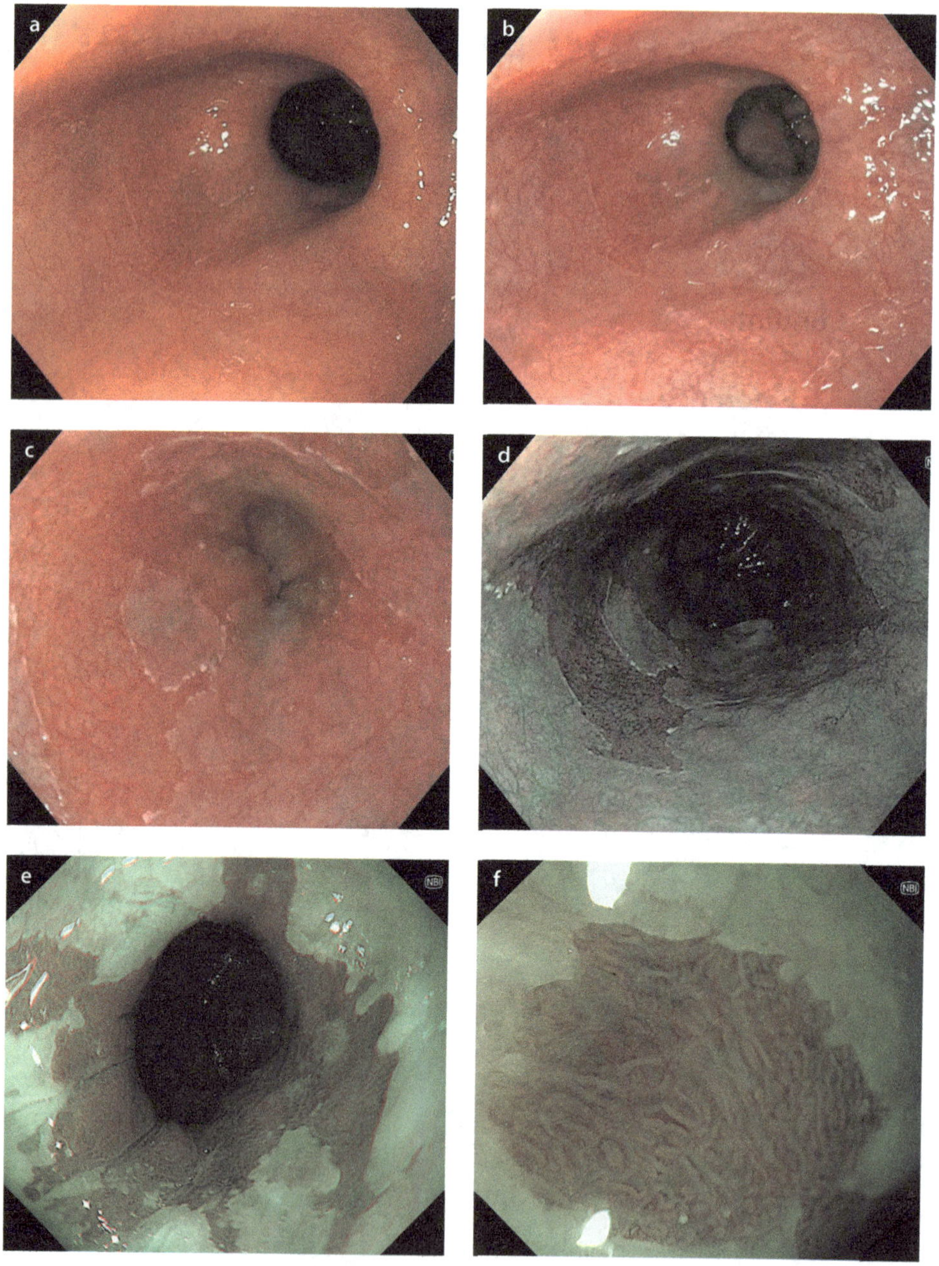
a
b
c
d
e
NBI
f
NBI

▪ Pathological Diagnosis

Barrett's esophagus

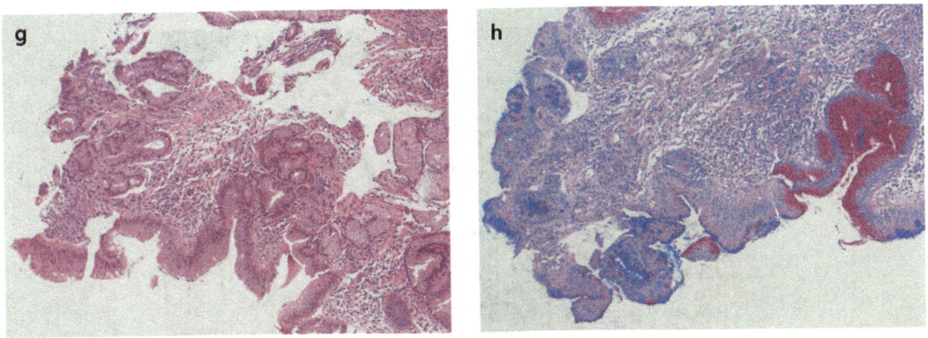

▪ Explanation

Barrett's esophagus is diagnosed when the normal distal squamous epithelial lining has been replaced by metaplastic columnar epithelium, which is clearly visible endoscopically (≥1 cm) above the gastroesophageal junction and confirmed histopathologically from esophageal biopsies. Endoscopic reporting should describe the length using the Prague criteria (circumferential extent (C), maximum extent (M) of endoscopically visible columnar-lined esophagus in centimeters). A maximum extent less than 3 cm is defined as short-segment Barrett's esophagus, while a length 3 cm and longer is defined as long-segment Barrett's esophagus. Using non-magnifying WLI, columnar-lined epithelium consistent with Barrett's esophagus can be seen in the distal esophagus, with maximum extent more than 3 cm, meeting criterion for long-segment Barrett's esophagus (Fig. a). TXI (mode 1) enhances the margin and color difference between the Barrett's epithelium and normal squamous epithelium lesion more effectively than WLI (Figs. b and c). Switching to non-magnifying NBI mode, the long-segment Barrett's mucosa is clearly demarcated from the normal squamous mucosa (Fig. d). Non-magnifying NBI observation shows short-segment Barrett's esophagus which is defined as the maximum length of columnar-lined epithelium being less than 3 cm (Fig. e). Switching to magnifying NBI mode, regular mucosal patterns are observed, consistent with non-dysplastic Barrett's mucosa (Fig. f). Histology image of hematoxylin and eosin-stained biopsy of Barrett's esophagus is characterized by columnar lined mucosa. No dysplasia is seen (Fig. g). Histology image of Barrett's mucosa with Alcian blue stain highlighting goblet cells represents intestinal metaplasia (Fig. h).

Tips

The top of the visible gastric folds are identified as the anatomic landmark of the gastroesophageal junction, the position from which measurements of visible columnar tissue should begin. The Prague classification requires endoscopists to define the location of the top of the gastric folds, the proximal extent of both circumferential Barrett's esophagus segments (C), as well as the maximal extent of any tongues (M) or islands. NBI, especially with optical magnification, will allow better visualization of the mucosa pit pattern for detection of dysplasia [1].

■ Video

A short video is available as ◘ Fig. 1.

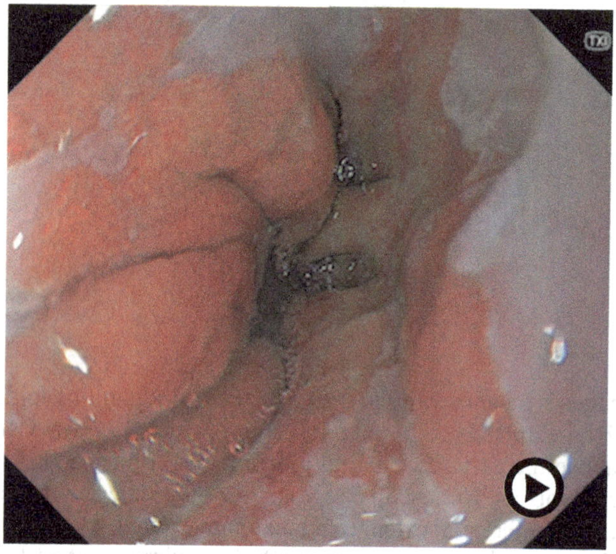

◘ **Fig. 1** WLI, TXI, non-magnifying and magnifying NBI views of SSBE without dysplasia (▶ https://doi.org/10.1007/000-c86)

Reference

1. Everson MA, Ragunath K, Bhandari P, Lovat L, Haidry R. How to perform a high-quality examination in patients with Barrett's esophagus. Gastroenterology. 2018;154(5):1222–6.

Barrett's Esophagus with Low-Grade Dysplasia

Stefan Seewald and Tiing Leong Ang

Content

Supplementary Information The online version contains supplementary material available at https://doi.org/10.1007/978-981-97-2732-2_29. The videos can be accessed individually by clicking the DOI link in the accompanying figure caption or by scanning this link with the SN More Media App.

Y. Sano et al. (eds.), *Atlas of Advanced Endoscopy*, https://doi.org/10.1007/978-981-97-2732-2_29

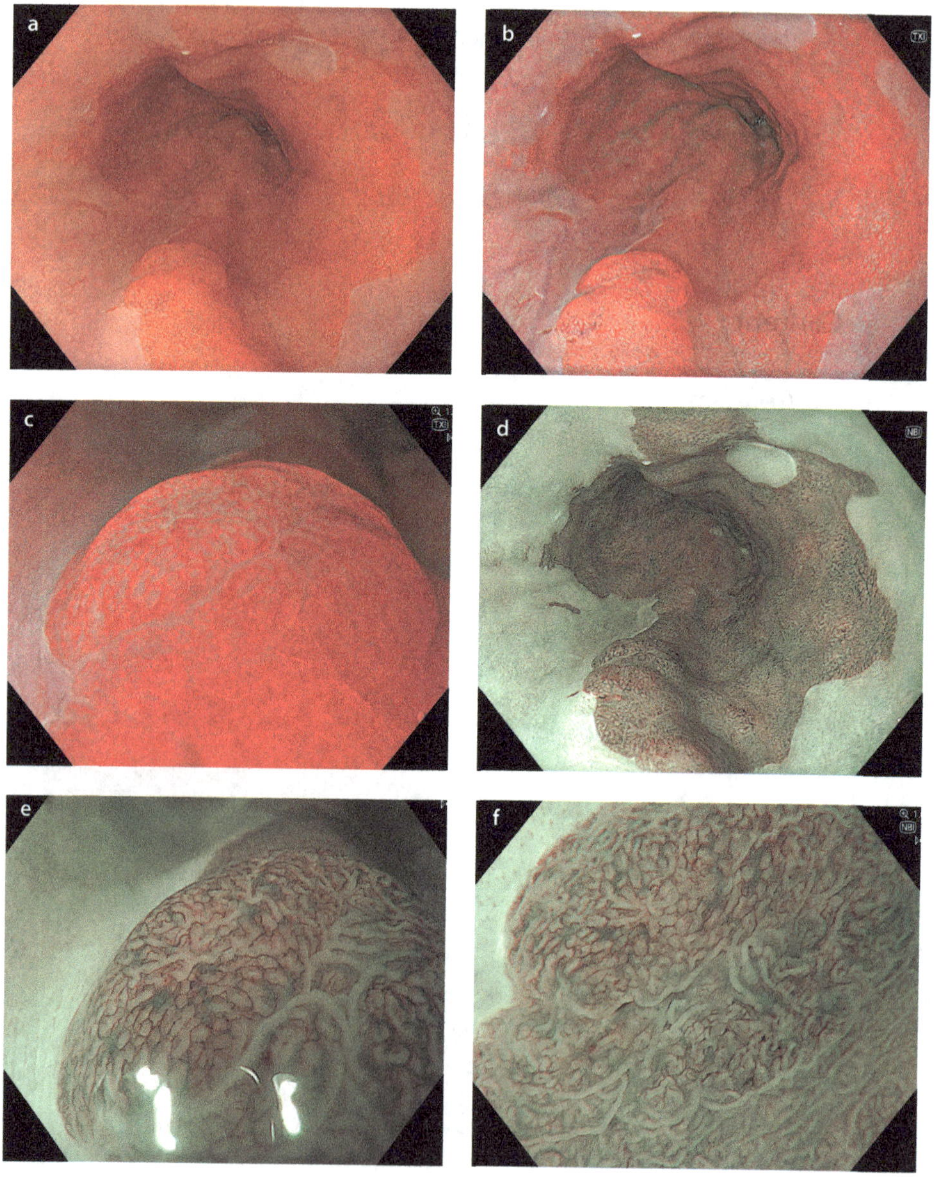
a
b
TXI
c
TXI
d
NBI
e
f
NBI

▪ Pathological Diagnosis

Barrett's Esophagus with Low-Grade Dysplasia

g

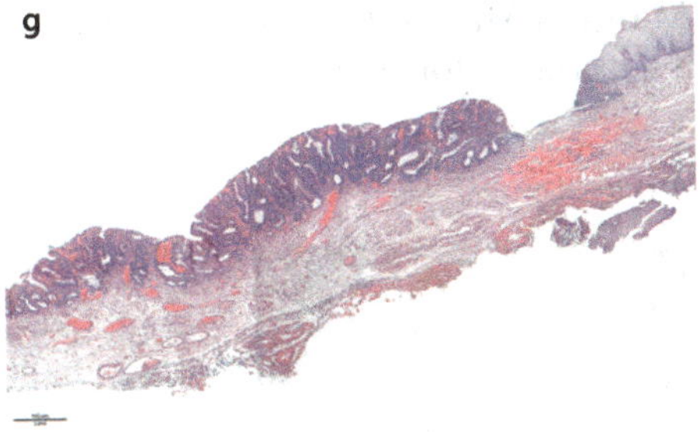

h

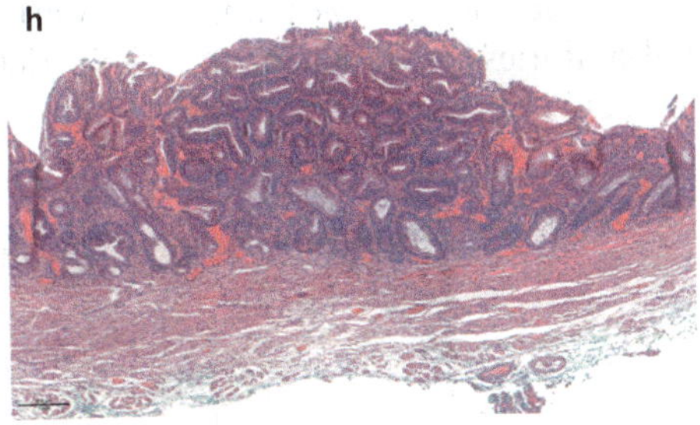

▪ Explanation

Barrett's esophagus (BE) with low-grade dysplasia (LGD) is usually considered to be endoscopically invisible, and the endoscopic features are not well. However, there is now awareness that it may be possible to detect BE with LGD endoscopically. It has been reported that the endoscopic features of BE with visible LGD included both elevated (Paris 0-IIa) and flat (Paris 0-IIb) morphology, combined with abnormal mucosal and/or vascular patterns, and with clear demarcation from the normal regular flat BE [1]. Using non-magnifying WLI, columnar-lined epithelium consistent with BE was observed. There was a Paris 0-IIa lesion at the proximal end of the Barrett's mucosa at 6 o'clock position (Fig. a). TXI (mode 1) further enhanced the color of the lesion, and a clear demarcation line that separated it from the normal Barrett's mucosa could be seen (Fig. b). TXI combined with magnification highlighted the mucosa patterns and microvessels (Fig. c). Switching to non-magnifying NBI mode, the demarcation line and thickened villous pattern of the lesion could be seen more clearly (Fig. d). The magnifying NBI mode enabled clear visualization of the demarcation line, as well as thickened villous pattern and irregular microvessels within the lesion (Figs. e and f). The LGD was treated by endoscopic mucosal resection (EMR). Histology (Figs. g and h) confirmed the diagnosis of BE with LGD.

Tips

BE with LGD may be recognized endoscopically as discrete Paris 0-IIa or 0-IIb lesions with a clear demarcation line, as well as abnormal mucosal and/or vascular patterns.

▪ Video

A short video is available as ◘ Fig. 1

Acknowledgments We thank Dr. Ewerton Marques Maggio from Medica Medical Laboratories Dr. F. Kaeppeli AG, Zurich, Switzerland, for providing the histology images.

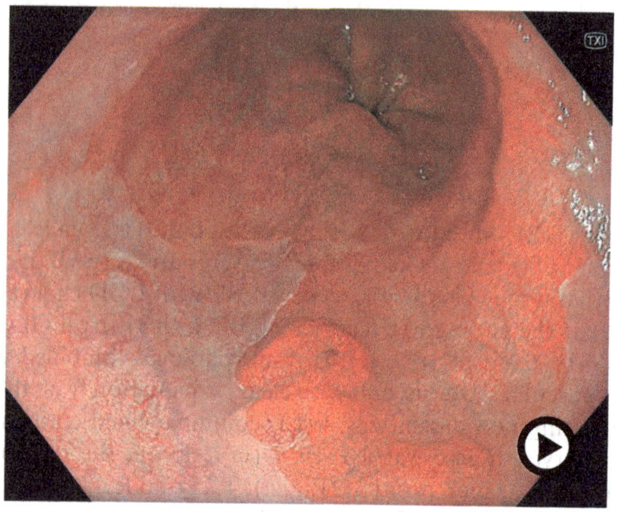

◘ **Fig. 1** WLI, TXI, non-magnifying and magnifying NBI views of BE with LGD (▶ https://doi.org/10.1007/000-c87)

Reference

1. He T, Iyer KG, Lai M, et al. Endoscopic features of low-grade dysplastic Barrett's. Endosc Int Open. 2023;11(8):E736–42.

Barrett's Esophagus with High-Grade Dysplasia

Stefan Seewald and Tiing Leong Ang

Content

Y. Sano et al. (eds.), *Atlas of Advanced Endoscopy*, https://doi.org/10.1007/978-981-97-2732-2_30

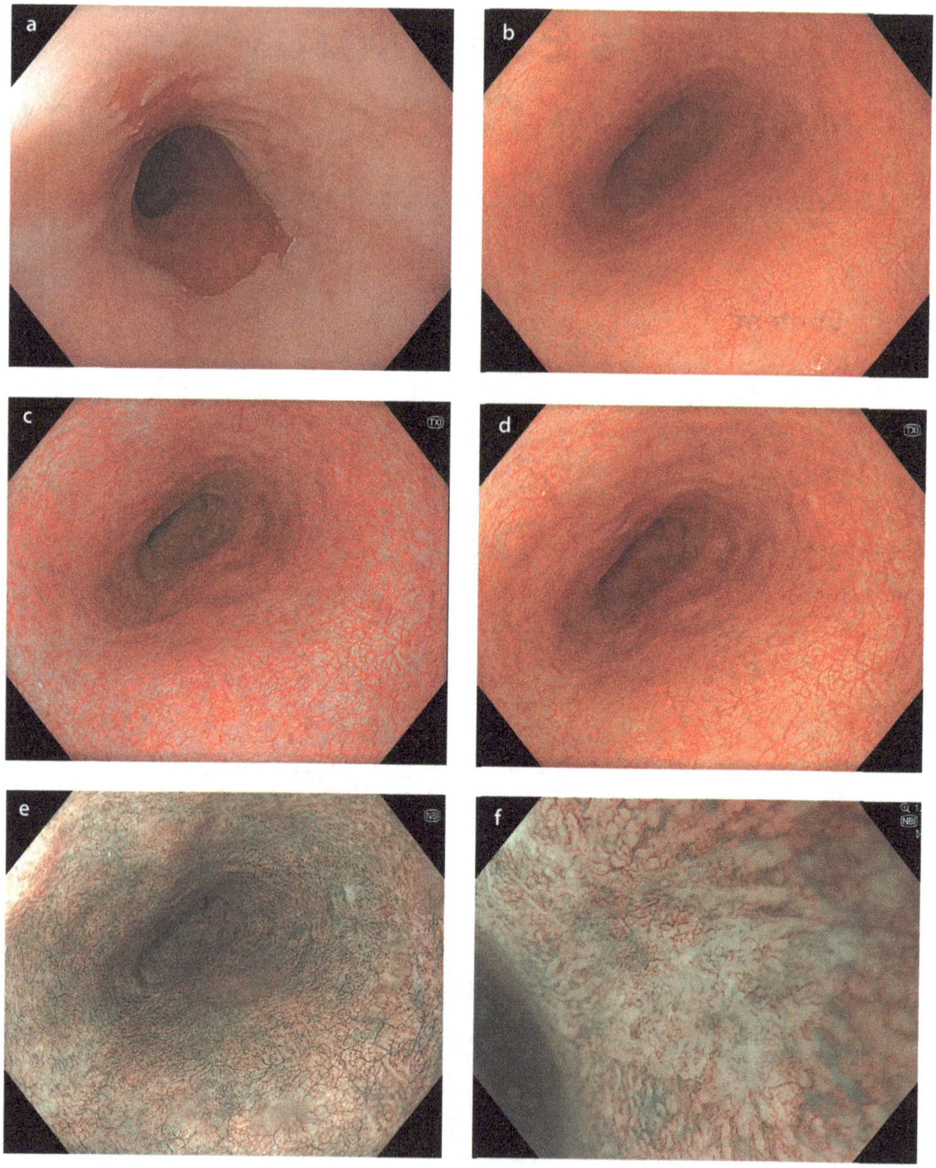
a
b
c
TXI
d
TXI
e
NBI
f
NBI

Pathological Diagnosis

Barrett's esophagus with high-grade dysplasia

g

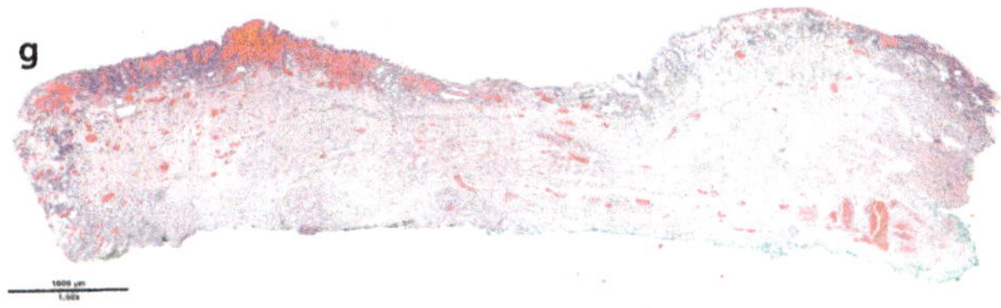

h

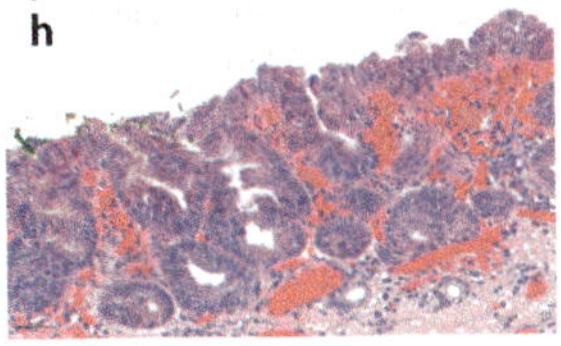

Explanation

BE with HGD may be recognized endoscopically as discrete lesions with irregular/distorted mucosal or microvessel patterns. In this case example, non-magnifying WLI was performed. The proximal margin of the long-segment BE was visualized (Fig. a). Within the long segment of Barrett's mucosa, using non-magnifying WLI, a subtle Paris 0-IIb lesion was noted at 1 o'clock position (Fig. b). TXI (mode 1) further enhanced the color of the lesion (Fig. c), while TXI (mode 2) enhanced the texture of the lesion (Fig. d). Switching to non-magnifying NBI mode, the dysplastic lesion was more obvious (Fig. e). With magnifying NBI, irregular mucosal and microvessel patterns were observed (Fig. f). Endoscopic mucosal resection (EMR) with en bloc resection was performed. Histology (Figs. g and h) confirmed the presence of HGD.

Tips

BE with HGD may be recognized endoscopically as discrete lesions with irregular/distorted mucosal or microvessel patterns. The appearance is enhanced with the use of image-enhanced endoscopy techniques such as TXI and NBI. Magnifying NBI is important for visualizing the irregular mucosal and microvessel patterns of BE with HGD [1].

Acknowledgments We thank Dr. Ewerton Marques Maggio from Medica Medical Laboratories Dr. F. Kaeppeli AG, Zurich, Switzerland, for providing the histology images.

Reference

1. Silva FB, Dinis-Ribeiro M, Vieth M, et al. Endoscopic assessment and grading of Barrett's esophagus using magnification endoscopy and narrow-band imaging: accuracy and interobserver agreement of different classification systems (with videos). Gastrointest Endosc. 2011;73(1):7–14.

Barrett's Dysplasia and Focal Adenocarcinoma

Hon Chi Yip and Rajvinder Singh

Supplementary Information The online version contains supplementary material available at https://doi.org/10.1007/978-981-97-2732-2_31. The videos can be accessed individually by clicking the DOI link in the accompanying figure caption or by scanning this link with the SN More Media App.

Y. Sano et al. (eds.), *Atlas of Advanced Endoscopy*, https://doi.org/10.1007/978-981-97-2732-2_31

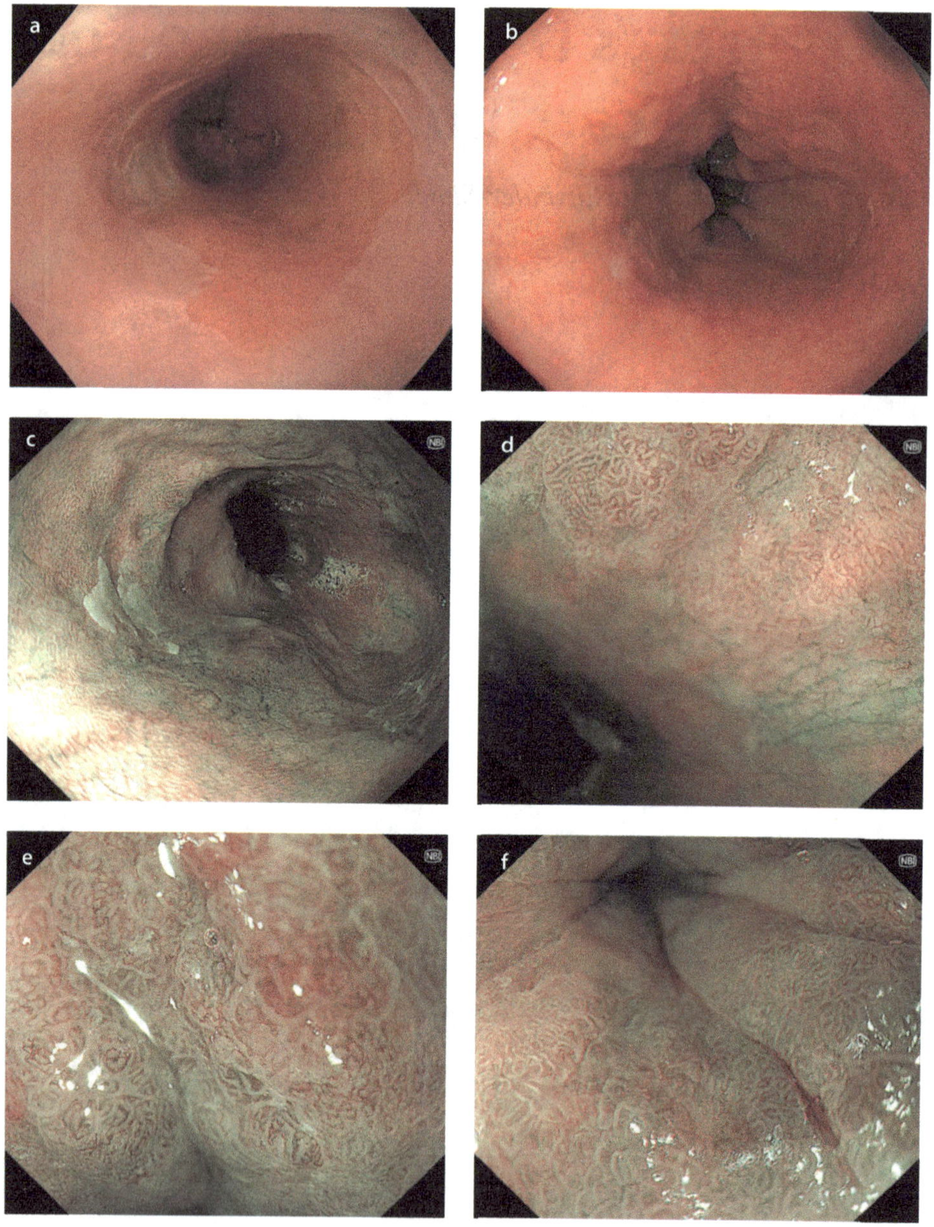
a
b
c
NBI
d
NBI
e
NBI
f
NBI

■ Pathological Diagnosis

High-grade dysplasia, focal intramucosal adenocarcinoma within background long-segment Barrett's esophagus

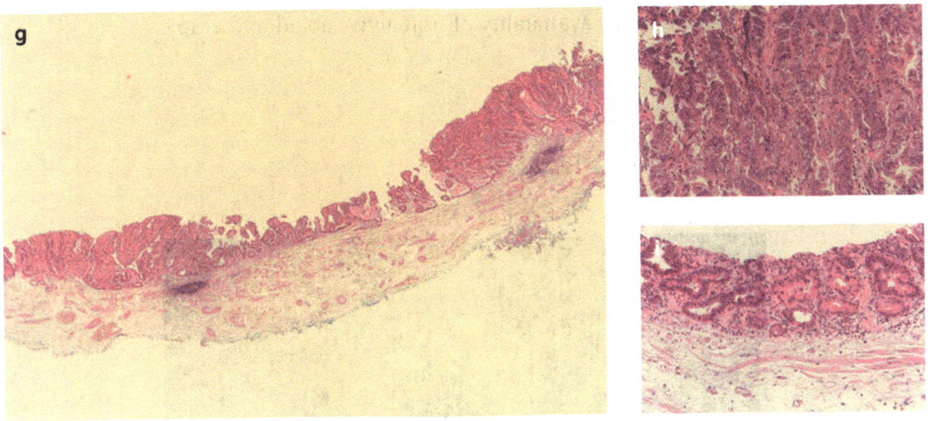

■ Explanation

On WLI (Fig. a), salmon pink mucosa could be observed in the distal esophagus, spanning circumferentially for 3 cm and extending up to 4 cm above the gastroesophageal junction (GEJ), compatible with C3M4 Barrett's esophagus. It was also associated with a sliding type hiatus hernia of 2 cm in length. Multiple raised red nodular components were identified in the anterior and right wall within the distal part of the Barrett's mucosa (Fig. b). On low magnification NBI image, metaplastic columnar epithelium could be easily observed, with small islands of squamous mucosa seen on the left side (Fig. c). In addition, surface pattern over the mucosa at the posterior wall also appeared to be irregular. In moderate power magnification NBI examination, irregularly branched glandular structures (non-pit-type) could be seen in the nodules over anterior wall, while the adjacent mucosa appeared flat and atrophic (Fig. d). Distal esophagus palisading veins (DEPV) could be identified that extended down to the GEJ. Upon high-power magnification NBI, in addition to irregularly arranged surface pattern, the underlying subepithelial capillary network also appeared irregular with dilatation and nonuniform appearance, some forming open loops while others in coiled shapes (Fig. e). Multifocal dysplasia was suspected as more lesions were detected in the posterior wall with irregular surface and vascular pattern (Fig. f). En bloc circumferential endoscopic submucosal dissection (ESD) was performed to resect all dysplastic lesions. Pathology of ESD specimen showed multiple high-grade dysplasia in the background of Barrett's mucosa and a foci of intramucosal adenocarcinoma (Figs. g, h, i).

Tips

During endoscopic surveillance of long-segment Barrett's esophagus, it is imperative that detailed mucosal examination is carried out with image enhancement technique (such as NBI). Neoplastic lesions often occur in a multifocal manner without distinct demarcation, making detection particularly challenging. Biopsies should be taken liberally in long-segment disease at suspicious sites to ensure adequate diagnosis of all lesions. Selection of the optimal endoscopic resection strategies should take into account the ability of en bloc margin negative resection, the risk of post-resection esophageal stricture, and the availability of adjunctive ablation therapy.

▪ Video

A short video is available as Fig. 1.

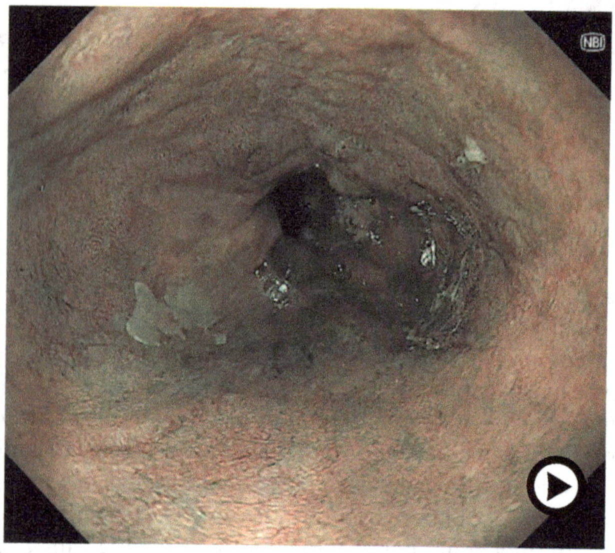

Fig. 1 Practical observation and diagnosis of Barrett's dysplasia and focal adenocarcinoma (▶ https://doi.org/10.1007/000-c88)

Barrett's Adenocarcinoma (0-IIa) (1)

Rajvinder Singh and Amol Bapaye

Supplementary Information The online version contains supplementary material available at https://doi.org/10.1007/978-981-97-2732-2_32. The videos can be accessed individually by clicking the DOI link in the accompanying figure caption or by scanning this link with the SN More Media App.

Y. Sano et al. (eds.), *Atlas of Advanced Endoscopy*, https://doi.org/10.1007/978-981-97-2732-2_32

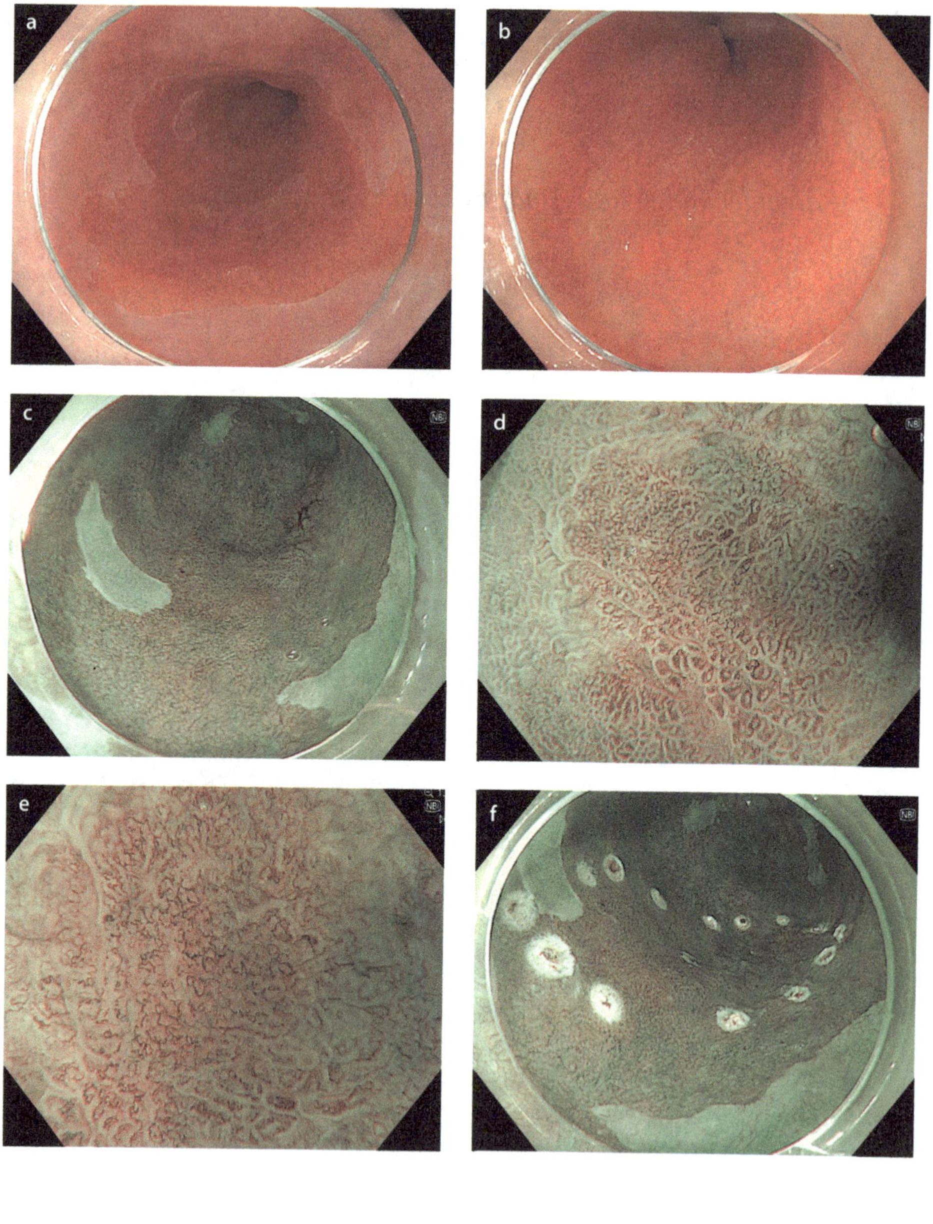

▪ Pathological Diagnosis

Well-differentiated adenocarcinoma: 23 × 12 mm, 0-IIa, pT1a(M)

g

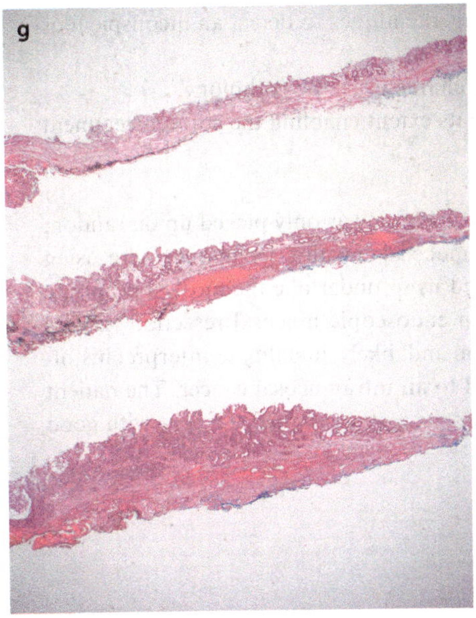

h

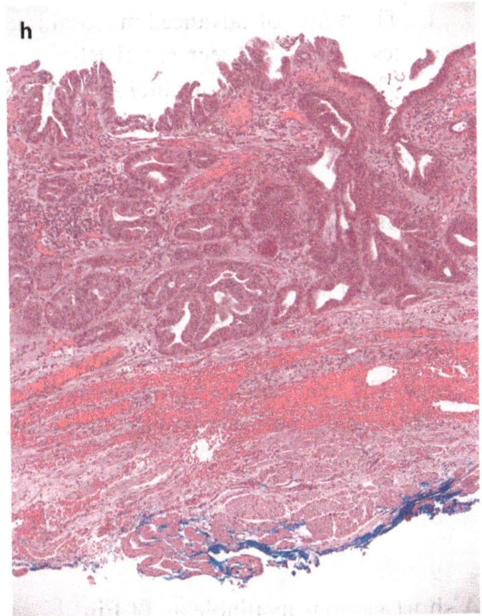

▪ Explanation

A 78-year-old man with a past medical history of stable Crohn's colitis on biologics, ischemic heart disease on dual antiplatelet agents, and type 2 diabetes mellitus presented with inconspicuous low-grade dysplasia which was detected incidentally on random biopsies in a long-segment (C7:M8) Barrett's esophagus (BE).

Figure a shows the proximal extension of the lesion with a few squamous patches. A lesion in the middle of the BE segment was vaguely visible on further careful observation with white light endoscopy (Fig. b). Narrow band imaging (NBI) was used to further accentuate the area of concern (Fig. c) and appears to have a more proximal extension compared to the white light image. Using NBI with low magnification and the underwater technique, the margins of the lesion were then clearly defined (Fig. d). A demarcation line was evident, and distorted pit and vascular pattern were visualized within the lesion itself. This was further accentuated on NBI and high magnification (Fig. e). Using NBI and magnification (Fig. f), the lesion measuring 2.5 cm was marked prior to resection. A decision was made to resect the lesion in a single piece using the endoscopic submucosal dissection (ESD) technique.

In Fig. g (low magnification), glandular mucosa in keeping with Barrett's esophagus can be seen. Within the glands, architectural complexity in the form of budding and early cribriform change can be seen. Atypical glands can be seen lying between disorganized smooth muscle, consistent with reduplicated muscularis mucosae. In Fig. h (high magnification), epithelial cells show marked atypia characterized by nuclear enlargement; nuclear hyperchromatic, cytoplasmic mucin depletion, and an increase in mitotic figures. The features are of high-grade dysplasia amounting to intramucosal adenocarcinoma. The underlying muscularis mucosae remains intact and no submucosal invasion is seen.

Tips

This case clearly highlights three key learning points.

1. The utility of advanced mucosal imaging techniques to detect an inconspicuous lesion in a long segment of BE
2. The ability of the technology to further characterize the pathology
3. The clear demarcation line determining its extent enabling the correct treatment strategy

The initial diagnosis of a low-grade dysplastic lesion was only picked up on random biopsies. The examination with advanced mucosal imaging techniques here using NBI with and without magnification enabled us to undertake the most appropriate treatment strategy (ESD), in contrast to an endoscopic mucosal resection (EMR) which would have led to a piecemeal resection and, likely, inability to interpret histology. The final histopathology was upgraded to an intramucosal cancer. The patient subsequently underwent ablation therapy for the rest for the BE segment with good outcomes.

Video

A short video is available as Fig. 1.

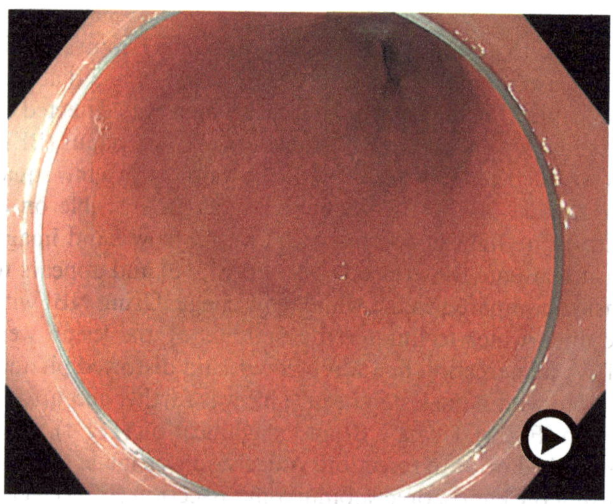

Fig. 1 Practical observation and diagnosis of flat inconspicuous Barrett's esophageal intramucosal cancer with WLI, TXI, and magnifying NBI (▶ https://doi.org/10.1007/000-c89)

Barrett's Adenocarcinoma (0-IIa) (2)

Edward Young and Rajvinder Singh

Supplementary Information The online version contains supplementary material available at https://doi.org/10.1007/978-981-97-2732-2_33. The videos can be accessed individually by clicking the DOI link in the accompanying figure caption or by scanning this link with the SN More Media App.

Y. Sano et al. (eds.), *Atlas of Advanced Endoscopy*, https://doi.org/10.1007/978-981-97-2732-2_33

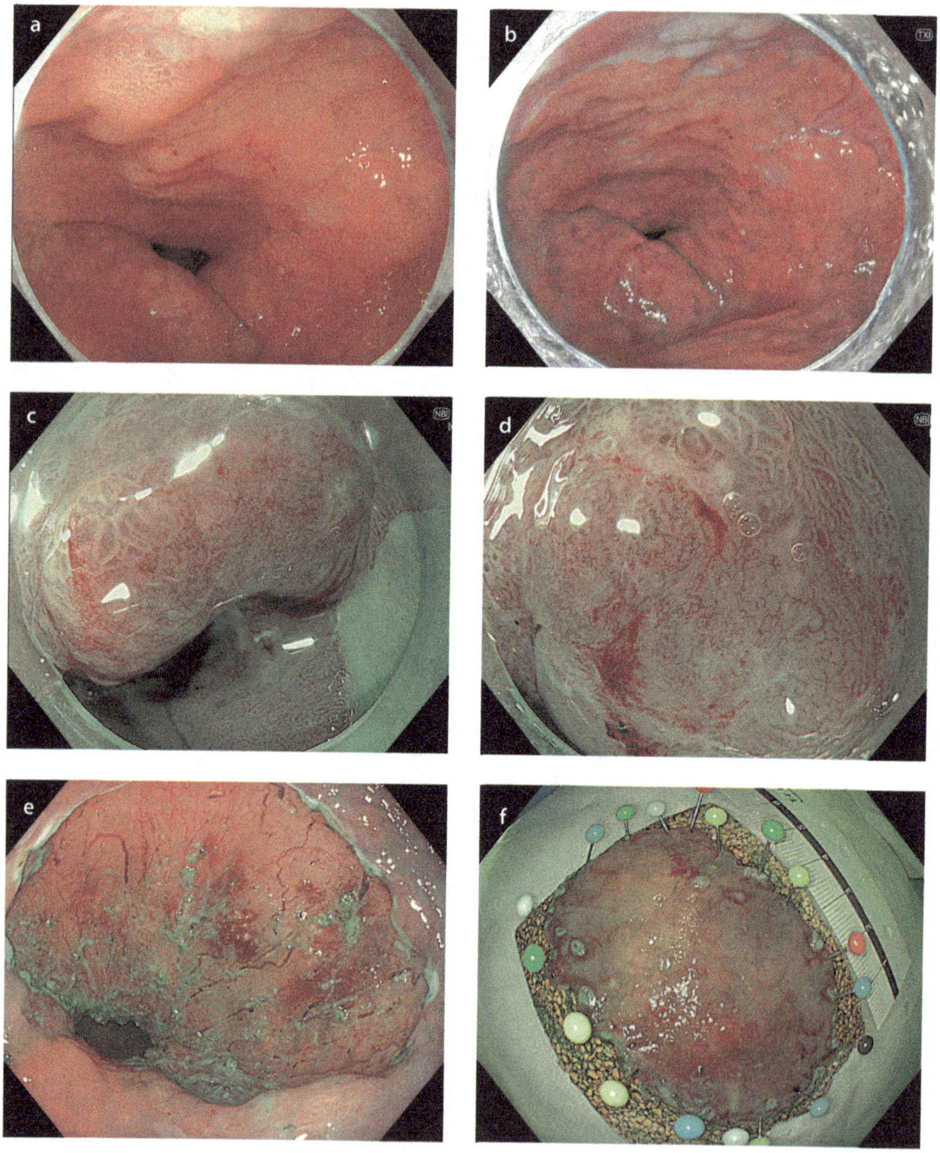
a
b
TXI
c
NBI
d
NBI
e
f

▪ Pathological Diagnosis

Moderately differentiated adenocarcinoma, pT1b, SM2 (800 µm)

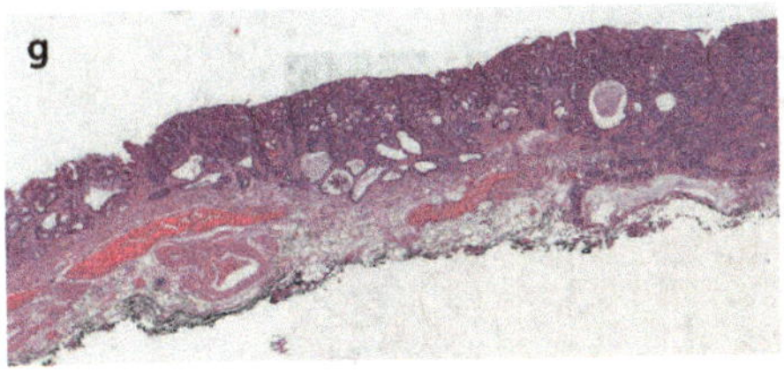

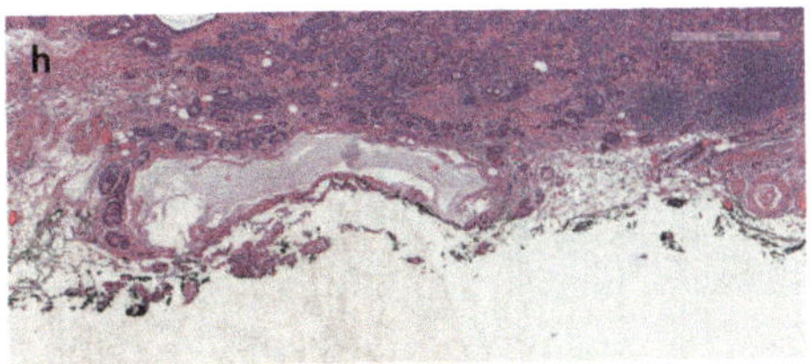

▪ Explanation

A 79-year-old man with a 55 pack-year smoking history and a background of hypertension presented with nodular Barrett's esophagus, biopsies of which were in keeping with an adenocarcinoma.

Figures a (seen on WLI) and b (seen on TXI) reveal a slightly elevated nodular-appearing lesion measuring about 4 cm in size which was straddling the squamocolumnar junction. CT and PET staging did not reveal any evidence of metastatic disease although there was abnormal FDG uptake at the lower esophagus and gastroesophageal junction. The patient underwent a staging endoscopic submucosal dissection (ESD). Advanced mucosal imaging with narrowband imaging prior to the ESD demonstrated two distinct patterns: (i) on NBI and low magnification (Fig. c) depicting the upper margin of the lesion where the pit pattern is totally distorted and a faint demarcation line is seen and (ii) NBI and high magnification (Fig. d) at the lateral flat margin of the lesion where a clear demarcation line and abnormal vascular pattern can be clearly differentiated from regular pit pattern. The lesion was carefully assessed and marked and an ESD performed (Fig. e). The specimen measuring about 6.5 cm was then pinned down on a cork (Fig. f) and submitted for histopathology. The specimen (Fig. g) revealed a moderately differentiated adenocarcinoma with deep submucosal invasion up to 800 µm (pT1b Sm2). The nearest radial margin was 2 mm clear; however, there was the presence of lymphovascular invasion (Fig. h). The en bloc nature of the resection therefore allowed upstaging of the lesion histologically. The patient subsequently underwent adjuvant chemoradiotherapy and has been disease-free for 6 months.

Tips

This case highlights the importance of performing a thorough evaluation with NBI and high magnification to demarcate the lateral extent of the lesion prior to undertaking an ESD. Although there was deep submucosal invasion demonstrated on final histopathology, the ability to remove the lesion en bloc allowed proper staging and adjunctive additional therapy leading to good overall outcomes. The patient avoided an unnecessary surgical procedure which is associated with significant morbidity and mortality.

▪ Video

A short video is available as ◘ Fig. 1.

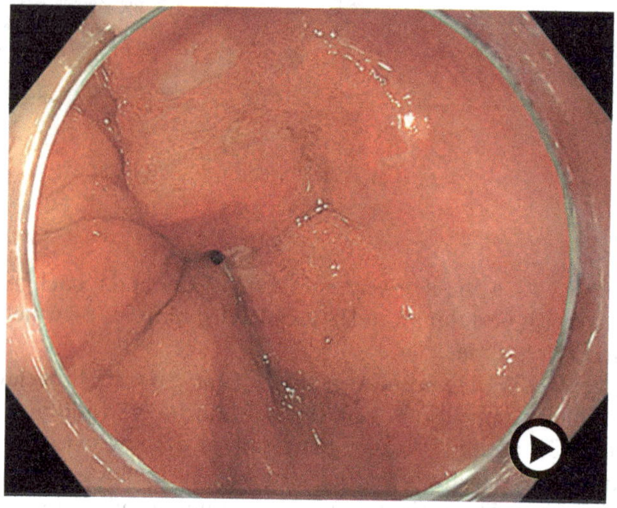

◘ **Fig. 1** Practical observation and diagnosis of Barrett's cancer with WLI and magnifying NBI. The lesion was resected using the ESD technique (▶ https://doi.org/10.1007/000-c8a)

Barrett's Adenocarcinoma (0-IIc)

Yasuhiko Mizuguchi and Seiichiro Abe

Supplementary Information The online version contains supplementary material available at https://doi.org/10.1007/978-981-97-2732-2_34. The videos can be accessed individually by clicking the DOI link in the accompanying figure caption or by scanning this link with the SN More Media App.

Y. Sano et al. (eds.), *Atlas of Advanced Endoscopy*, https://doi.org/10.1007/978-981-97-2732-2_34

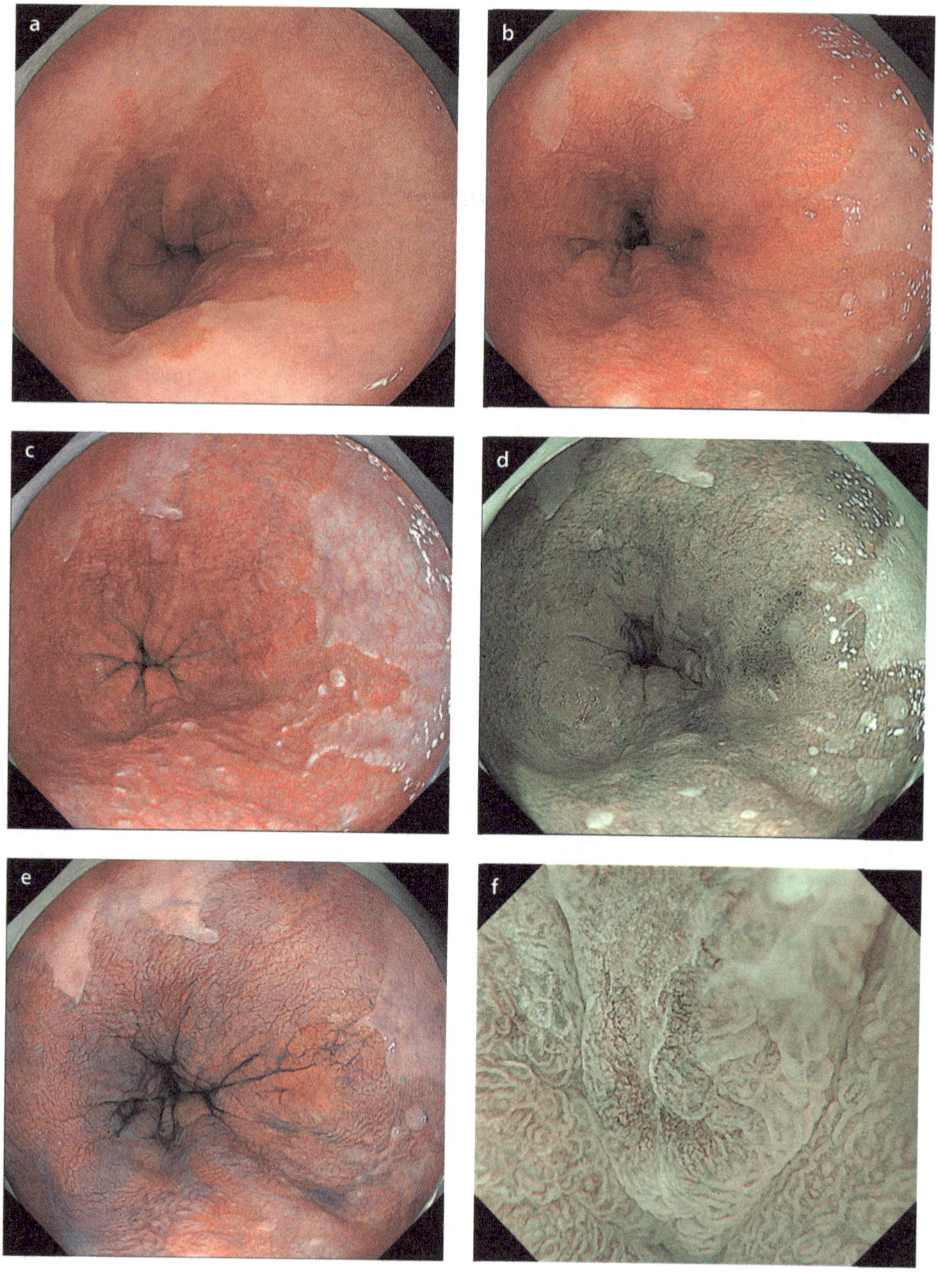
a
b
c
d
e
f

▪ Pathological Diagnosis

Tubular adenocarcinoma, moderately differentiated type: 17 × 9 mm, 0-IIc, pT1a (M)

g

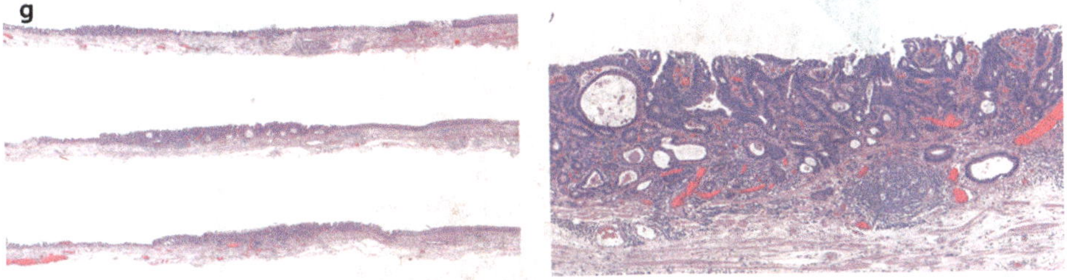

▪ Explanation

Endoscopically, the esophagogastric junction (EGJ) can be identified as the lower limit of the esophageal palisade vessels or, if it cannot be identified, the upper limit of the gastric longitudinal folds (Fig. a). By measuring the distance between the EGJ and the squamocolumnar junction, a diagnosis of short-segment Barrett's esophagus may be conferred (Prague's classification C1M2). Within the Barrett's mucosa, squamous islands were observed, indicating Barrett's esophagus. An erythematous depression was identified at 3 o'clock position in Barrett's mucosa, but the margin is indistinct under white light endoscopy (Fig. b). Using TXI mode 1, the distinction in coloration and texture between the lesion and the adjacent mucosa was highlighted, rendering the presence of the lesion more apparent (Fig. c). Under non-magnifying NBI, the lesion was seen as a well-demarcated brownish area (Fig. d). After 0.6% acetic acid followed by indigo-carmine spraying, the lesion was further delineated as an area with a distinctive pattern compared to the surrounding area (Fig. e). Magnifying NBI on the anal side of the lesion revealed a well-demarcated area with dilated, dense, and irregular vessels inside (Fig. f). A biopsy from the lesion revealed tubular adenocarcinoma (well to moderately differentiated type), and the lateral extent of the lesion was determined together with four negative biopsies around the lesion. The lesion was resected en bloc by the ESD technique. The lesion was histologically moderately differentiated adenocarcinoma, confined to the mucosal layer with partial invasion to the muscularis mucosae. Barrett's epithelium was adjacent to the lesion (Fig. g).

Tips

The diagnosis of the extent of Barrett's esophageal adenocarcinoma is challenging because it may involve a lateral extension that is macroscopically unrecognizable. Acetic acid has been reported to reversibly organize cytoplasmic proteins and improve the visibility of mucosal microscopic patterns. Combining acetic acid spray with indigo carmine may help to better define the boundaries of the lesion because the indigo carmine is repelled at the lesion. In order to achieve en bloc endoscopic resection, a careful diagnosis of the extent of the lesion combined with peripheral biopsy is required.

Video

A short video is available as Fig. 1.

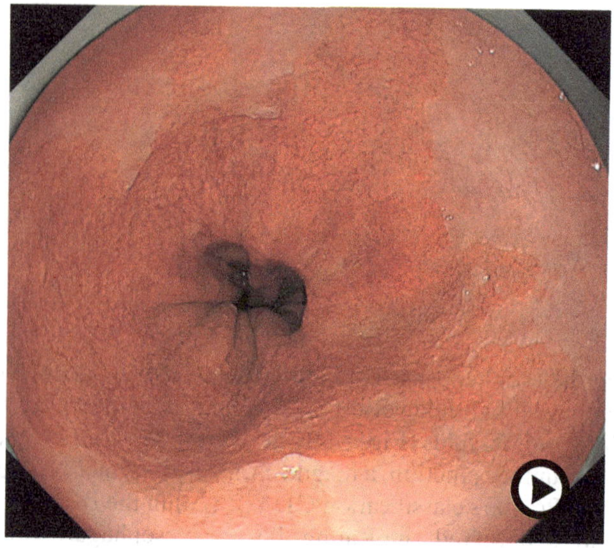

Fig. 1 Practical observation and diagnosis of Barrett's adenocarcinoma by WLI, TXI, NBI, and chromoendoscopy (0.6% acetic acid with 0.2% indigo carmine) (▶ https://doi.org/10.1007/000-c8b)

Stomach

Contents

Gastric Intestinal Metaplasia

Noriya Uedo

Content

Supplementary Information The online version contains supplementary material available at https://doi.org/10.1007/978-981-97-2732-2_35. The videos can be accessed individually by clicking the DOI link in the accompanying figure caption or by scanning this link with the SN More Media App.

Y. Sano et al. (eds.), *Atlas of Advanced Endoscopy*, https://doi.org/10.1007/978-981-97-2732-2_35

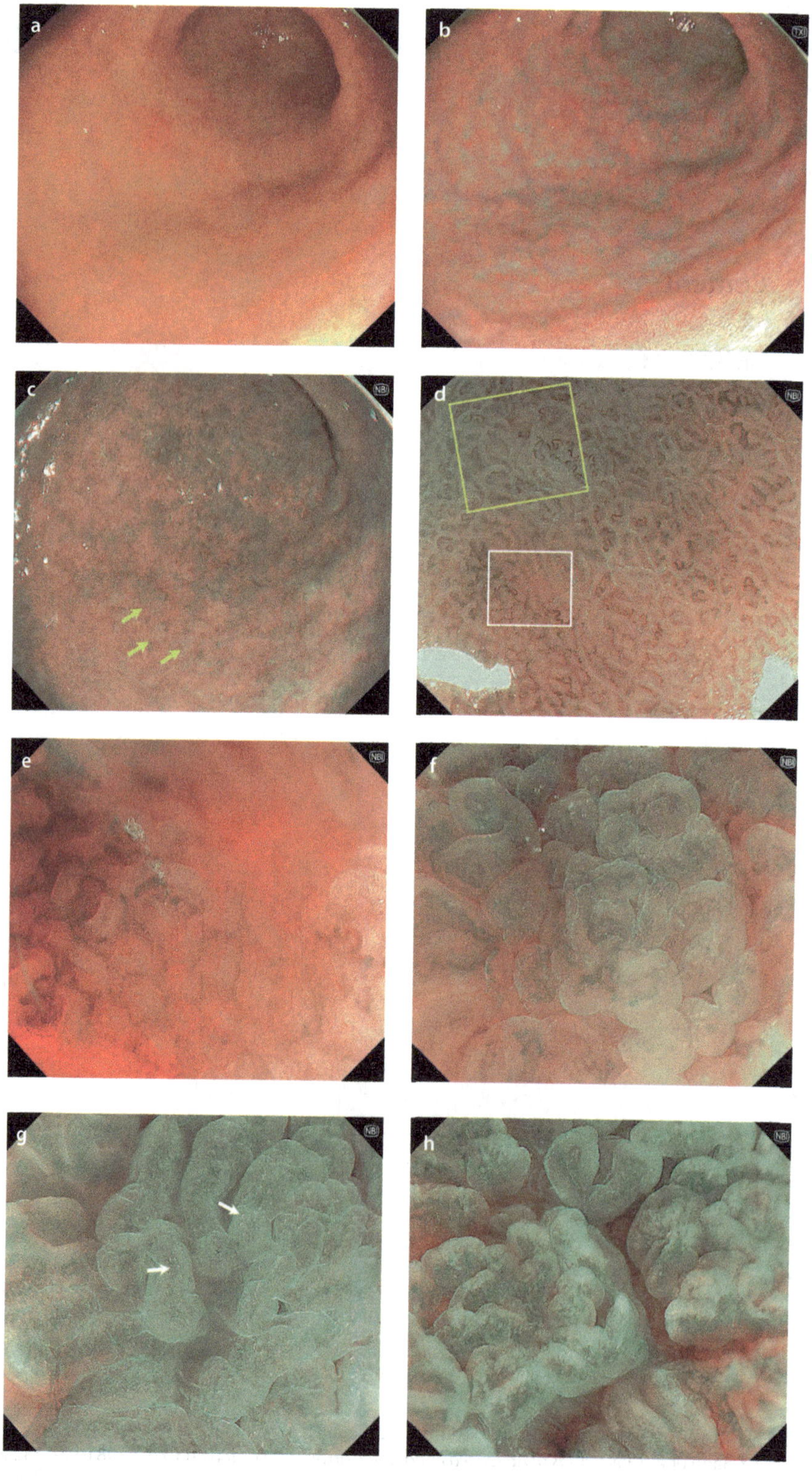
a
b
TXI
c
NBI
d
NBI
e
NBI
f
NBI
g
NBI
h
NBI

Explanation

A patient with *H. pylori*-associated chronic atrophic gastritis received gastroscopic examination. The antral mucosa looked having a mosaic of red and white colors (Fig. a). Texture and color enhancement imaging (TXI) mode enhanced the difference in colors (Fig. b). Narrowband imaging (NBI) revealed multiple whitish patches (Fig. c, yellow arrows), indicating the presence of gastric intestinal metaplasia (GIM). Magnifying NBI image distinguished an area with the GIM, in which the light blue crest (LBC) is seen (yellow box in Fig. d), from an area of the non-metaplastic mucosa (white box in Fig. d). Ultra-magnifying NBI image of the non-metaplastic mucosa showed homogeneous cellular structure in the marginal crypt epithelium with smooth margin (Fig. e). Ultra-magnifying NBI image of the GIM showed the marginal crypt epithelium with rough margin and the LBC (Fig. f); heterogeneous cellular structure in the marginal crypt epithelium including sporadic large cells, which is speculated as goblet cells (Fig. g, white arrows); and white opaque substance (WOS) in the marginal crypt epithelium (Fig. h) [1].

Tips

An ultrahigh magnification zoom videoendoscope (EVIS-H290EC, Olympus Co. Ltd., Tokyo, Japan) is used with an image processor (CV-1500, Olympus). A soft black cap (MAJ-1989, Olympus) was mounted on the tip of the endoscope. The protruding length of the cap is adjusted to make the long dimension of the endoscopic image ~2 mm. Structural enhancement was set at B8 for NBI. Brightness control was set at average mode. GIM distributes mosaic pattern in the gastric mucosa. NBI represents areas with GIM as whitish patches. The GIM usually shows ridged/villous microsurface structure. GIM has two indicative endoscopic signs: the LBC and the WOS. LBC corresponds with histological microvilli (brush border), and WOS corresponds with lipid droplets absorbed in the marginal crypt epithelium.

■ Video

A short video is available as ◘ Fig. 1.

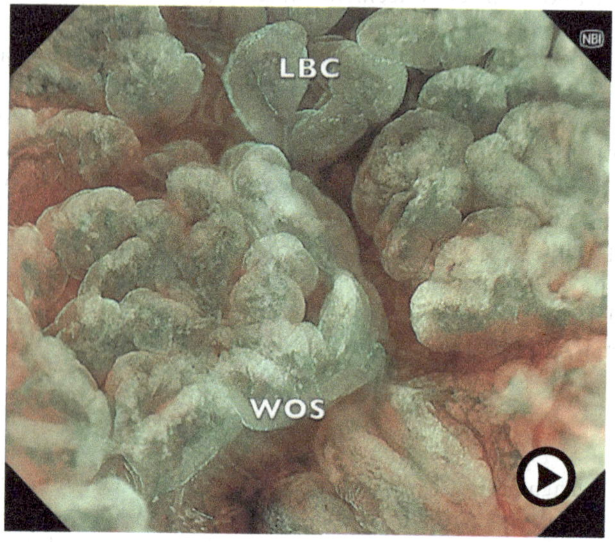

◘ **Fig. 1** Practical observation and diagnosis of gastric intestinal metaplasia with TXI and NBI (▶ https://doi.org/10.1007/000-c8c)

Reference

1. Iwagami H, Uedo N, Yip HC, et al. Ultra-magnifying narrow-band imaging for endoscopic diagnosis of gastric intestinal metaplasia: a pilot image analysis study. Endosc Int Open. 2021;9:E522–9.

Gastric Hyperplastic Polyp with Differentiated Intramucosal Adenocarcinoma

Yoichi Akazawa and Hiroya Ueyama

Content

Supplementary Information The online version contains supplementary material available at https://doi.org/10.1007/978-981-97-2732-2_36. The videos can be accessed individually by clicking the DOI link in the accompanying figure caption or by scanning this link with the SN More Media App.

Y. Sano et al. (eds.), *Atlas of Advanced Endoscopy*, https://doi.org/10.1007/978-981-97-2732-2_36

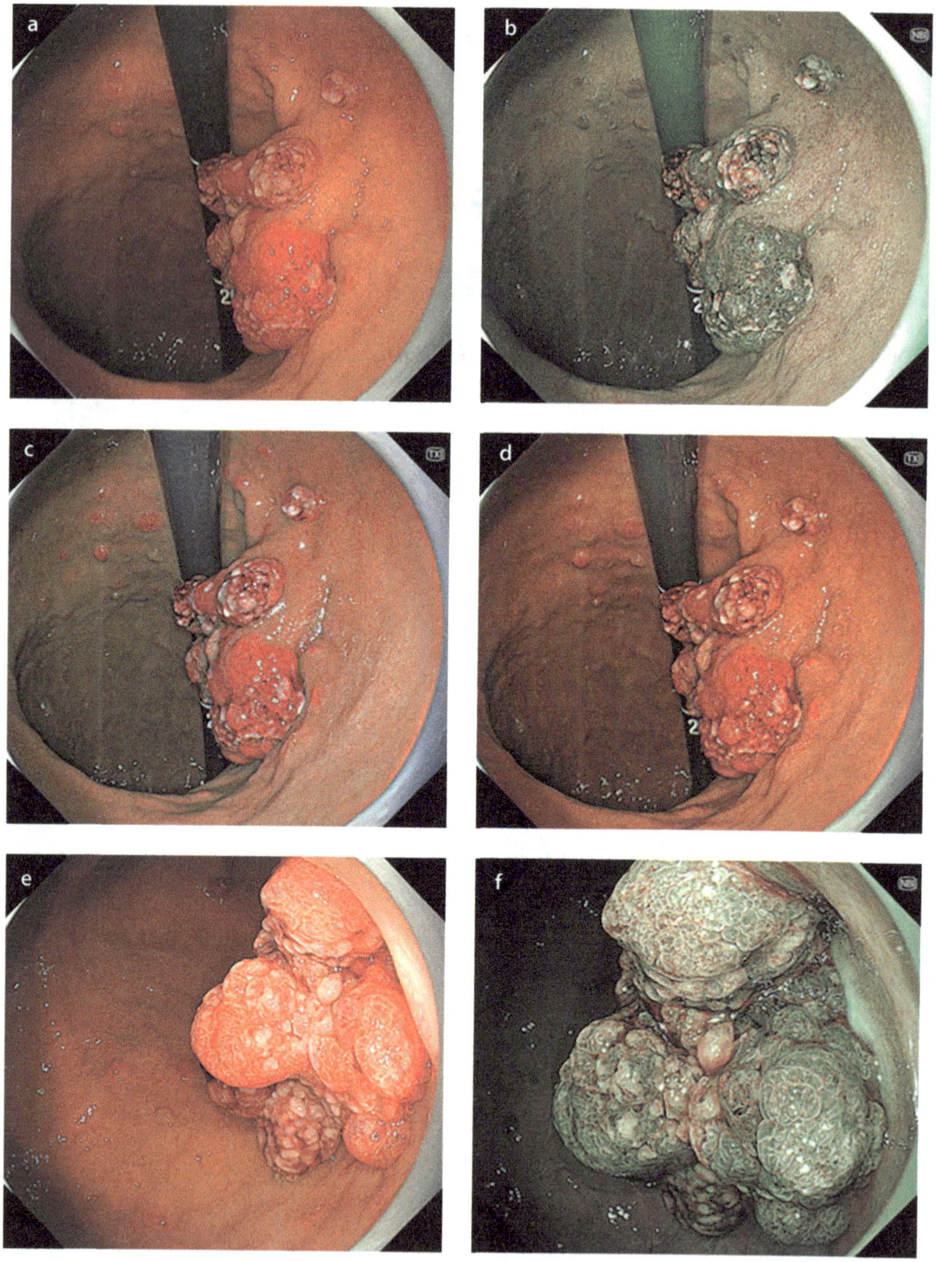
a
b
c
d
e
f

▪ Pathological Diagnosis

Well-differentiated intramucosal adenocarcinoma arising from hyperplastic polyp

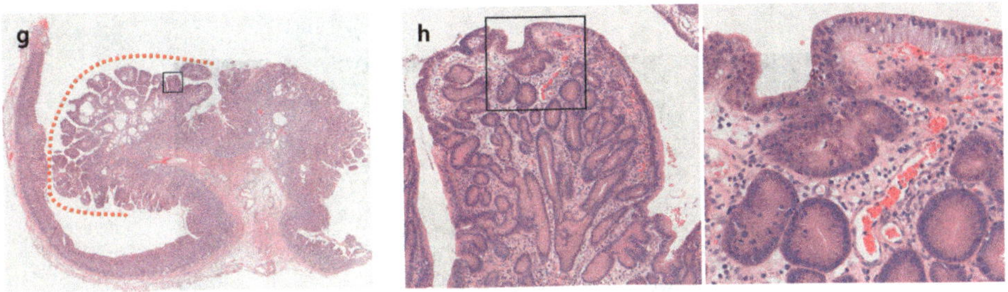

▪ Explanation

In retroflexed view, a bright reddish protruded lesion approximately 30 mm was located at the lesser curvature of the middle third of the stomach. The apex of the lesion showed a lobular and granular appearance with whitish exudate and was easily bleeding which seems to be a giant hyperplastic polyp (Fig. a). Narrow Band Imaging (NBI) enhanced the margin of the lesion more clearly than white light imaging (WLI) (Fig. b). Texture and color enhancement imaging (TXI) mode 1/mode 2 enhanced the color contrast of the red and whitish areas (Figs. c and d). In a forward view close to the lesion, WLI showed that the surface structure was uneven and irregular and the mucosa was partially white in color (Fig. e). NBI emphasized the surface structure showing nonuniform granular structures of different sizes (Fig. f). Based on these endoscopic findings, the lesion was suspected to be an intramucosal adenocarcinoma arising from a hyperplastic polyp and resected by endoscopic submucosal dissection. Histopathologically, the specimen showed a pedunculated polyp with lobular surface structures (Fig. g). Histopathological findings for the most part were non-neoplastic hyperplastic polyps. At the surface area of the tumor (*red dotted line in* Fig. h), irregular branching ducts with enlarged round nuclei were observed and diagnosed as well-differentiated adenocarcinoma arising from a hyperplastic polyp (Fig. h).

Tips

Hyperplastic polyps are the most frequent type of gastric polyps and occur with chronic inflammation such as *Helicobacter pylori*-related gastritis or autoimmune gastritis. Gastric hyperplastic polyps are usually small, flat, or sessile dome-shaped lesions with smooth surfaces and sometimes have erosions on their surface. A small number (<2%–6%) of gastric hyperplastic polyps, especially the larger ones (>1–2 cm), coexist with focal dysplasia or adenocarcinoma [1, 2]. Generally, the cancerous area in hyperplastic polyps shows mucosal surface irregularity and mucosal whitening in WLI. Magnifying endoscopy with NBI is useful to diagnose the cancerous area with irregular microvascular or microsurface pattern. Therefore, in large gastric hyperplastic polyps, it is important to perform target biopsies in areas where cancer is suspected referring to such endoscopic findings [3]. Even if the biopsy results do not reveal cancer in a large hyperplastic polyp, endoscopic resection should be considered as a total biopsy.

■ Video

A short video is available as ◘ Fig. 1.

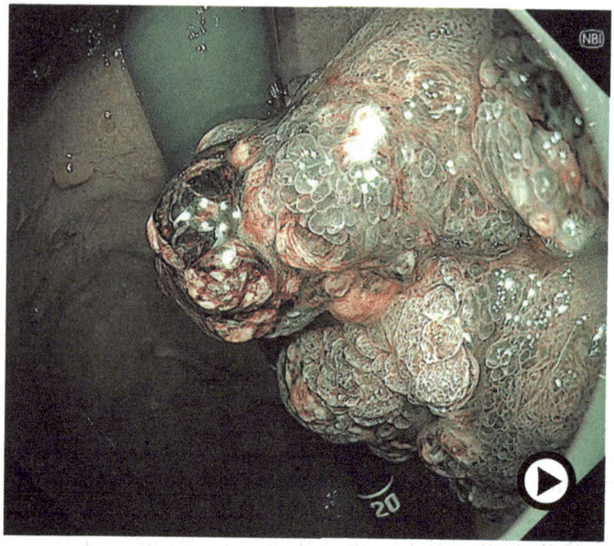

◘ **Fig. 1** Practical observation and diagnosis of gastric hyperplastic polyp with differentiated intramucosal adenocarcinoma with TXI and NBI (▶ https://doi.org/10.1007/000-c8d)

References

1. Yoshida K, Mikami T, Oyama T, Sato Y, Saito T, Mikami T, et al. Signet-ring cell carcinoma in hyperplastic polyp of the stomach. Intern Med. 2019;58:3531–5.
2. Markowski AR, Markowska A, Guzinska-Ustymowicz K. Pathophysiological and clinical aspects of gastric hyperplastic polyps. World J Gastroenterol. 2016;22:8883–91.
3. Horiuchi H, Kaise M, Inomata H, Yoshida Y, Kato M, Toyoizumi H, et al. Magnifying endoscopy combined with narrow band imaging may help to predict neoplasia coexisting with gastric hyperplastic polyps. Scand J Gastroenterol. 2013;48:626–32.

Gastric Adenoma

Takashi Kanesaka and Shigenori Nagata

Supplementary Information The online version contains supplementary material available at https://doi.org/10.1007/978-981-97-2732-2_37. The videos can be accessed individually by clicking the DOI link in the accompanying figure caption or by scanning this link with the SN More Media App.

Y. Sano et al. (eds.), *Atlas of Advanced Endoscopy*, https://doi.org/10.1007/978-981-97-2732-2_37

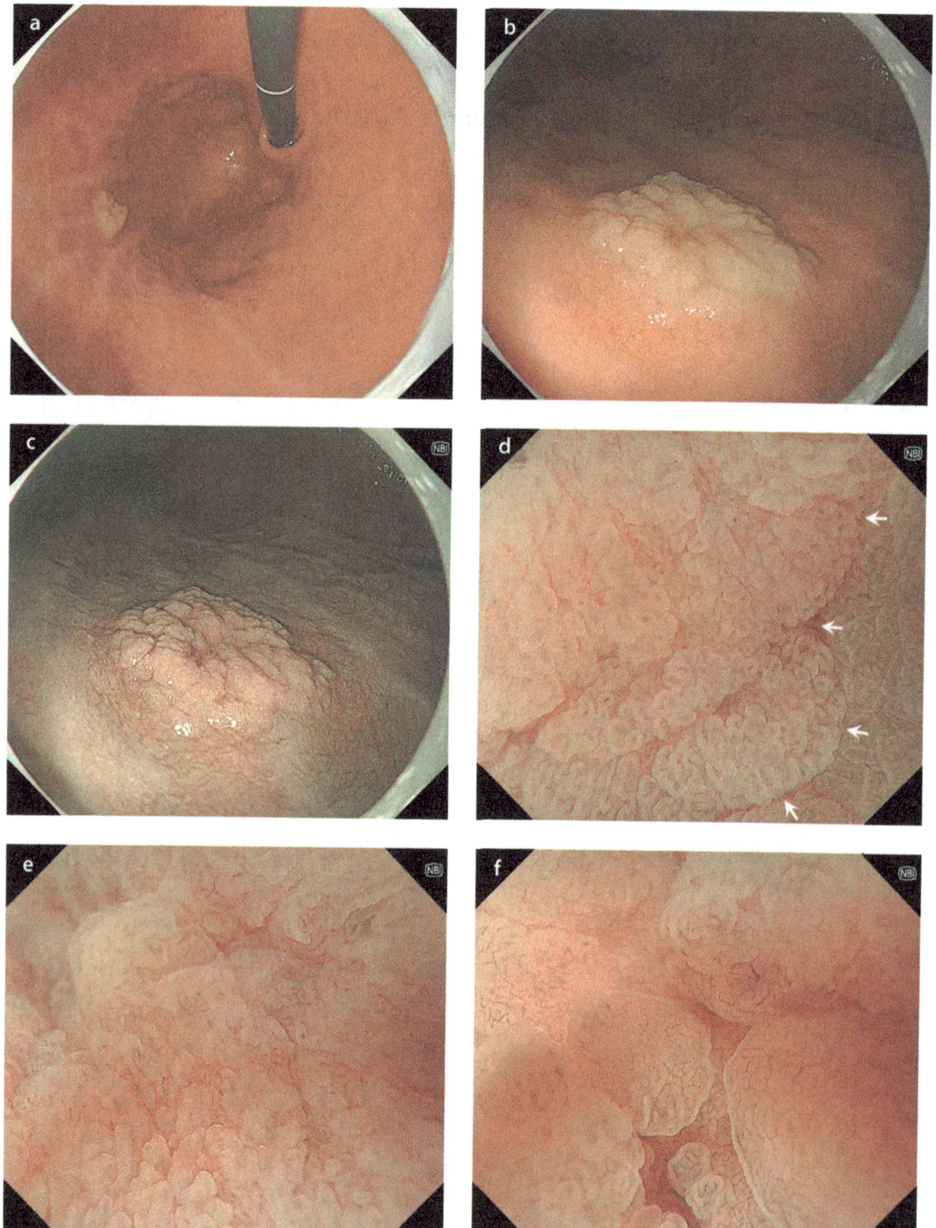
a
b
c
NBI
d
NBI
e
NBI
f
NBI

■ Pathological Diagnosis

Pyloric gland-type adenoma

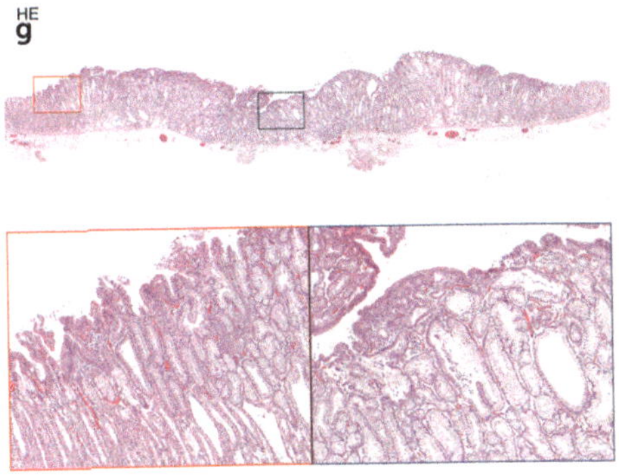

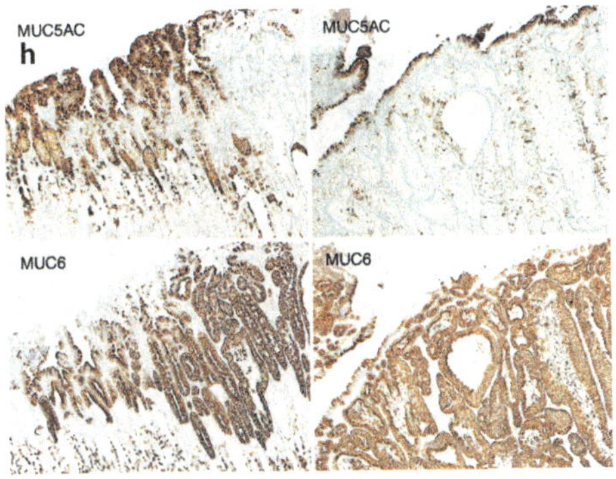

■ Explanation

A 20-mm superficial elevated lesion (0-IIa) has been detected in the greater curvature of the upper gastric body (Fig. a). The background mucosa shows the normal corpus mucosa without atrophic change. (The patient has no history of *H. pylori* eradication, and the serological findings indicate the absence of *H. pylori* infection.) The lesion has a whitish color and coarse granular surface, and the extent of the lesion is clearly identified even when observed with non-magnifying WLI (Fig. b). Non-magnifying NBI clearly shows the surface structure of the lesion (Fig. c). When switching to high-magnification NBI, in the marginal part of the lesion, the microvessels are slightly shown within epithelial circle (Fig. d). A demarcation line between the lesion and the background mucosa is clearly identified as shown by arrows. Although the evaluation of morphology of the microvessels is impossible in this part, the microsurface pattern is considered to be regular because similar-sized round epithelium was densely shown. However, in the central part of the lesion, groove-like depressions are observed (Figs. e and f). In this part, irregular polygonal microvessels are clearly observed (Fig. f). This lesion was resected using the ESD technique. Histopathological examination revealed a pyloric gland-type adenoma (Figs. g and h). Tumor glands are strongly positive for MUC6 and the superficial layer expresses MUC5AC. Pyloric gland-type adenoma was reported first in 1990. It occurs more frequently in women. The predilection site in the stomach is the corpus mucosa. In 30% of pyloric gland-type adenomas, transition to differentiated adenocarcinoma of predominantly gastric phenotype has been noted.

Tips

As the infection rate of *H. pylori* is declining, attention should be paid to tumors arising from non-atrophic mucosa. This case can be easily detected by observation after cleansing the stomach, and it is relatively easy to diagnose a gastric-type tumor because the findings of WLI are characteristic, i.e., a whitish, granular, superficial elevated lesion arising from a non-atrophic gastric mucosa. However, in this case, it is difficult to distinguish between cancer and adenoma even with magnifying NBI.

Video

A short video is available as Fig. 1.

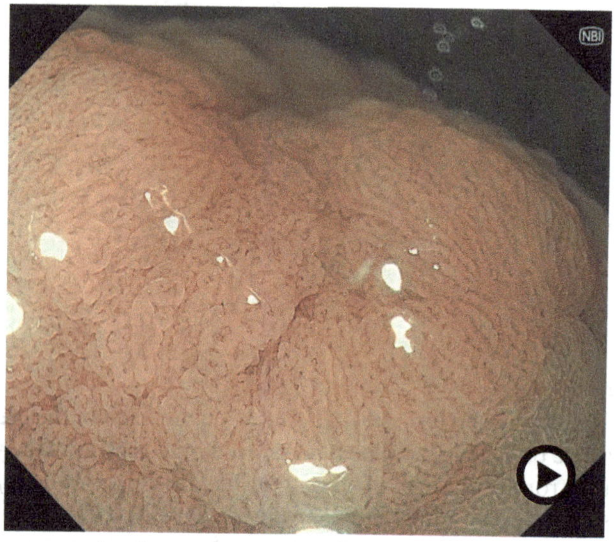

Fig. 1 Practical observation and diagnosis of gastric adenoma with magnifying NBI (▶ https://doi.org/10.1007/000-c8e)

Early Gastric Cancer (0-I, Differentiated Type)

Kenshi Yao and Takao Kanemitsu

Y. Sano et al. (eds.), *Atlas of Advanced Endoscopy*, https://doi.org/10.1007/978-981-97-2732-2_38

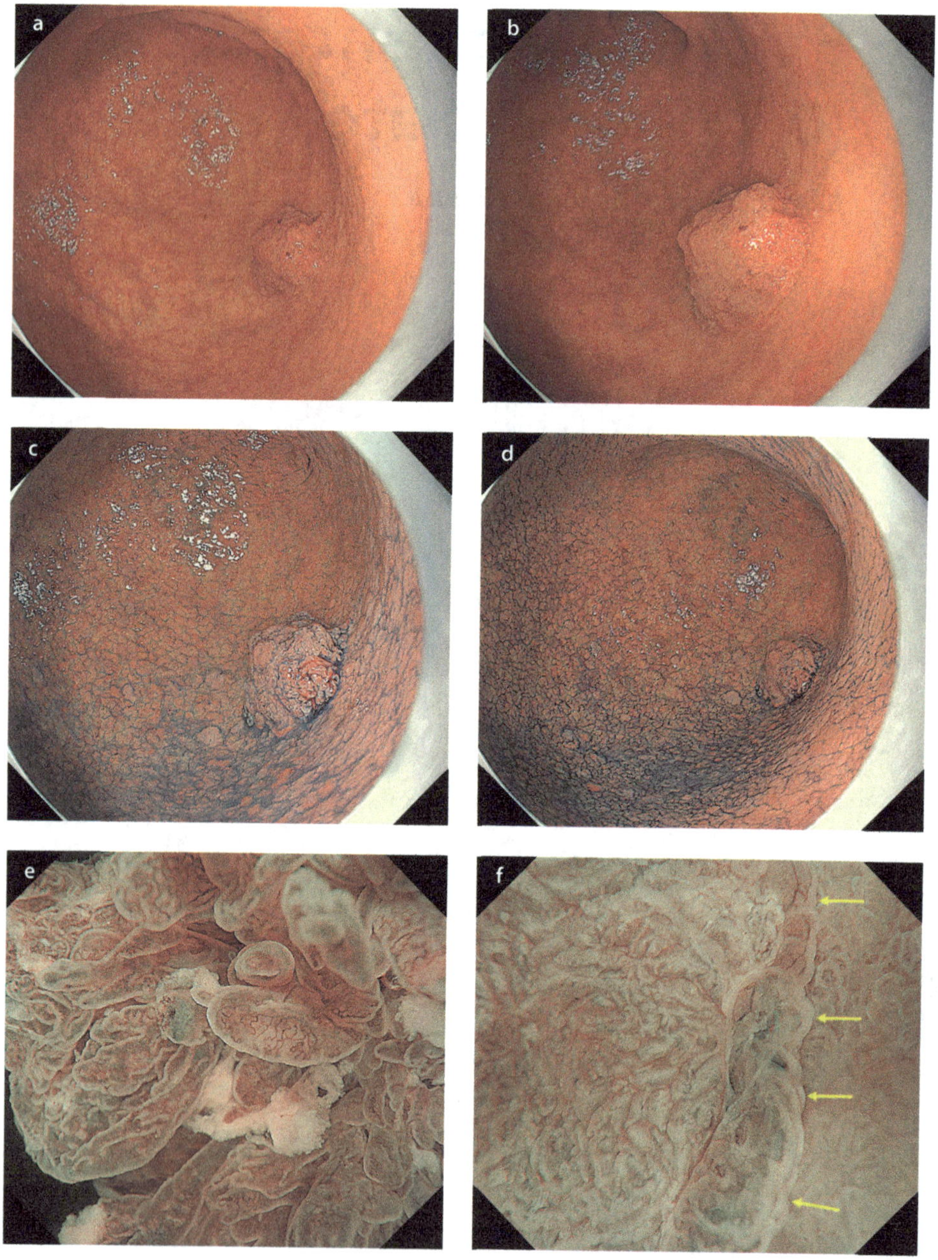

Pathological Diagnosis

Well-differentiated adenocarcinoma, pT1a(MM), Ly1, V0, CAT 1, SAT 1, INFa, HM0, VM0 (Since it was a non-curative resection, additional surgery was explained, but the patient wanted to be followed up. The patient has been recurrence-free for 2 years postoperatively.)

g

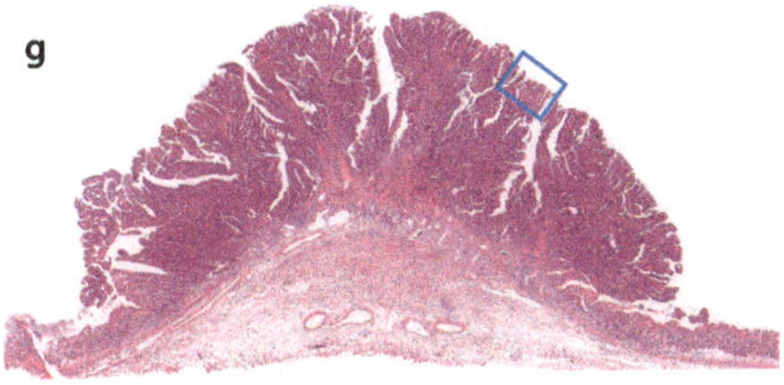

h

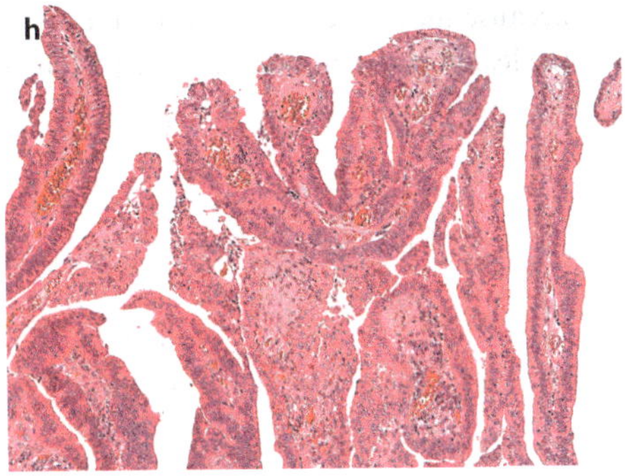

Explanation

White light observation: A 20-mm-sized, semipedunculated lesion is seen on the posterior wall of the greater curvature of the gastric angle. The background mucosa is markedly atrophic (Fig. a). The lesion has a mixture of erythematous and whitish coloration, and its surface is rough (Fig. b).

Indigo carmine dye spray observation: The surface of the lesion is irregularly granular with irregular sizes (Fig. c). Since there was no obvious non-extension sign on the observation of the gastric wall under strong extension, this lesion is diagnosed as intramucosal carcinoma (Fig. d).

Magnifying endoscopy with NBI (at the apex of the lesion) (Fig. e): The lesion was judged to have an irregular microvascular (MV) pattern because of the heterogeneity of the shape of individual microvessels. The microsurface (MS) pattern of the lesion was determined to be irregular MS pattern, because the intervening parts were irregularly shaped and irregularly arranged with a villi-like morphology.

Magnifying endoscopy with NBI (at the base of the lesion) (Fig. f): A clear demarcation line between the lesion and the background mucosa (indicated by yellow arrows). The MV pattern was judged as absent MV pattern because it could not be evaluated due to the presence of white opaque substance. The MS pattern of the lesion is judged as irregular MS pattern, because the individual WOS are irregularly mottled, nonuniform in shape, and irregularly arranged. Based on the above enlarged endoscopic findings, a diagnosis of differentiated carcinoma is possible.

Histopathological findings of ESD specimens (H&E stain): The lesion consists of tall, atypical glandular ducts with narrow interstitial. The lesion is an intramucosal carcinoma (Fig. g). Nuclear atypia and structural atypia are seen in the ducts of the tumor, which is a finding of well-differentiated tubular adenocarcinoma (Fig. h).

Tips

Magnification of type 0-I lesions in the stomach is often difficult to observe in detail because the field of view is difficult to fix and often bleeds. It is important to observe the lesion under flooding, at a short distance, and with a rather weak magnification.

Early Gastric Cancer (0-IIa + IIc, Differentiated Type)

Kunihisa Uchita

Supplementary Information The online version contains supplementary material available at https://doi.org/10.1007/978-981-97-2732-2_39. The videos can be accessed individually by clicking the DOI link in the accompanying figure caption or by scanning this link with the SN More Media App.

Y. Sano et al. (eds.), *Atlas of Advanced Endoscopy*, https://doi.org/10.1007/978-981-97-2732-2_39

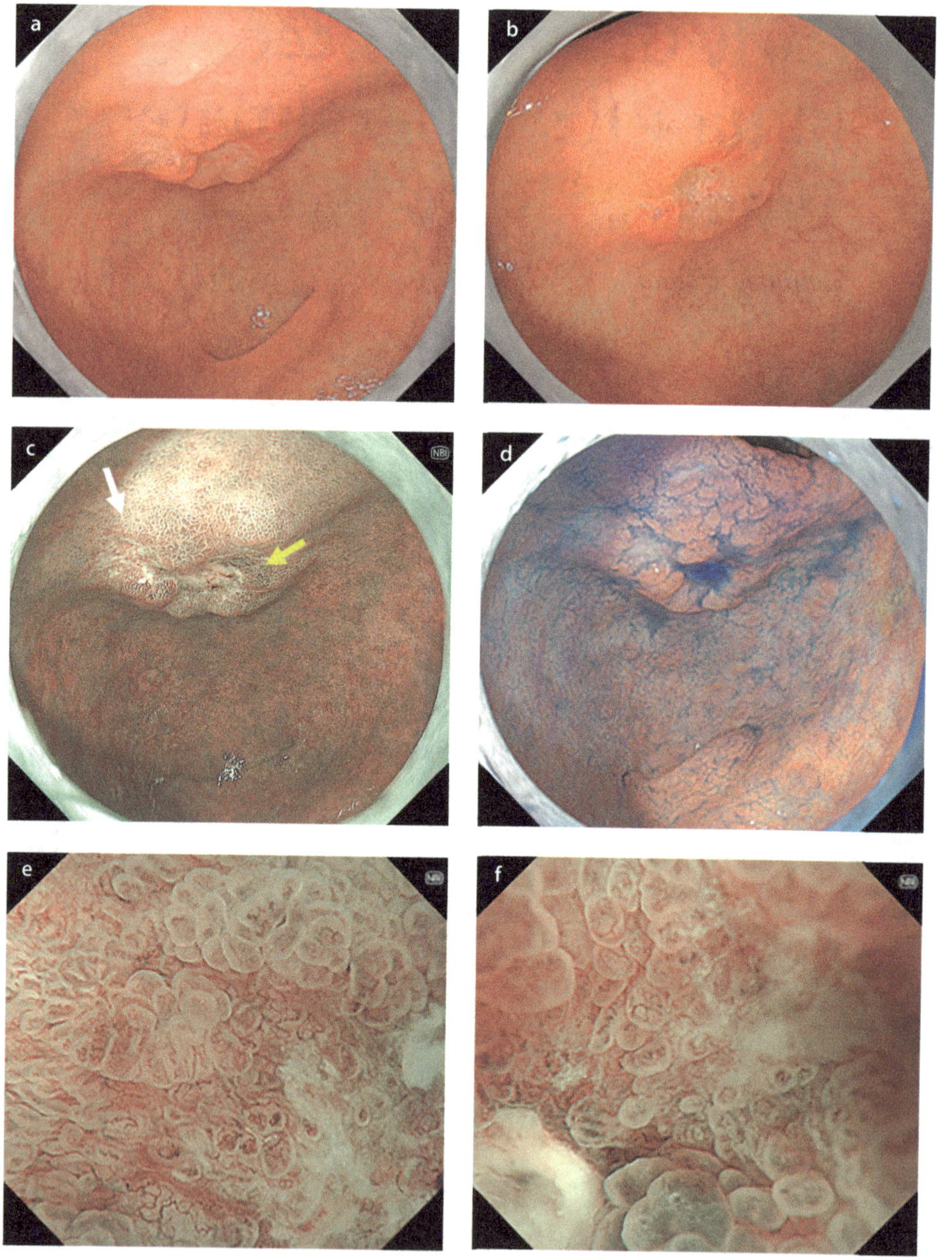
a
b
c
NBI
d
e
f

▪ Pathological Diagnosis

Moderately differentiated tubular adenocarcinoma invading the submucosa (tub2 > tub1 > por2, pT1b1 (SM330μm), pUL1, Ly0, V0)

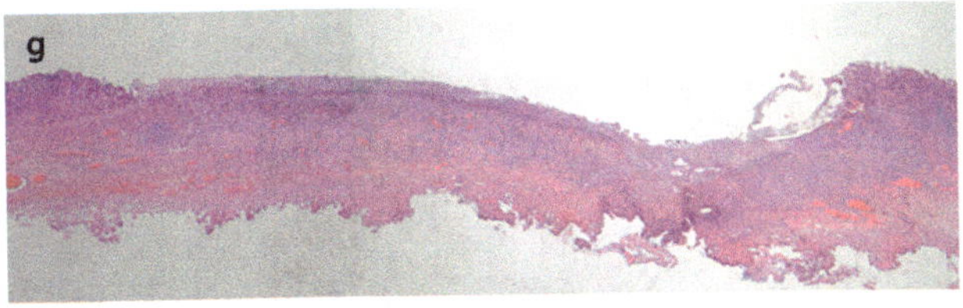

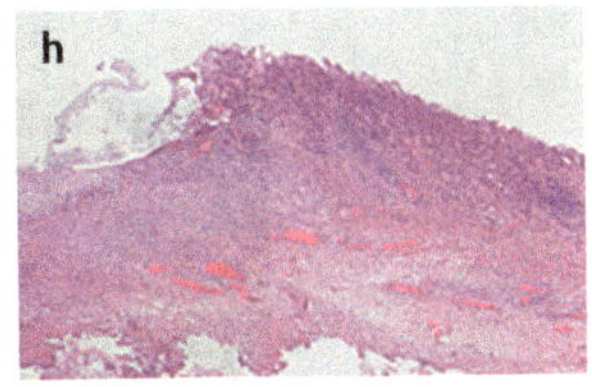

▪ Explanation

The conventional endoscopy revealed a 25-mm elevated lesion with a depression at the lesser curvature of the gastric antrum (Fig. a). Part of the depression area was covered with white coat. The surrounding mucosa rose smoothly like a submucosal tumor (SMT), and the appearance of the overlying mucosa was also smooth (Fig. b). These findings suggest the possibility of submucosal invasion by the tumor. Under non-magnifying NBI (Fig. c), the microsurface pattern of the elevated area appeared regular. On the other hand, chromoendoscopy made the depression part more evident, and the margin appeared irregular, indicating possible encroachment (Fig. d). Figures e and f show the findings of the magnifying endoscopy with NBI (ME-NBI) in the areas indicated by the white and yellow arrows in Fig. c, respectively. The lesion was slightly depressed, with a clear demarcation line, an unequal microstructure, and capillaries of varied shapes (Fig. e). In the lower left of Fig. f, white coat was recognized, along with an unequal and papillary structure known as the "vec" pattern. The microvascular pattern was also irregular around white coat. Histopathological examination revealed a moderately differentiated tubular adenocarcinoma invading the submucosa (tub2 > tub1 > por2, pT1b1 (SM330μm), pUL1, Ly0, V0). At the borderline, cancer cells extended under the noncancerous epithelium, presenting SMT-like elevations (Figs. g and h).

Tips

Based on the irregular depression with ulceration seen in conventional chromoendoscopy and the irregular microsurface and microvascular pattern observed with ME-NBI, this lesion can be confidently diagnosed as gastric cancer. The smooth raising of the surrounding mucosa like an SMT and the regular microstructure are important indicators of SM invasion.

Video

A short video is available as Fig. 1.

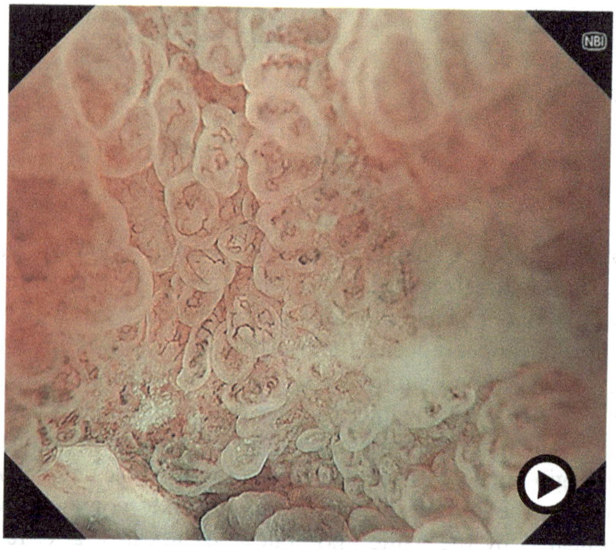

Fig. 1 Practical observation and diagnosis of early gastric cancer with NBI and chromoendoscopy (▶ https://doi.org/10.1007/000-c8f)

Early Gastric Cancer (0-IIc, Differentiated Type) (1)

Noriya Uedo and James Weiquan Li

Content

Supplementary Information The online version contains supplementary material available at https://doi.org/10.1007/978-981-97-2732-2_40. The videos can be accessed individually by clicking the DOI link in the accompanying figure caption or by scanning this link with the SN More Media App.

Y. Sano et al. (eds.), *Atlas of Advanced Endoscopy*, https://doi.org/10.1007/978-981-97-2732-2_40

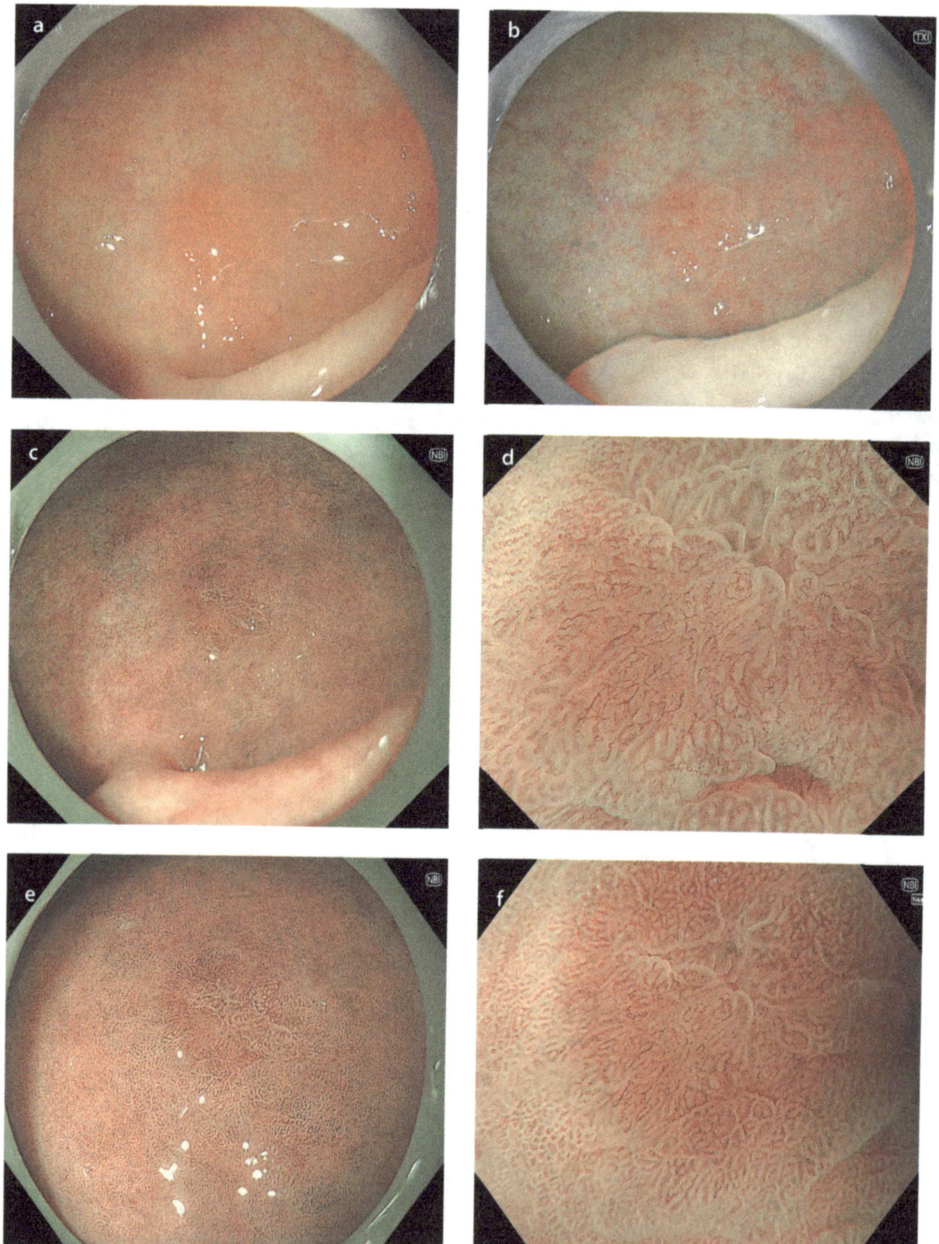
a
b
TXI
c
NBI
d
NBI
e
NBI
f
NBI

Pathological Diagnosis

Well-differentiated tubular adenocarcinoma, 5 × 5 mm, intramucosal (pT1a), lymphovascular invasion negative, no ulcerative findings, horizontal margin negative, and vertical margin negative

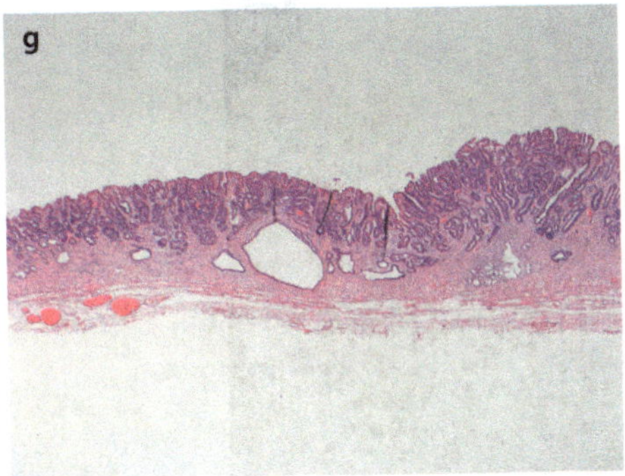

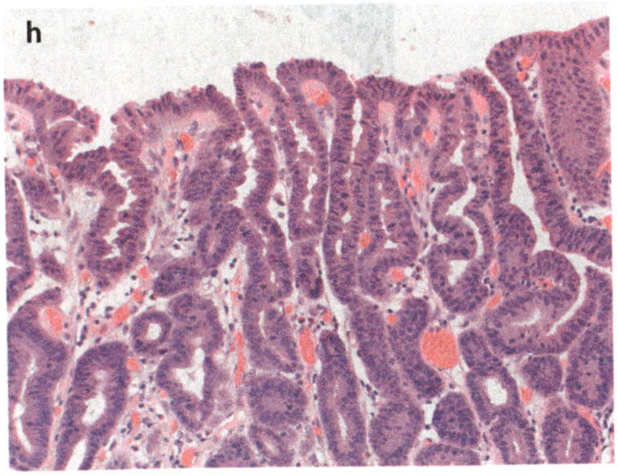

Explanation

A woman in her 80s presented for evaluation of small early gastric cancer in the antrum. A localized redness was found as a suspicious lesion in white light image (WLI, Fig. a). Texture and color enhancement imaging (TXI, mode 1) enhanced reddish color of the lesion (Fig. b). Narrowband imaging (NBI) showed the lesion as a brownish area (Fig. c). Magnifying NBI image using water-immersion technique clearly revealed irregular microvessel pattern within the demarcation line, indicating the lesion as an early gastric cancer (Fig. d). In the center of the lesion (upper middle in the Fig. d), there was a regenerative epithelium caused by biopsy performed by a referral endoscopist. The lesion was re-examined by near-focus video endoscope with the expanded-depth-of-focus (EDOF) function (Fig. e). Because of the EDOF function, panoramic view of the entire lesion was obtained without fine adjustment of the focus with magnification (Fig. f). The lesion was removed by endoscopic submucosal dissection technique, and histopathological examination of the resected specimen was well-differentiated tubular adenocarcinoma (Figs. g and h).

Tips

Initially, the mucosa is carefully observed to first find localized irregularity (unevenness) of color or surface in WLI. When the suspicious lesion for neoplasm is found, the lesion is inspected with magnifying NBI to make differential diagnosis between carcinoma and non-carcinoma [1]. A soft black cap (MAJ-1989; Olympus Co. Ltd., Tokyo, Japan) is mandatory to stabilize endoscopic images for magnifying observation. The tip of the cap is touched on the proximal non-neoplastic mucosa to avoid contact bleeding. Distance between the object (mucosal surface) and the lens is manually adjusted by fine movement of scope push-pull maneuver and air deflation. In addition, the magnification lever on the scope control head is used for adjusting the focus. Water-immersion technique avoid catch light at the mucosal surface and achieve slight natural magnification effect by the different reflection ratio of the water.

■ Video

A short video is available as ◘ Fig. 1.

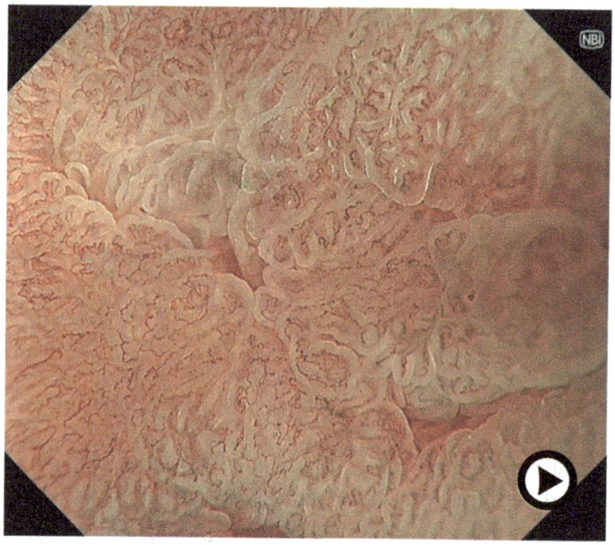

◘ **Fig. 1** Practical observation and diagnosis of early gastric cancer with TXI and NBI. A zoom video-endoscope (EVIS-XZ1200, Olympus) and near-focus video-endoscope with EDOF function (EVIS-EZ1500; Olympus) are used with the new image processor (CV-1500; Olympus) for inspection. Structure enhancement mode was set at B5 for WLI and B8 for NBI. The soft black cap (MAJ-1989, Olympus) was mounted on the tip of the endoscopes. The protruding length of the cap is adjusted to make the long dimension of the endoscopic image to be 4.5 mm for XZ1200. After evaluation of microvessel and microsurface pattern of the lesion using XZ1200, the lesion is reassessed with the EZ1500 (▶ https://doi.org/10.1007/000-c8g)

Reference

1. Muto M, Yao K, Kaise M, et al. Magnifying endoscopy simple diagnostic algorithm for early gastric cancer (MESDA-G). Dig Endosc. 2016;28:379–93.

Early Gastric Cancer (0-IIc, Differentiated Type) (2)

Jun Liang Teh and Hon Chi Yip

Supplementary Information The online version contains supplementary material available at https://doi.org/10.1007/978-981-97-2732-2_41. The videos can be accessed individually by clicking the DOI link in the accompanying figure caption or by scanning this link with the SN More Media App.

Y. Sano et al. (eds.), *Atlas of Advanced Endoscopy*, https://doi.org/10.1007/978-981-97-2732-2_41

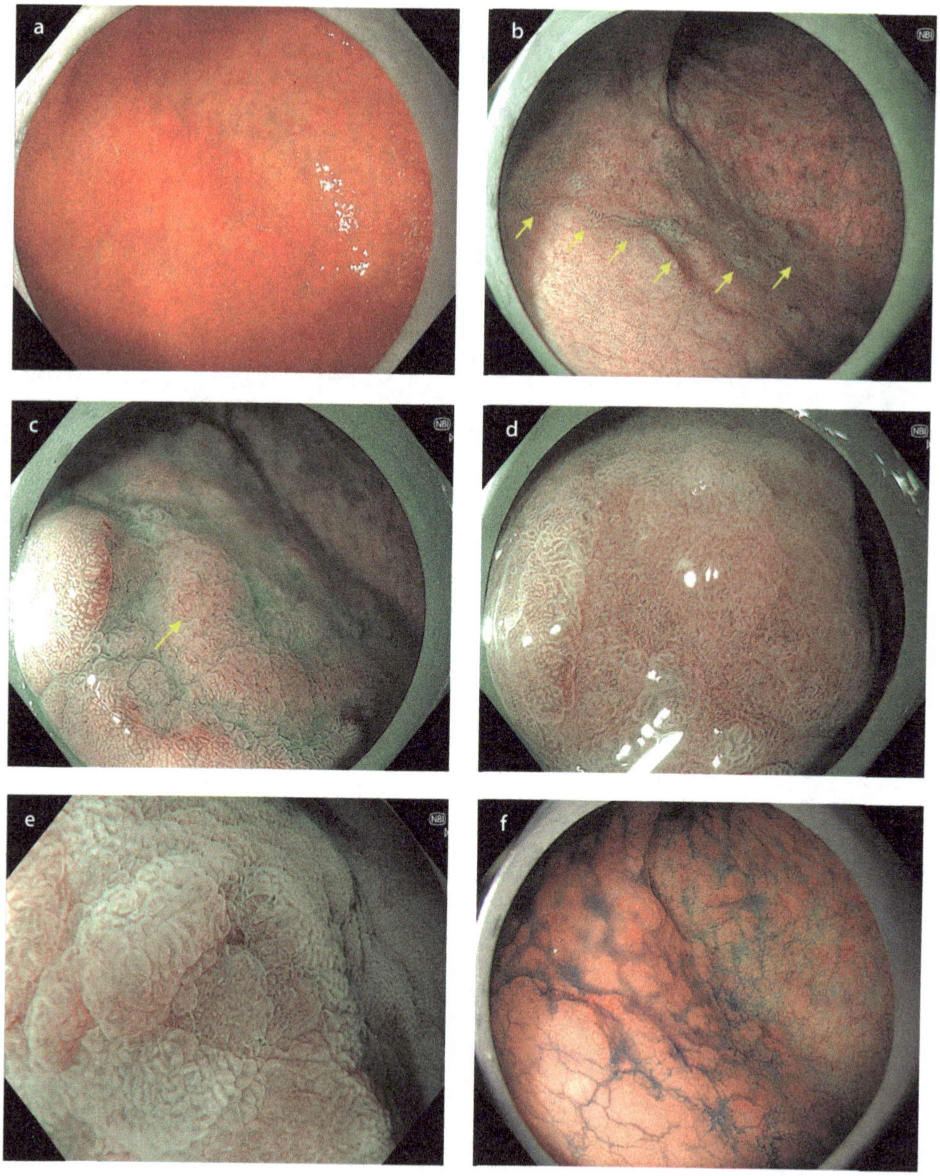
a
b
NBI
c
NBI
d
NBI
e
NBI
f

■ Pathological Diagnosis

Moderately differentiated tubular adenocarcinoma (T1a).

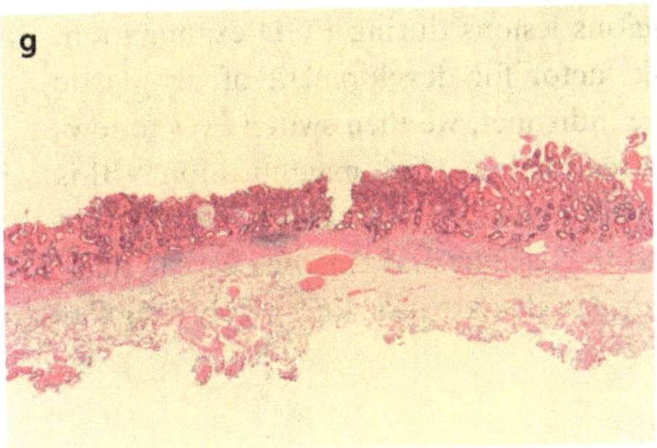

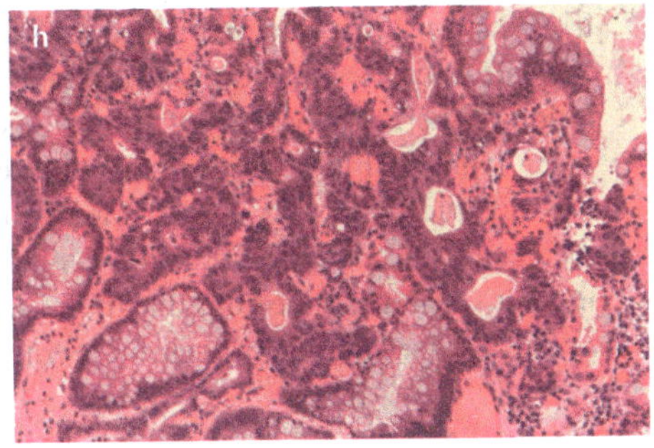

■ Explanation

Non-magnifying WLI observation with GIF-EZ1500 endoscope revealed a 2.5-cm Paris 0-IIb + IIc reddened lesion in the anterior wall of the proximal antrum on a background of chronic atrophic gastritis. The lesion appeared demarcated with irregularity in surface and color, associated with nodularity and depressed areas (Fig. a). On NBI observation, demarcation line can be appreciated clearly (yellow arrow) (Fig. b). Using high magnification with NBI near-focus mode, a clear demarcation on the oral side of the lesion can again be appreciated. The subepithelial capillary network (SECN) appeared uncoiled with irregular polarity suggestive of irregular microvascular pattern (IMVP) (yellow arrow in Fig. c). Within the demarcation line (Fig. d), the marginal crypt epithelium demonstrated nonuniform morphology with varied directionality, representing irregular microsurface pattern (IMSP). Proliferation of the SECN is also observed with varied morphology (open-loop, linear) and differing diameters. At the inferior edge of the image, irregular microvascular pattern is seen to encroach into the noncancerous epithelium, suggesting the presence of intraepithelium microinvasion (IEMI). In the adjacent normal mucosa, the presence of light blue crest as well as white opaque substance suggested the presence of intestinal metaplasia (Fig. e). Chromoendoscopy with 0.04% indigo carmine also delineated the macroscopic margin of the lesion (Fig. f). Non-extension sign is absent. The endoscopic diagnosis is 2.5-cm intramucosal differentiated-type adenocarcinoma without ulceration, fulfilling indication for endoscopic resection. This lesion was resected by endoscopic submucosal dissection (ESD). In the ESD specimen, the lesion showed a moderately differentiated tubular adenocarcinoma confined to the mucosa, with negative lymphovascular invasion and negative margins, indicating a curative resection (Figs. g and h).

Tips

With non-magnifying white light endoscopy, we first look for demarcated areas with irregular surface or color to identify suspicious lesions during EGD examination. The presence of atrophic gastritis is a risk factor for development of neoplastic lesions. Where the margins of the lesions are indistinct, we then switch over to low-magnifying NBI to better delineate the demarcation line. High-magnification NBI is used to study the microsurface glandular pattern and microvascular pattern within the demarcation line, the transition zone, and the background mucosa. We can diagnose this lesion as cancerous due to the presence of demarcation line, highly irregular microvascular and microsurface pattern.

Video

A short video is available as Fig. 1.

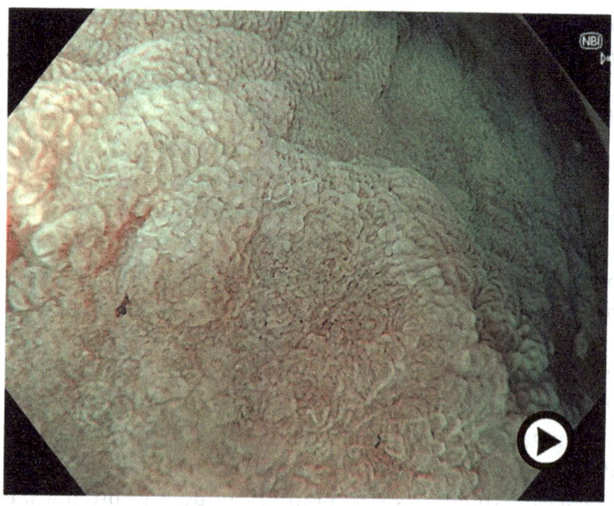

Fig. 1 Practical observation and diagnosis of early gastric cancer with NBI and chromoendoscopy (▶ https://doi.org/10.1007/000-c8h)

Early Gastric Cancer (0-IIc, Differentiated Type) (3)

Takeshi Uozumi and Seiichiro Abe

Content

Supplementary Information The online version contains supplementary material available at https://doi.org/10.1007/978-981-97-2732-2_42. The videos can be accessed individually by clicking the DOI link in the accompanying figure caption or by scanning this link with the SN More Media App.

Y. Sano et al. (eds.), *Atlas of Advanced Endoscopy*, https://doi.org/10.1007/978-981-97-2732-2_42

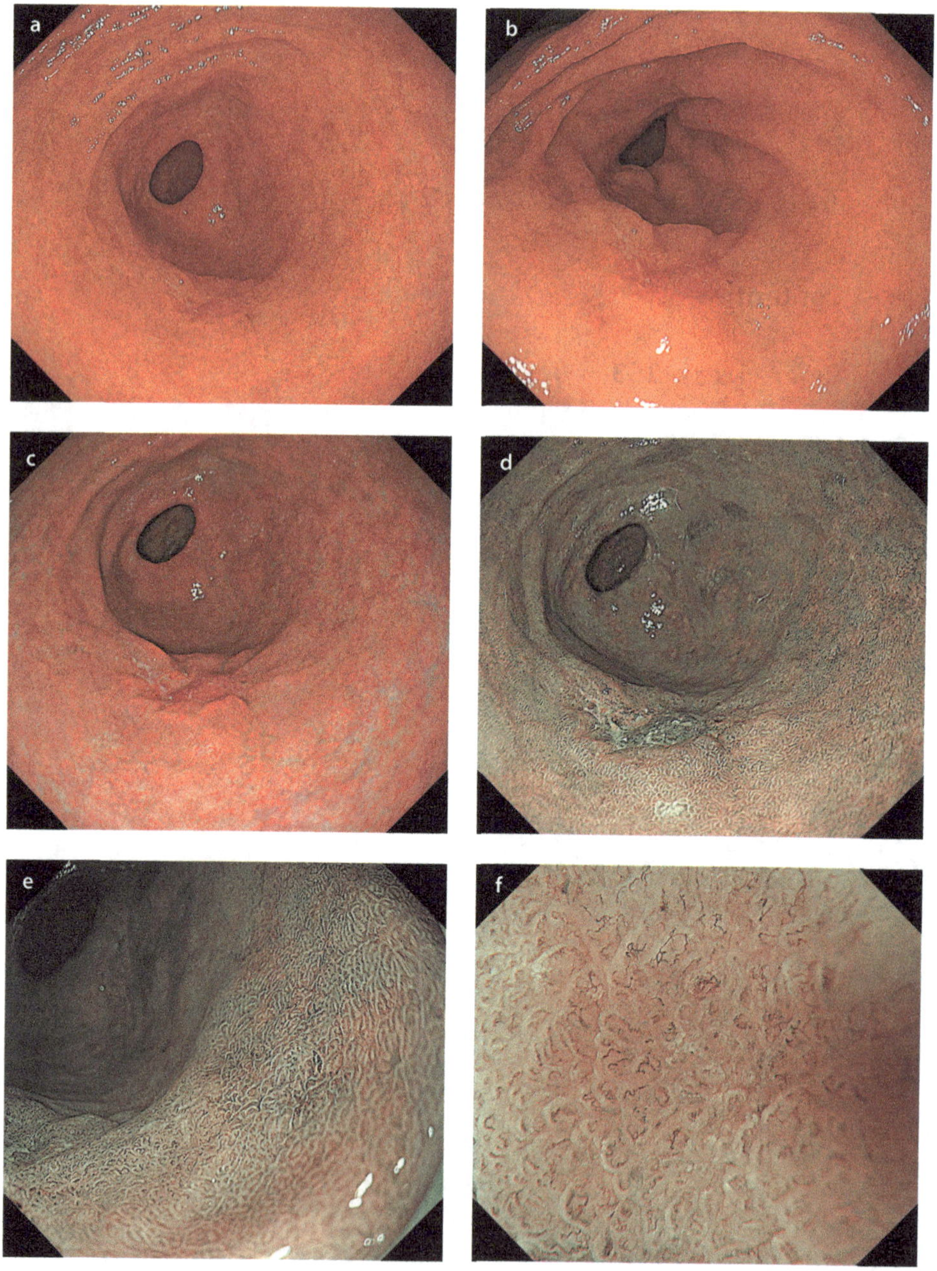
a
b
c
d
e
f

■ Pathological Diagnosis

Tubular adenocarcinoma, moderately differentiated type: 21 × 20 mm, 0-IIc, cT1b(SM).

g

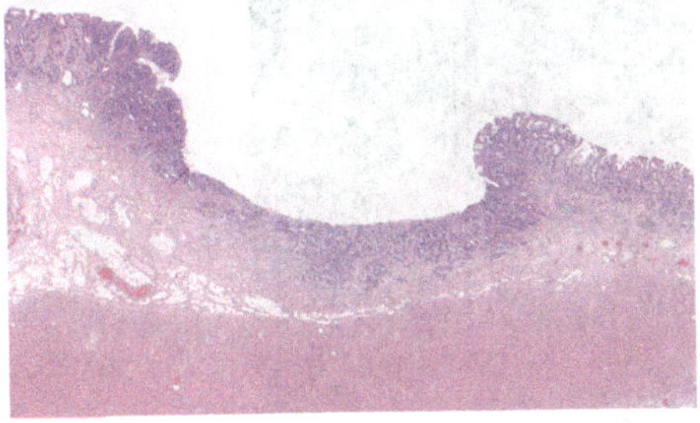

h

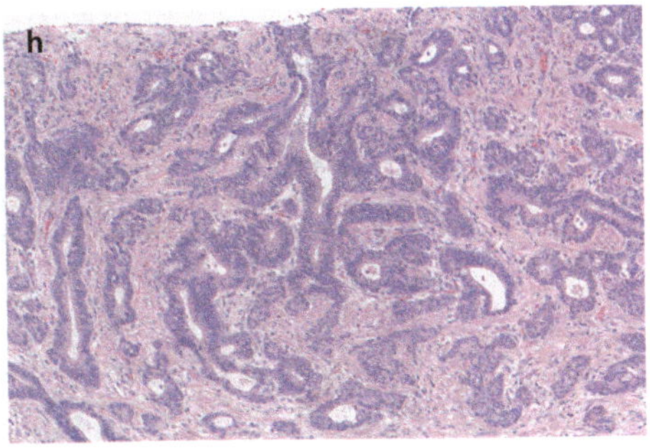

■ Explanation

A 25-mm reddish depressed lesion was identified in the greater curvature of the gastric antrum. This lesion was characterized by asymmetrical remarkable redness and mucosal fold convergence with an irregular margin, which indicated a malignant neoplasm. The non-extension sign was observed with a large volume of air insufflation (Fig. a). Marginal elevation and stiffness of the depressed area stood out with a moderate volume of air insufflation (Fig. b). These endoscopic findings with WLI indicated deep submucosal invasive gastric cancer. TXI (mode 1) enhanced the color contrast and surface structure (Fig. c). Non-magnifying NBI highlighted the lesion as a brownish area, and the margin of the lesion became obvious (Fig. d). Low-power magnification was useful for precisely delineating the demarcation line (Fig. e). High-power magnification revealed irregular microvascular and microsurface pattern (Fig. f). A biopsy from the lesion revealed tubular adenocarcinoma, well to moderately differentiated type. The clinical diagnosis was a 25-mm tubular adenocarcinoma in the gastric antrum, suspicious of submucosal invasion, and distal gastrectomy was performed. Histopathological examination revealed a moderately differentiated tubular adenocarcinoma invading the submucosal layer (900 μm) without peptic ulcer scar, lymphovascular invasion, or lymph node metastasis. The proximal and distal margin of the resected specimen was negative (Figs. g and h).

Tips

Endoscopic resection for early gastric cancer is a minimally invasive treatment, and the indication is described in the JGES guideline [1]. It is important to distinguish the deep submucosal invasive cancer (≥500 μm) from the other shallower one to properly decide on a treatment strategy. A simple scoring model for predicting the depth of invasion is reported [2]. Marginal elevation, tumor size ≥30 mm, remarkable redness, and uneven surface are significantly associated with deep submucosal invasion [2]. A "non-extension sign" is also useful for diagnosing deep submucosal invasive cancer [3]. At the submucosal invasive area, a trapezoid elevation or mucosal folds convergence with elevation can be observed with a large air volume insufflation.

Video

A short video is available as Fig. 1.

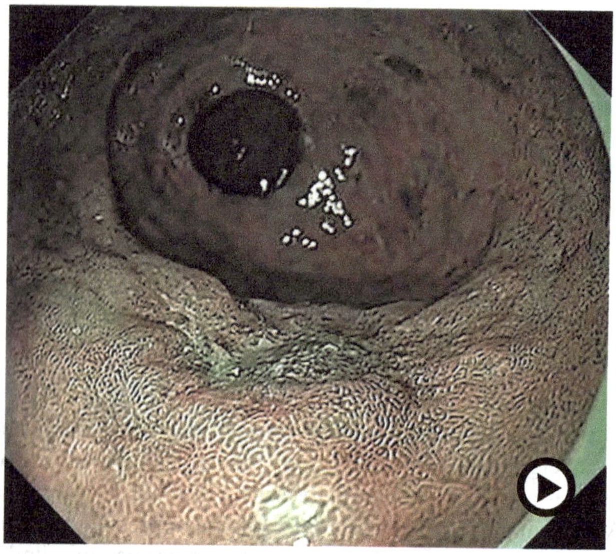

Fig. 1 Practical observation and diagnosis of early gastric cancer with TXI and NBI (▶ https://doi.org/10.1007/000-c8j)

Acknowledgments The authors would like to thank S. Sekine for reviewing the pathological findings.

References

1. Ono H, Yao K, Fujishiro M, et al. Guidelines for endoscopic submucosal dissection and endoscopic mucosal resection for early gastric cancer. Dig Endosc. 2021;33:4–20.
2. Abe S, Oda I, Shimazu T, et al. Depth-predicting score for differentiated early gastric cancer. Gastric Cancer. 2011;14:35–40.
3. Nagahama T, Yao K, Imamura K, et al. Diagnostic performance of conventional endoscopy in the identification of submucosal invasion by early gastric cancer: the "non-extension sign" as a simple diagnostic marker. Gastric Cancer. 2017;20:304–13.

Early Gastric Cancer (0-IIc, Undifferentiated Type) (1)

Toshiaki Hirasawa and Wataru Kurihara

Content

Y. Sano et al. (eds.), *Atlas of Advanced Endoscopy*, https://doi.org/10.1007/978-981-97-2732-2_43

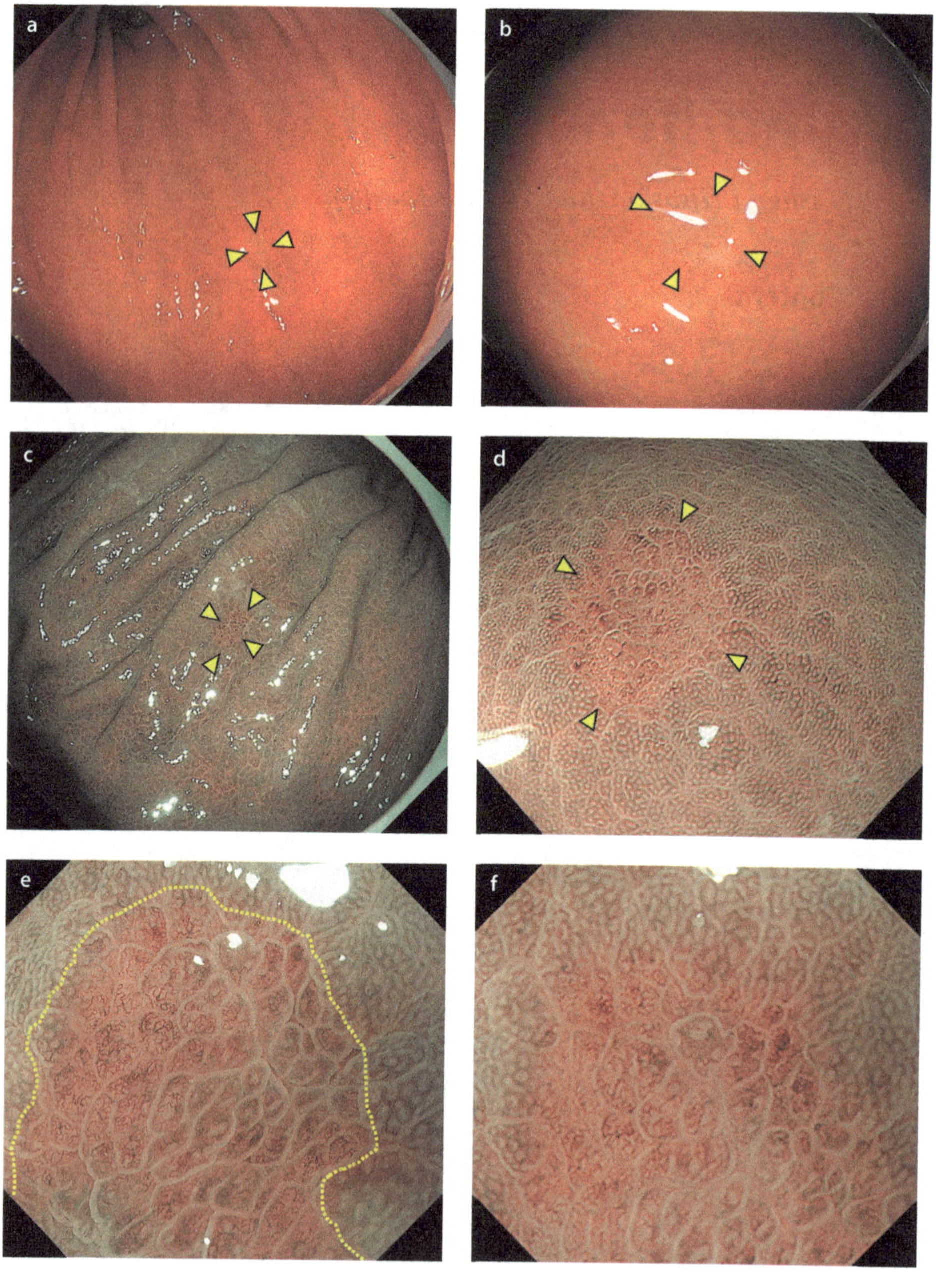
a
b
c
d
e
f

Pathological Diagnosis

Adenocarcinoma, sig > por.

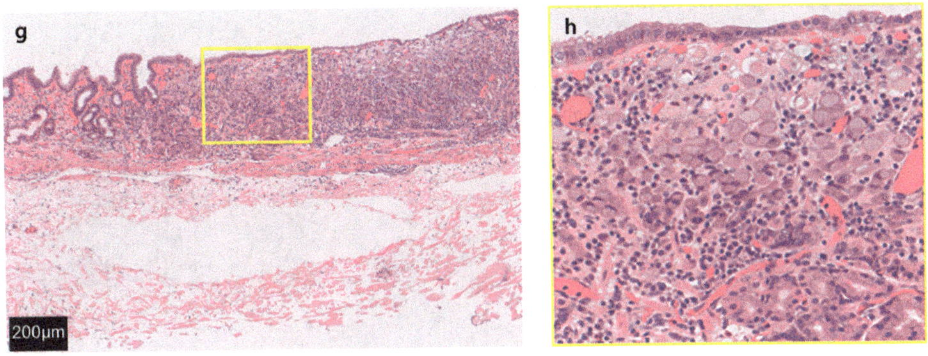

Explanation

WLI reveals a 5-mm-diameter pale and oval lesion in the greater curve of the remnant stomach body. No atrophic changes are observed in the background gastric mucosa (Figs. a and b). NBI shows the lesion as a brownish area (Fig. c). Magnifying endoscopy with NBI (ME-NBI) reveals a slightly depressed lesion with a clear demarcation line of the brownish area (Fig. d). At higher magnification using ME-NBI, expanded intervening parts and wavy microvessels are observed (Figs. e and f). The demarcation line is highlighted by the yellow line. Histologically, signet ring cells are located at whole lamina propria layer and spreading laterally. The proper gastric glands remain at the bottom of the mucosal layer. At the surface, nonneoplastic one-layered foveolar epithelium covers the lesion (Figs. g and h).

Tips

Undifferentiated-type early gastric cancer often occurs in young people without gender differences. WLI enables the identification of a pale lesion particularly in the fundic gland region of the middle and low parts of the stomach. The lesion is often observed as type 0-IIb in the early stage, progressing to type 0-IIc. Using NBI, the pale lesion appears as a brownish area. With ME-NBI, expanded intervening parts, wavy microvessels, and corkscrew patterns can be observed. "Wavy microvessels" refer to dilated and tortuous vessels observed within the surface microstructure, while "corkscrew patterns" refer to dilated and tortuous vessels observed in the obscured surface microstructure. Histopathologically, carcinoma cells confined to the proliferative zone are visualized as expanded intervening parts using ME-NBI [1]. Carcinoma cells extending from the proliferative zone into the superficial layer are visualized as wavy-micro vessels using ME-NBI, while carcinoma cells involving the whole layer of the mucosa are visualized as corkscrew patterns using ME-NBI [1].

Acknowledgments We would like to express our gratitude to Dr. Manabu Takamatsu for his valuable assistance as a pathologist.

Reference

1. Horiuchi Y, et al. Gastric Cancer. 2016;19:515–23.

Early Gastric Cancer (0-IIc, Undifferentiated Type) (2)

Kazuo Shiotsuki and Kohei Takizawa

Y. Sano et al. (eds.), *Atlas of Advanced Endoscopy*, https://doi.org/10.1007/978-981-97-2732-2_44

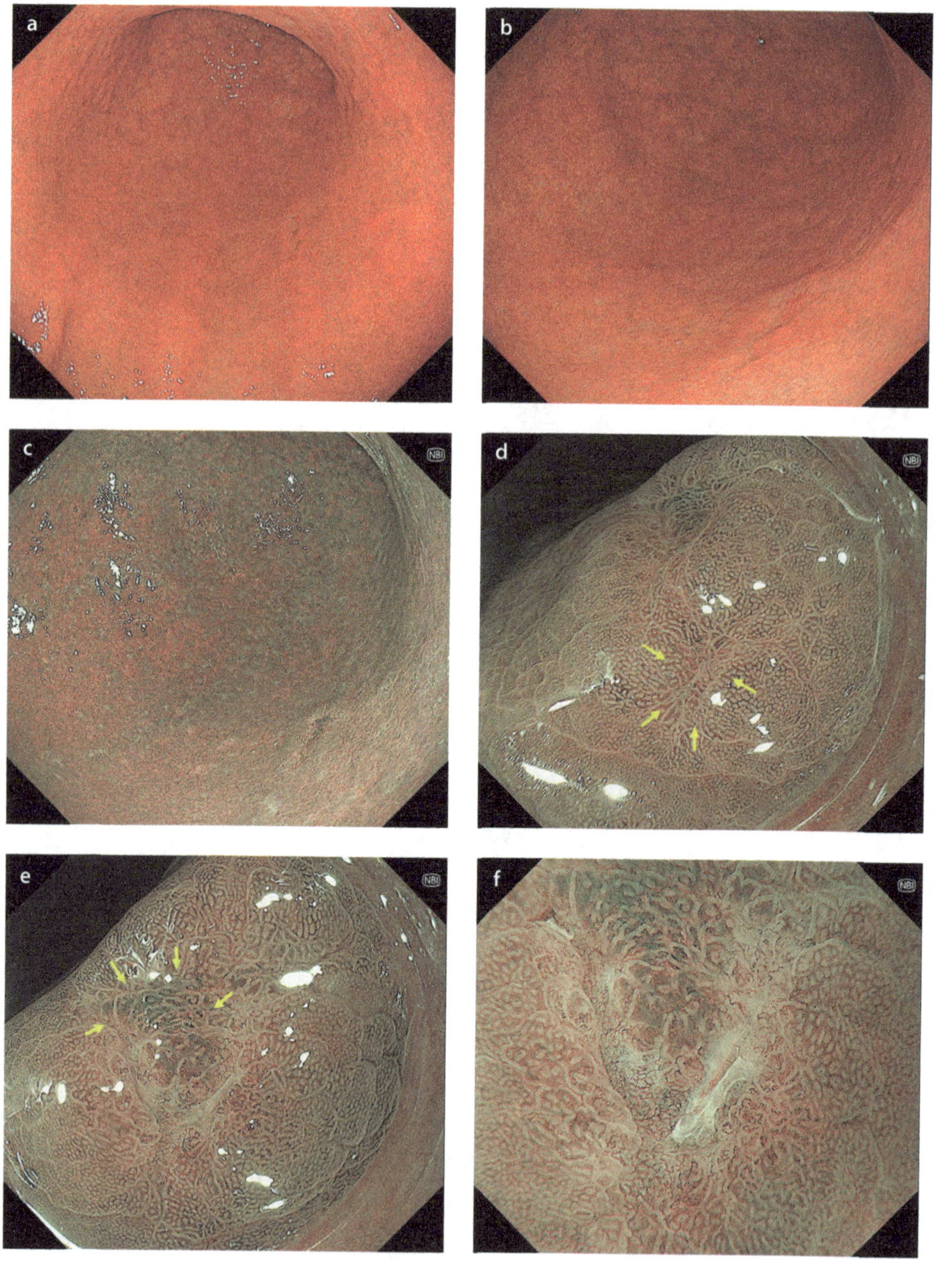
a
b
c
NBI
d
NBI
e
NBI
f
NBI

Pathological Diagnosis

Undifferentiated-type (por, sig) intramucosal early gastric cancer without any submucosal invasion or lymphovascular invasion.

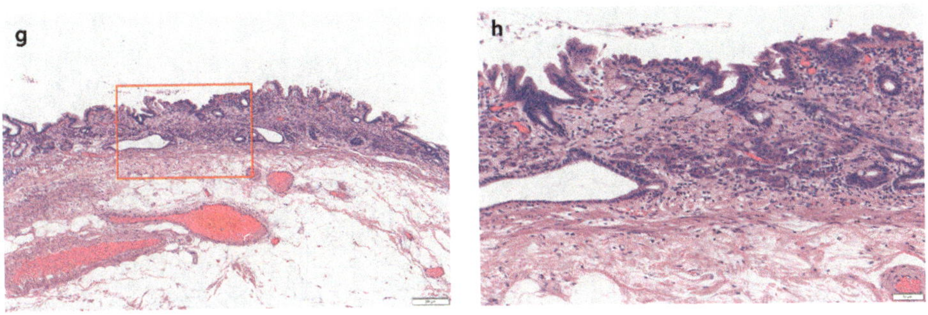

Explanation

A slightly depressed lesion measuring approximately 12 mm has been located at the greater curvature in the middle gastric body (Fig. a). The color of its entire lesion is discolored, but the center of the lesion is recognized as reddish when observed with non-magnifying WLI (Fig. b). Similar to appearance of WLI, under non-magnifying NBI, the lesion can be recognized as a discolored depression, which included in slightly brownish tone (Fig. c). Next, low-magnification NBI shows oral side as depressed with slightly irregular microvessels (Fig. d). The part of large intervening part is recognized at the center of the anal side of the depression, that is, a reactive change by performed previous biopsy (Fig. e). The high-magnification NBI under water condition shows irregular microvessels called cork screw patterns and wavy microvessels (Fig. f). Finally, based on these findings, the lesion was diagnosed as an undifferentiated-type intramucosal early gastric cancer without ulceration, ≤2 cm, and that was resect en bloc fashion by ESD. Histopathological examination revealed a poorly differentiated adenocarcinoma that includes signet ring cells proliferated in the mucosal layer without any submucosal invasion or lymphovascular invasion (Figs. g and h).

Tips

It is very important to detect an undifferentiated-type EGC; if the lesion is the undifferentiated-type intramucosal EGC without ulceration, ≤2 cm, it can be treated by ESD as curative resection. Because the undifferentiated-type EGC sometimes proliferates and expands laterally without any change in surface structure, so when the undifferentiated-type EGC will be removed by ESD, it should be performed negative biopsy around the lesion to confirm a cancer-free margin. It is hard to detect an undifferentiated-type EGC, because it is very small and less frequent than differentiated-type EGC. The tips of detection of an undifferentiated-type EGC are paid attention to observe the atrophic borders and the discolored lesions seen especially in the *Helicobacter pylori*-uninfected stomach.

Early Gastric Cancer After HP Eradication (1)

Masaaki Kobayashi

Supplementary Information The online version contains supplementary material available at https://doi.org/10.1007/978-981-97-2732-2_45. The videos can be accessed individually by clicking the DOI link in the accompanying figure caption or by scanning this link with the SN More Media App.

Y. Sano et al. (eds.), *Atlas of Advanced Endoscopy*, https://doi.org/10.1007/978-981-97-2732-2_45

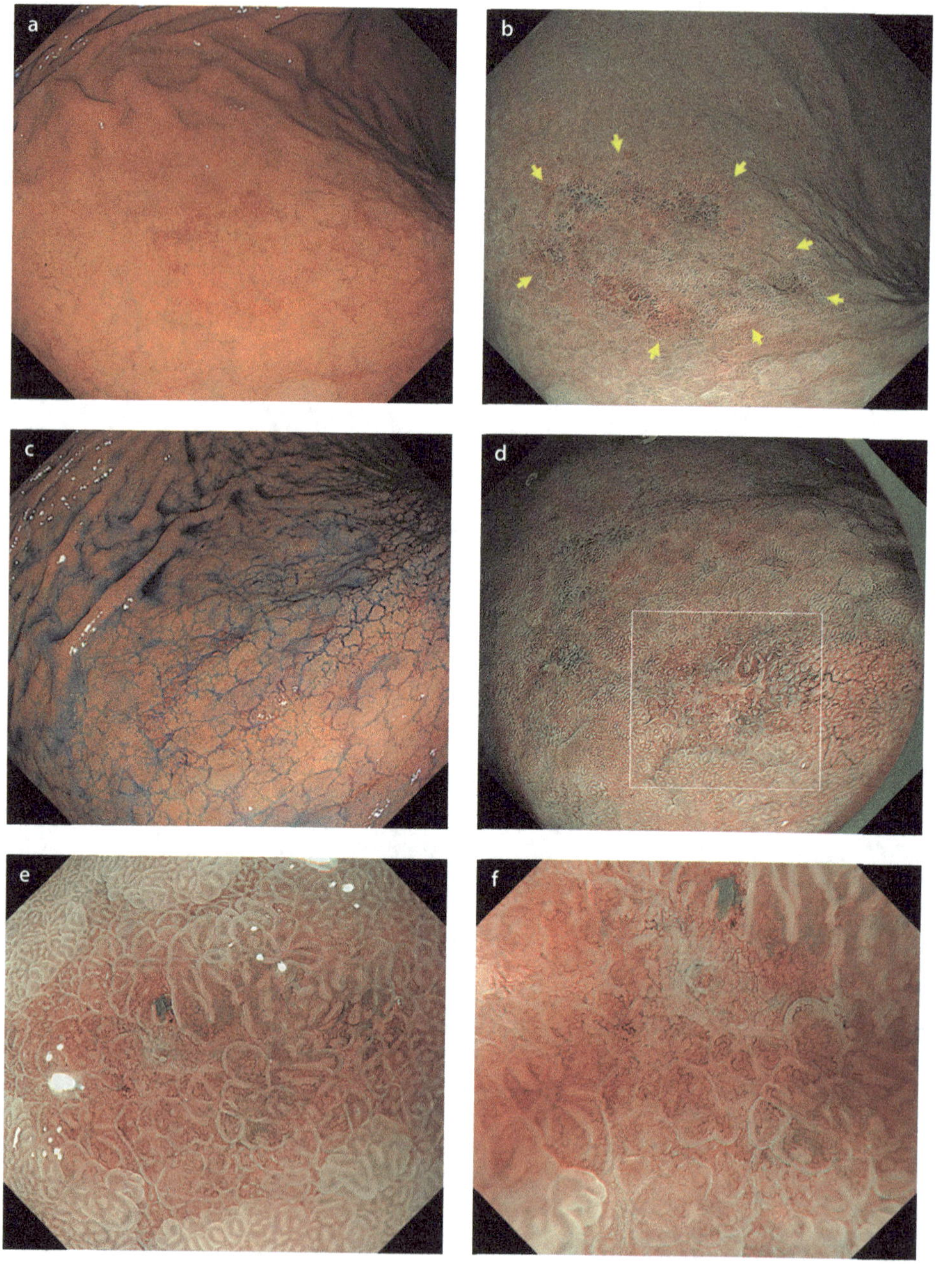
a
b
c
d
e
f

■ Pathological Diagnosis

M, Gre, type 0-IIc, 37 × 27 mm, adenocarcinoma (sig, por2), pT1a (M), pUL0, Ly0, V0.

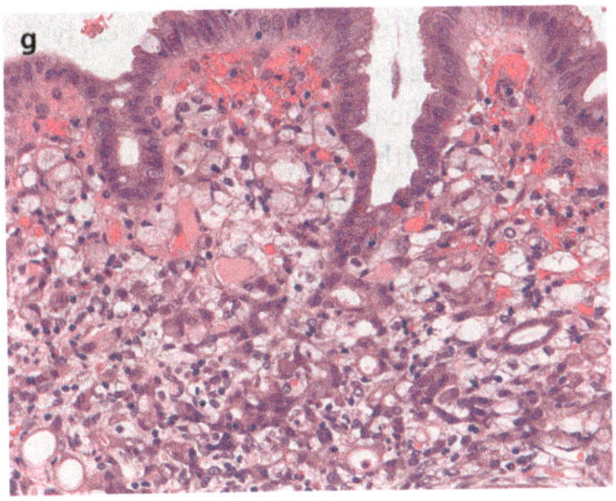

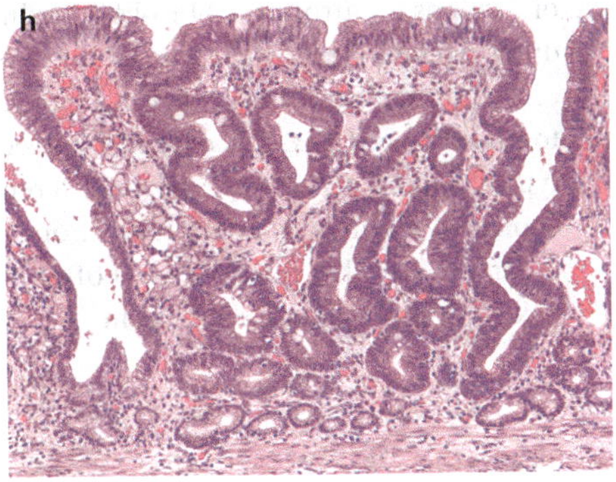

■ Explanation

A slightly depressed lesion measuring approximately 35 mm has been detected in the greater curvature of the upper gastric body. The lesion consists of multiple depressed components that resemble to map-like redness. The periphery of the lesion is slightly reddish, but it is difficult to recognize the border by color difference under WLI (Fig. a). Non-magnifying NBI enhances the border and color difference of the lesion more effectively (Fig. b, arrows) than WLI or chromoendoscopy using indigo carmine dye (Fig. c). Under half-magnifying NBI (Fig. d), the depressed lesion forms a mixture of swollen villi and crypts. Demarcation line can be drawn with adjacent noncancerous area, depending on differences in brownish color and surface structure. Under high-magnification NBI (Fig. e), the brownish area in the white box of Figure e reveals tortuous vessels within the swollen villi, exhibiting weave-like microvessels. Irregular microvascular images are more clearly observed under water high-magnification NBI (Fig. f). This irregular microvascular picture can easily diagnose gastric cancer according to MESDA-G, while the enlarged villous structures are continuous from the surrounding noncancerous area. LBC can also be observed both in cancer and noncancer villous structures. Based on these findings, the lesion was diagnosed as undifferentiated-type early gastric cancers and was resected en bloc using the ESD technique. Histopathological examination revealed adenocarcinoma (sig, por2), pT1a (M), 37 × 27 mm. The carcinoma is extending from the proliferative zone to the superficial layer, leaving the construction of non-neoplastic crypts with intestinal metaplasia (Figs. g and h).

Tips

This case underwent ESD for early gastric cancers followed by successful *H. pylori* eradication. Map-like redness is present after eradication, often in the border region of mucosal atrophy. Map-like redness requires differentiation from early differentiated-type gastric cancer, but this case had undifferentiated-type gastric cancer. Undifferentiated-type cancer larger than 20 mm is indicative for surgery, but due to his advanced age, this case is followed up after ESD with informed consent. The typical endoscopic image of undifferentiated-type gastric cancer is a discolored, depressed lesion even after eradication. However, undifferentiated-type cancers arising in the atrophic border region often show erythematous depressed lesions associated with regenerating epithelium.

Video

A short video is available as Fig. 1.

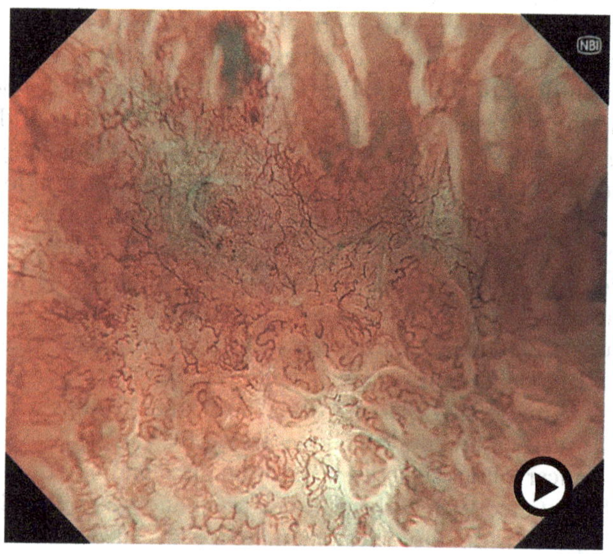

Fig. 1 Practical observation and diagnosis of early gastric cancer with NBI (▶ https://doi.org/10.1007/000-c8k)

Early Gastric Cancer After HP Eradication (2)

Masafumi Takatsuna and Manabu Takeuchi

Y. Sano et al. (eds.), *Atlas of Advanced Endoscopy*, https://doi.org/10.1007/978-981-97-2732-2_46

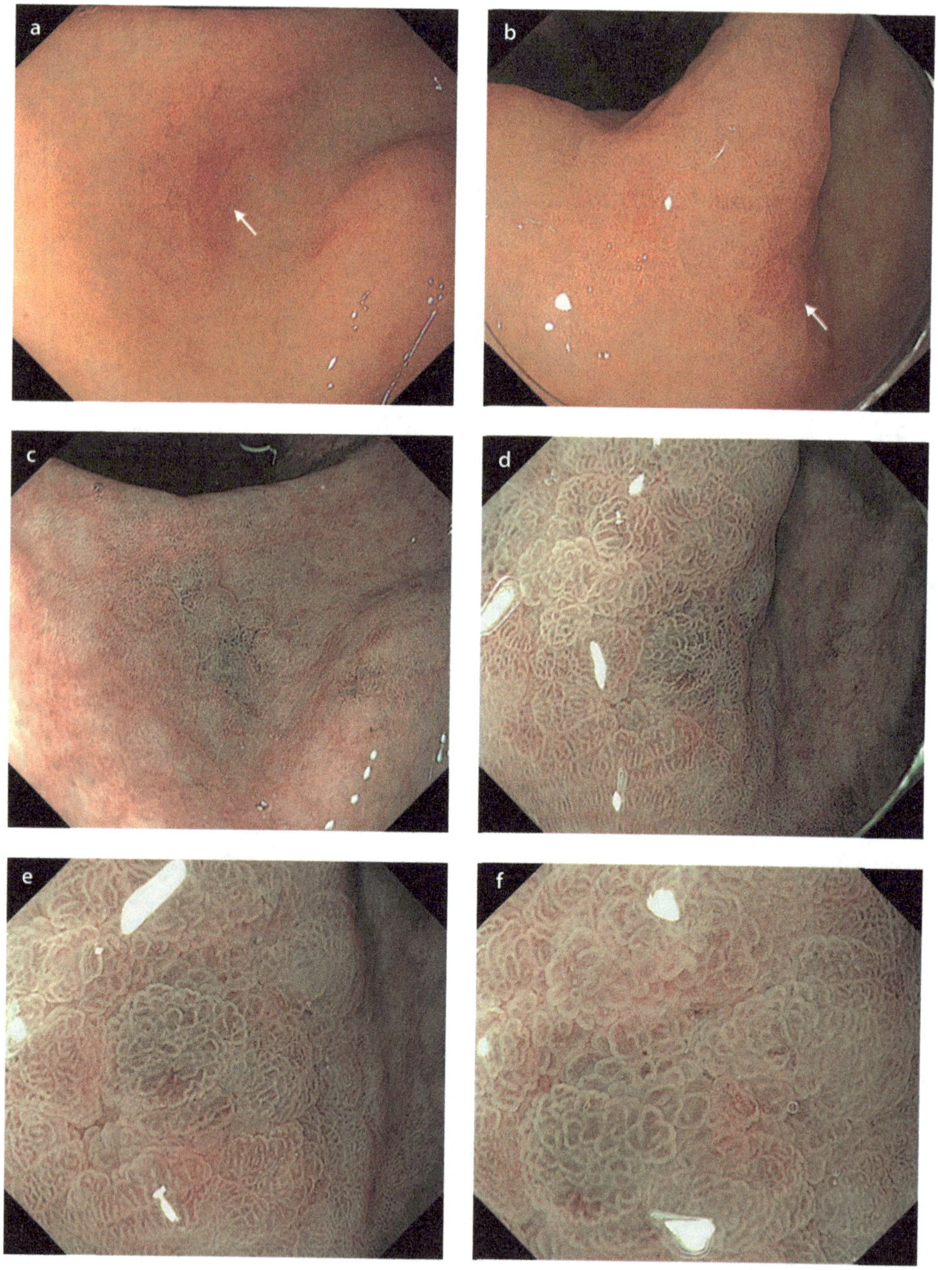
a
b
c
d
e
f

▪ Pathological Diagnosis

Poorly differentiated adenocarcinoma (pT1a(M), ly-, v-, UL-, pHM0, pVM0).

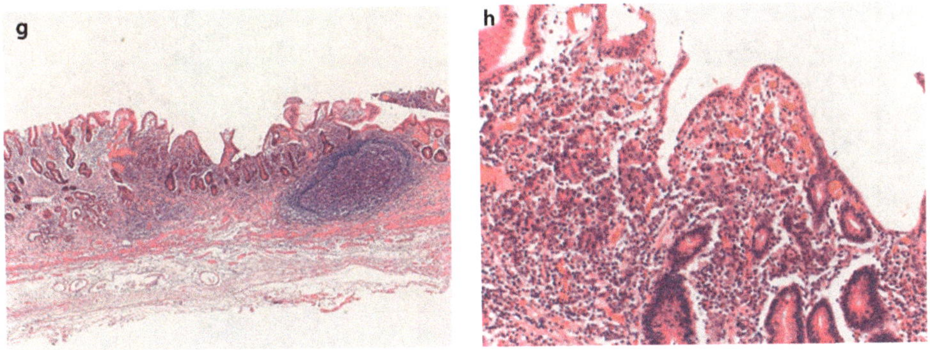

▪ Explanation

Non-magnifying white light imaging (WLI) reveals a 5-mm reddish depressed lesion in the lesser curvature of antrum (white arrow) (Fig. a). The redness of the lesion is stronger than the surrounding mucosa (Fig. b). After switching to non-magnifying narrowband imaging (NBI) with GIF-XZ1200 scope, the lesion shows a brownish color compared with its surrounding mucosa (Fig. c). In the low-magnifying NBI, the mucosal surface pattern shows a moderate clear demarcation line between the brownish lesion and the peripheral area (Fig. d). High-magnification NBI observation shows the lesion surface to have a papillary structure in which the white zone is uneven in shape compared to the surrounding area. However, no microvessel pattern is visible (Figs. e and f). Histopathological diagnosis after the lesion biopsy shows poorly differentiated adenocarcinoma; therefore, the lesion was removed by endoscopic submucosal dissection (ESD). Histopathological diagnosis after ESD confirms the diagnosis to be poorly differentiated adenocarcinoma (pT1a(M), ly-, v-, UL-, pHM0, pVM0) (Fig. g). Histopathological findings also reveal that the poorly differentiated adenocarcinoma cells are covered by non-atypical mucosa (Fig. h).

Tips

Poorly differentiated early gastric cancer that develops after the eradication of *Helicobacter pylori* is difficult to detect; however, it is important to rely on its reddish color and mild depression compared to the surrounding mucosa. In NBI magnified observation, the reddish depressed lesion shows a brownish color compared to the surrounding mucosa and a papillary structure in which the white zone is uneven in shape compared to the surrounding area. Therefore, careful observation is important, as these findings may be interpreted as inflammatory changes.

Raspberry-Type Gastric Adenocarcinoma

Daisuke Kikuchi

Y. Sano et al. (eds.), *Atlas of Advanced Endoscopy*, https://doi.org/10.1007/978-981-97-2732-2_47

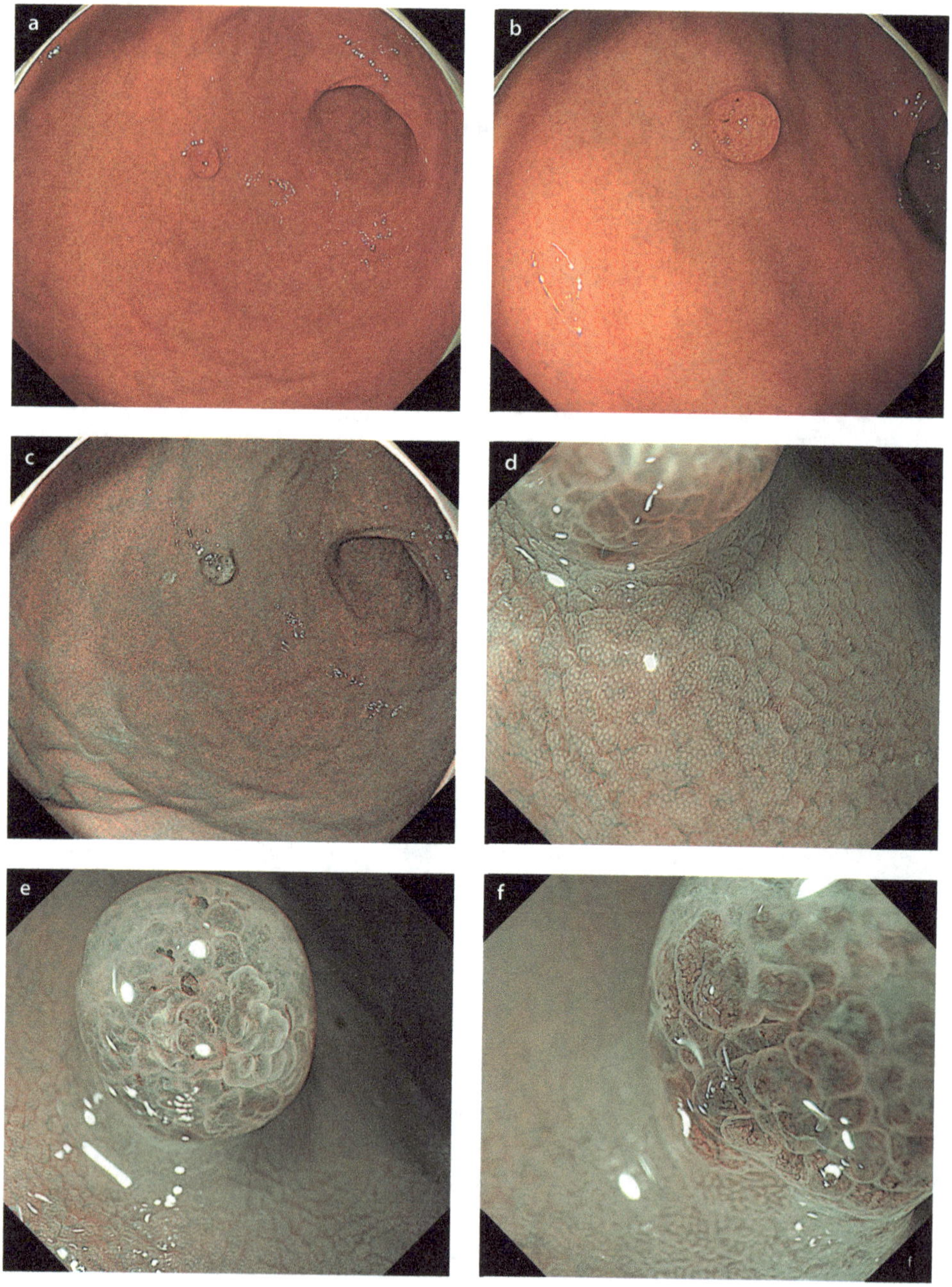
a
b
c
d
e
f

Pathological Diagnosis

The lesion was resected by EMR in an en bloc fashion. The final pathological diagnosis was foveolar-type adenocarcinoma, 5 × 5 mm, pT1a (M), ly0, v0, HM0, VM0.

g

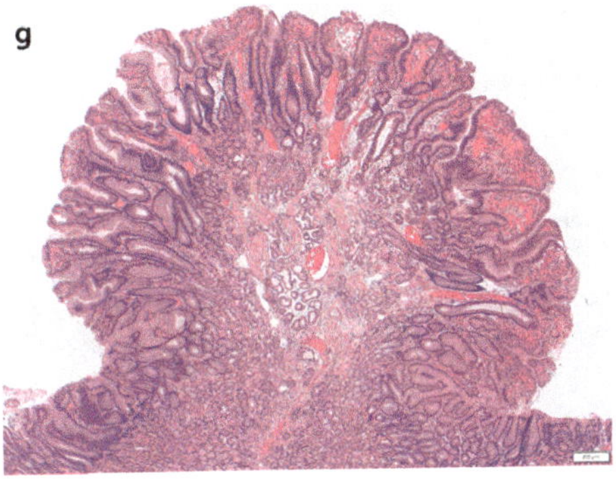

h

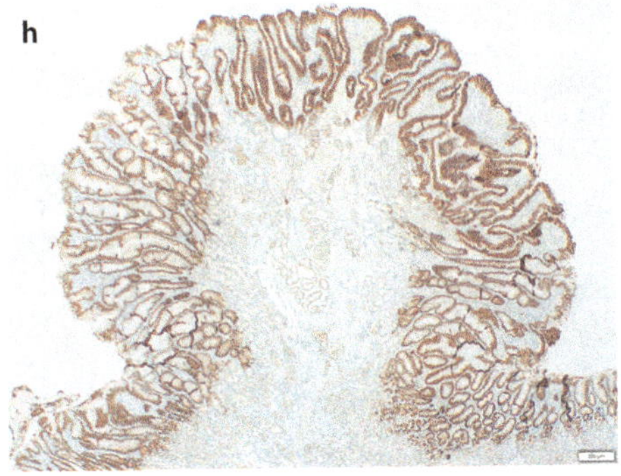

Explanation

White light endoscopy (WLE) revealed a 10-mm-sized reddish and protruded lesion in the greater curvature of the lower gastric body (Fig. a). Close observation revealed the presence of regular arrangement of collecting venules in the background mucosa (Fig. b). On narrowband imaging (NBI), this lesion appeared as a distinct brownish elevation (Fig. c). NBI magnification revealed that the background mucosa around the lesion had a regular array of dotted gland duct openings, indicating an *H. pylori*-uninfected and non-atrophic background mucosa (Fig. d). The patient had no history of *H. pylori* eradication, and both the urea breath test and fecal antigen test for *H. pylori* were negative, so it could be concluded that this patient was not infected with *H. pylori*. Using NBI magnification, the lesion exhibited enlarged granular microstructures, each surrounded by a white zone (Fig. e). The high-power magnification observation revealed irregularly dilated and tortuous microvessels within the microstructures (Fig. f). Histologically, the tumor exhibited a papillary growth pattern, and MUC5AC-positive cells were observed throughout the neoplastic lesion (Figs. g and h).

Tips

Raspberry-type gastric adenocarcinoma is one of the gastric neoplasms that occur in cases without *H. pylori* infection. The characteristic feature of raspberry-type gastric adenocarcinoma is noticeable reddish, subpedunculated protruded lesion, typically localized in the greater curvature of the gastric body. It is often accompanied by fundic gland polyps, which appear as similar small elevations, but differentiation is easily made by the difference in color tone in WLE and the pattern of microstructures in NBI. Another differential diagnosis is hyperplastic polyps. Unlike hyperplastic polyps, raspberry-type gastric adenocarcinoma arises in non-atrophic gastric mucosa, and under NBI magnification, the mucosal microstructure surrounded by a white zone appears nearly circular.

Gastric Adenocarcinoma of Fundic Gland Type

Hiroya Ueyama

Supplementary Information The online version contains supplementary material available at https://doi.org/10.1007/978-981-97-2732-2_48. The videos can be accessed individually by clicking the DOI link in the accompanying figure caption or by scanning this link with the SN More Media App.

Y. Sano et al. (eds.), *Atlas of Advanced Endoscopy*, https://doi.org/10.1007/978-981-97-2732-2_48

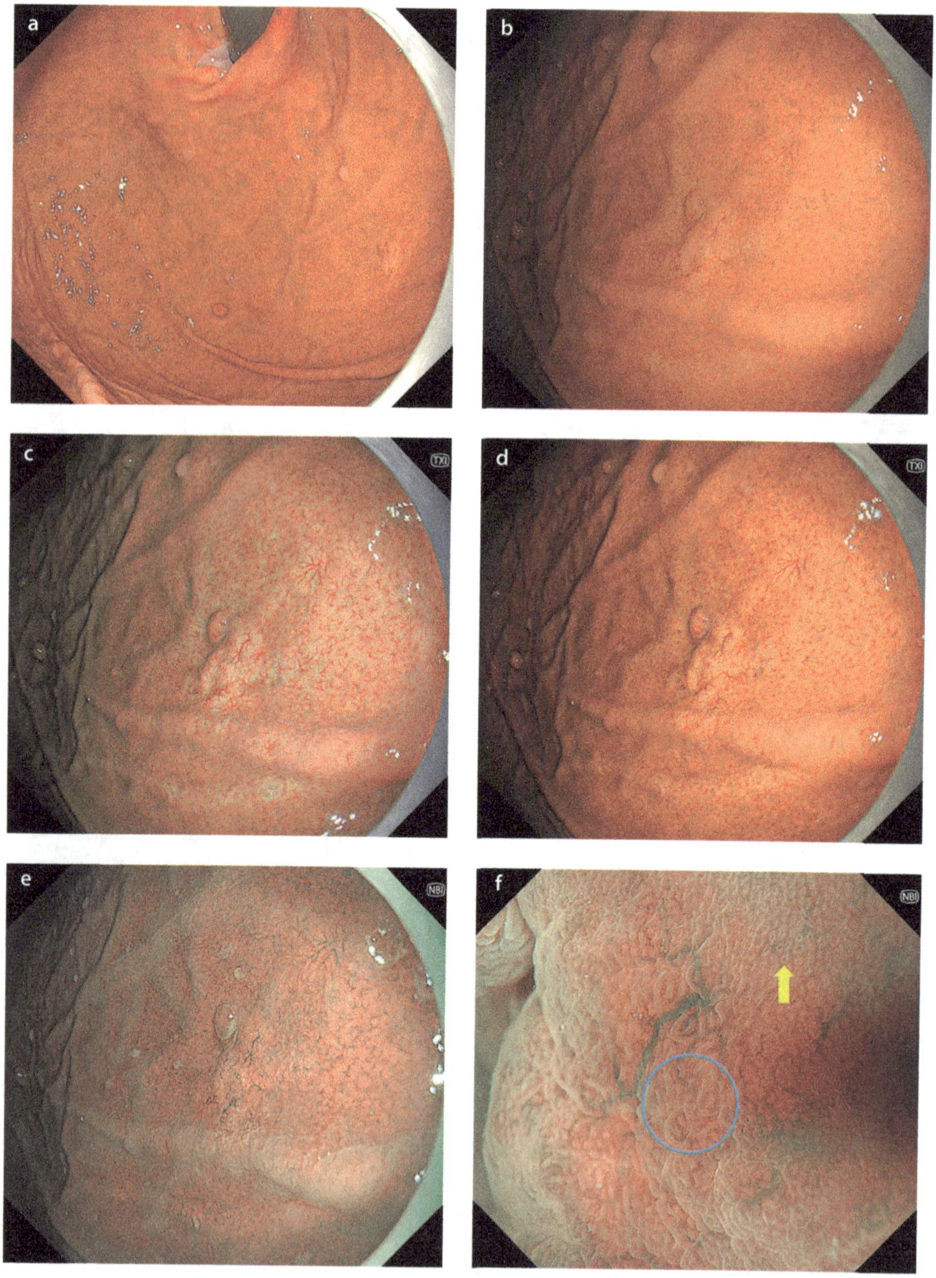
a
b
c
TXI
d
TXI
e
NBI
f
NBI

Pathological Diagnosis

Gastric adenocarcinoma of fundic gland type.

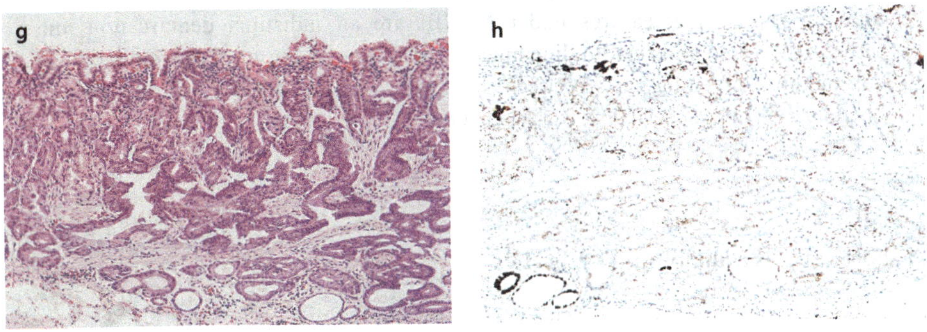

Explanation

An elevated lesion, 15 mm in size, was detected in the anterior wall of the upper body of the stomach. The lesion was slightly white and had the appearance of a subepithelial lesion and dilated vessels with branching architecture. There was no atrophy in the surrounding mucosa (Figs. a and b). Compared to WLI, TXI (mode 1 and mode 2) enhanced the margins, dilated vessels, and color difference between the lesion and the neighboring mucosa (Figs. c and d). NBI also enhanced the margins and dilated vessels more effectively (Fig. e). When switching to magnifying endoscopy with NBI (M-NBI), dilatation of the crypt openings (yellow arrowhead), dilatation of the intervening parts (blue circle), and microvessels without distinct irregularity were observed, but a distinct demarcation line was absent (Fig. f). Based on these findings, the lesion was endoscopically diagnosed as gastric adenocarcinoma of fundic gland type (GA-FG) and was resected en bloc with ESD. Histopathological examination revealed that the surface layer was covered by nonneoplastic mucosa, and tumor cells similar to chief cells were proliferating primarily in the deep mucosa, showing an irregularly branching structure and adhesions (Fig. g). Immunohistochemically, pepsinogen-I (a marker for chief cell) was diffusely positive (Fig. h), and the lesion was diagnosed as GA-FG. Final pathological diagnosis: U, 0-IIa, 9 mm, GA-FG, pT1b/SM2 (600 μm), UL0, Ly0, V0, HM0, VM0.

Tips

GA-FG is defined by a well-differentiated adenocarcinoma composed of pale gray-blue, basophilic columnar cells with mild nuclear atypia, resembling chief cells that are positive for immunohistochemical staining of pepsinogen-I and/or H^+/K^+-ATPase (J Gastroenterol 2021; 56: 814–828. WHO classification of tumours, fifth edn 2019). This is a novel subtype of gastric cancer that is not associated with *H. pylori* infection [Am J Surg Pathol. 2010; 34(5): 609–19.]. Endoscopically, the most common features with WLI are a submucosal tumor or subepithelial lesion shape, a slightly white appearance and dilated vessels with branching architecture, and a background mucosa without atrophy (Endoscopy 2014; 46: 153–157.). The four most common endoscopic features under M-NBI are an indistinct demarcation line between the lesion and surrounding mucosa, dilatation of the crypt openings, dilatation of the intervening parts between the crypts, and microvessels without distinct irregularity [Diagnostics 2022; 2; 12(11):2666.].

▪ Video

A short video is available as ◘ Fig. 1.

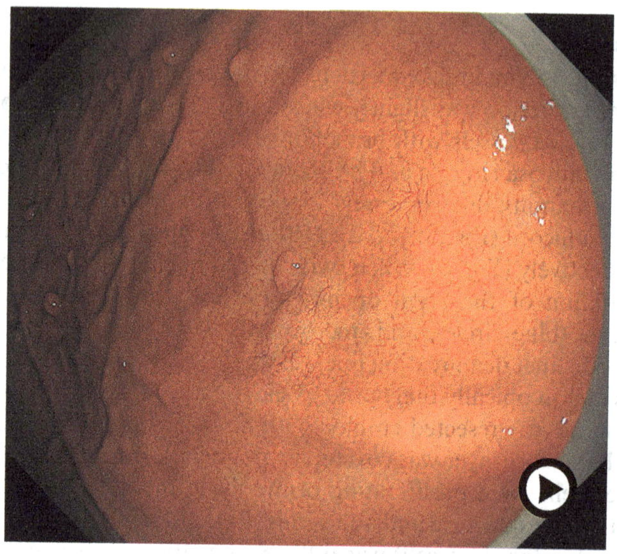

◘ **Fig. 1** Practical observation and diagnosis of gastric adenocarcinoma of fundic gland type with TXI and NBI (▶ https://doi.org/10.1007/000-c8m)

Gastric MALT Lymphoma

Masao Yoshida and Hiroyuki Ono

Y. Sano et al. (eds.), *Atlas of Advanced Endoscopy*, https://doi.org/10.1007/978-981-97-2732-2_49

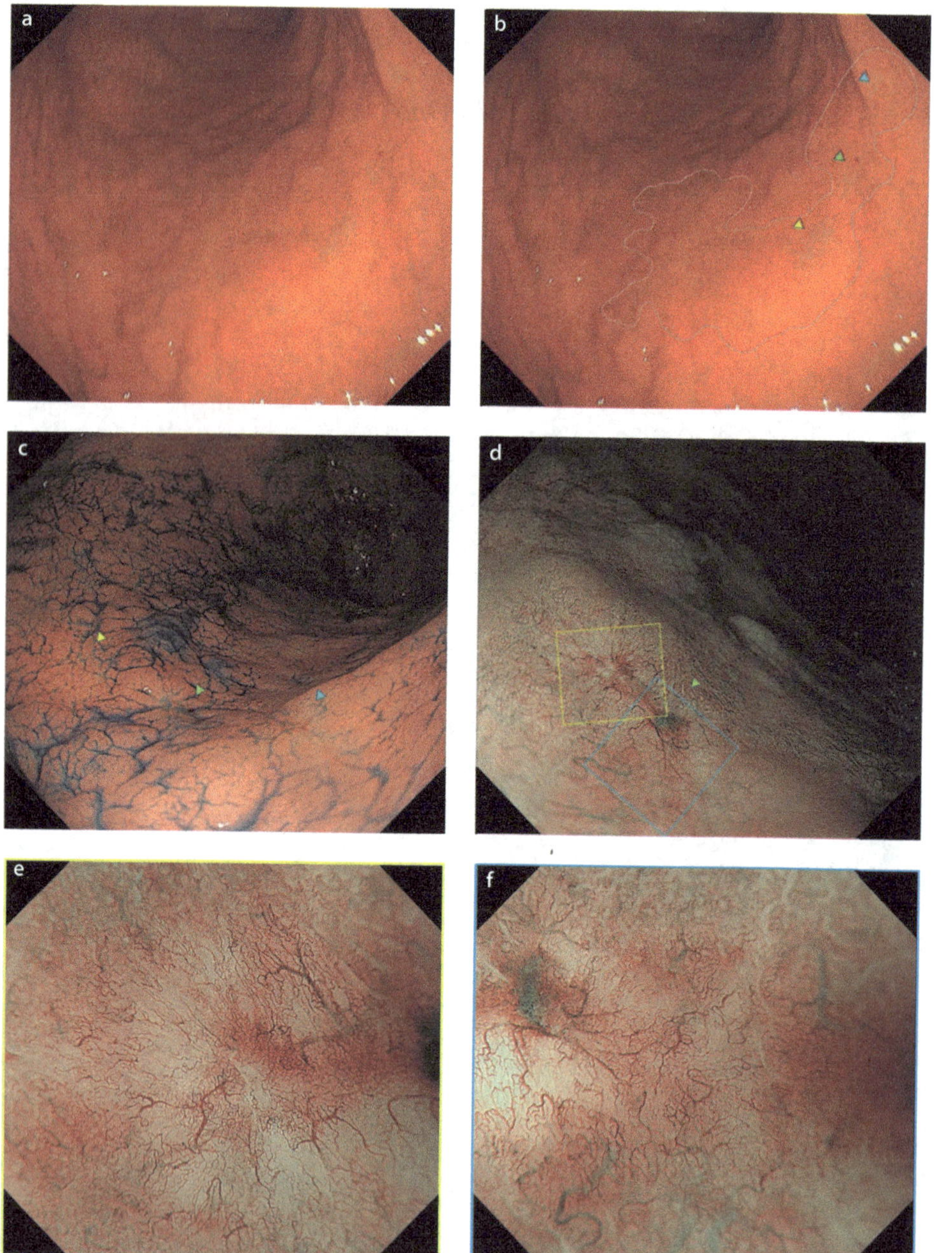
a
b
c
d
e
f

■ Pathological Diagnosis

MALT-type malignant lymphoma.

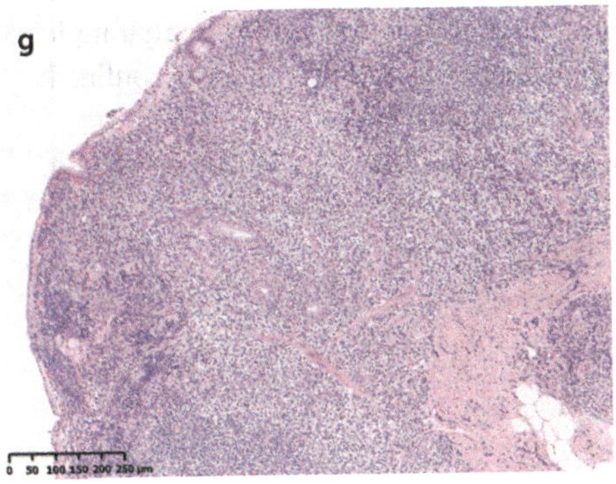

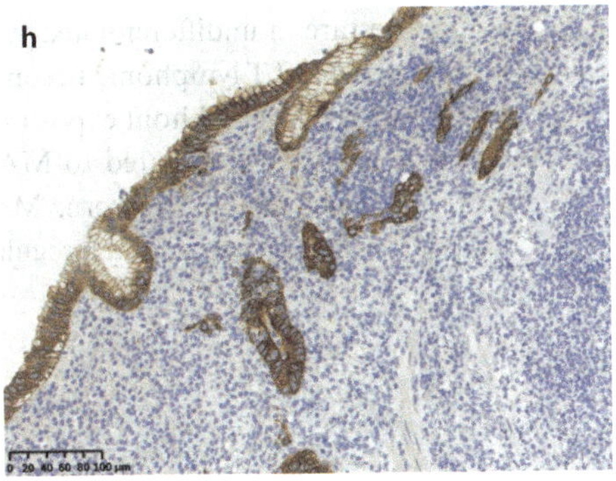

■ Explanation

Diffuse redness and mucosal swelling of the stomach suggest *Helicobacter pylori* infection (Fig. a). An unclear pale mucosal area of 50 mm is observed from the posterior wall to the greater curvature of the middle body (Fig. b, traced). Microvascular dilation is seen inside the pale mucosal areas on the posterior wall (Fig. b, blue and green arrows). Chromoendoscopy using indigo carmine dye highlights multiple slight depressions (Fig. c, arrows). NBI reveals that the foveola opening pattern of the background gastric fundic gland mucosa gradually becomes unclear near the depressed areas, making it challenging to establish a clear demarcation line between the lesion and the normal tissue (Fig. d). Magnifying NBI shows the disappearance of glandular structures and an extending, crooking dendritic vascular network on the mucosal surface, resembling a treelike appearance (TLA) due to the smooth tapering of the observed dendritic vessels (Figs. e and f). The following findings support a diagnosis of MALT lymphoma: (1) localized pale lesion, (2) multiple depressions, and (3) TLA. Histopathological examination of the biopsy confirms significant infiltration of atypical lymphocytes (Fig. g). Lymphoepithelial lesions (LEL) are confirmed using the pan-cytokeratin antibody (AE13) (Fig. h). Immunohistochemical stains yield positive results for CD20, CD79a, BCL-2, and CD21 while being negative for CD3, CD5, CD10, and cyclin D1.

Tips

Gastric MALT lymphoma is typically characterized by shallow, pale, and depressed lesions. Careful observation of the gastric mucosal surface after removing mucus and bubbles is essential to avoid overlooking any changes in color tone. The differential diagnosis of gastric MALT lymphoma includes undifferentiated-type gastric cancer and atrophic gastritis.

- Undifferentiated-type gastric cancer: Magnifying NBI observation often reveals corkscrew vasculature when cancer cells are exposed on the surface. Multiple lesions are rare in undifferentiated-type gastric cancer. However, differentiating it from gastric MALT lymphoma becomes challenging when cancer cells are confined to the mucosal layer without exposure to the mucosal surface.
- Atrophic gastritis: Compared to MALT lymphoma, atrophic gastritis exhibits a more indistinct regional pattern. Magnifying NBI observation shows a regular mucosal surface structure and a regular vascular network.

Duodenum

Contents

Adenoma (Papillary Region)

Shigetsugu Tsuji and Hisashi Doyama

Content

Y. Sano et al. (eds.), *Atlas of Advanced Endoscopy*, https://doi.org/10.1007/978-981-97-2732-2_50

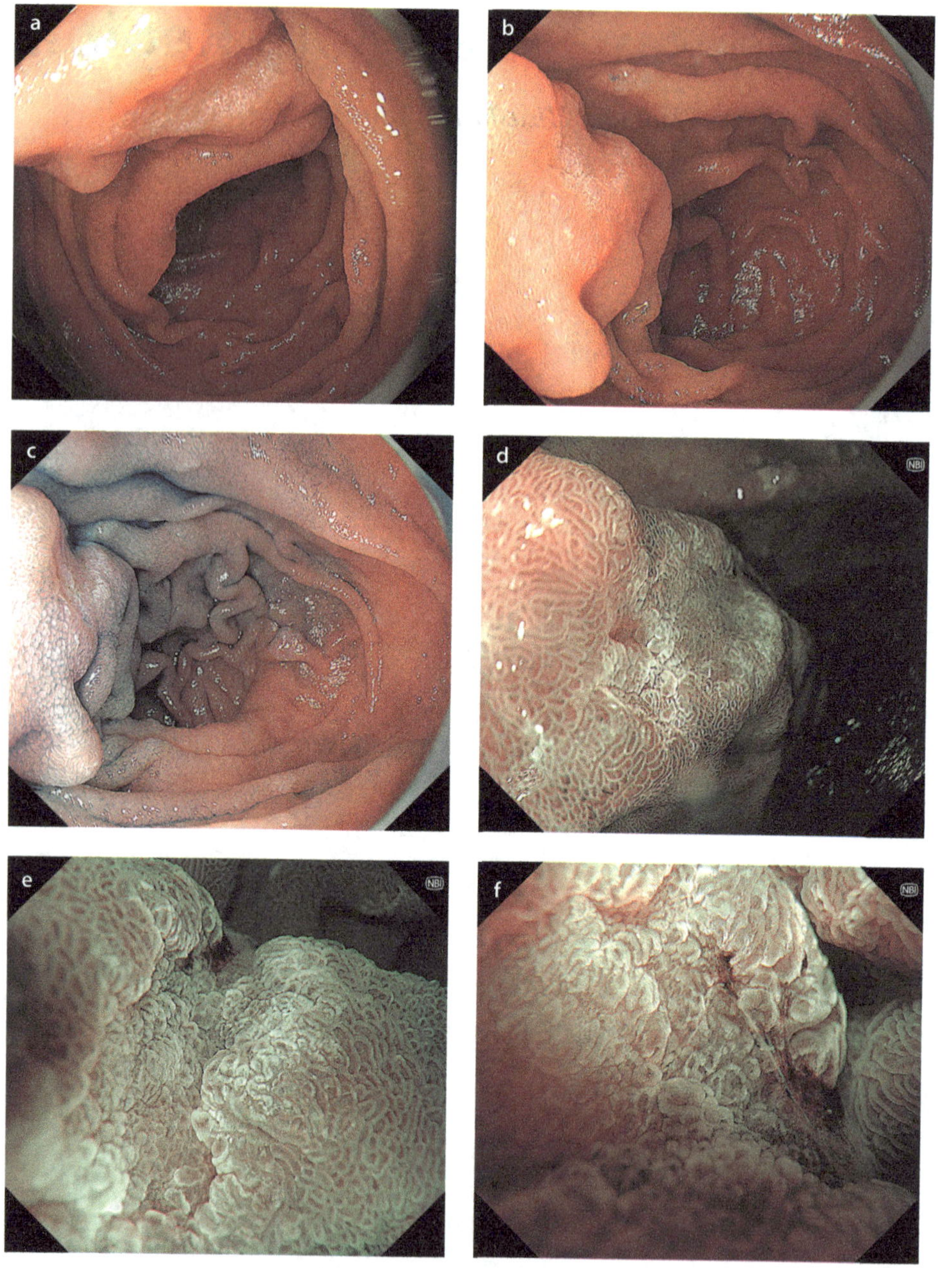

a, **b** Conventional endoscopy with white light imaging. **c** Dye spraying. **d** Non-magnifying endoscopy with narrowband imaging (NBI). **e**, **f** Magnifying endoscopy with NBI

■ Pathological Diagnosis

Histopathological findings of the biopsy specimen showed low-grade tubular adenoma.

g

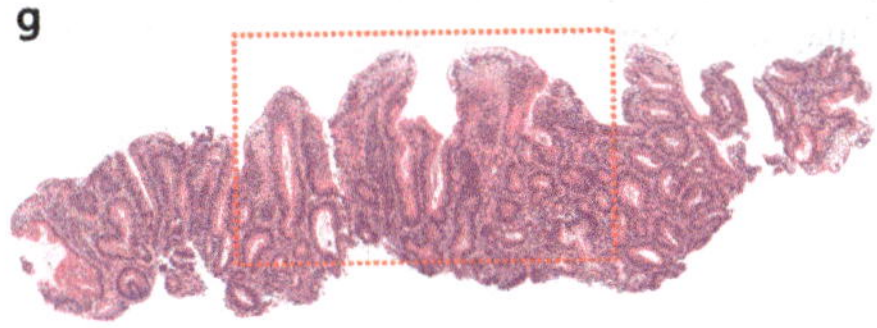

h

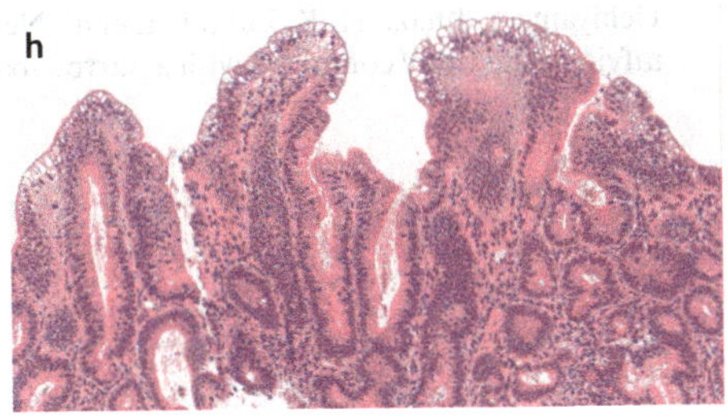

g, h Histopathological findings of the biopsy specimen

■ Explanation

Ampullary adenoma was found in a patient with familial adenomatous polyposis. Endoscopic examination using conventional endoscopy with white light imaging showed a whitish, slightly depressed lesion (10 mm in diameter) at the ampulla of Vater (Figs. a and b). The depressed margins were clear after dye spraying (Fig. c). White opaque substance (WOS) was more prevalent at the margins of the depressed area on non-magnifying endoscopy with narrowband imaging (NBI) (Fig. d). Magnifying endoscopy with NBI (M-NBI) revealed a clear demarcation line (DL) between the lesion and the surrounding normal mucosa, indicative of the change in the mucosal microsurface (MS) structure.

Vessel plus surface (VS) classification:

1. V: Because of the presence of the WOS, the morphology of the subepithelial microvessels could not be observed, making this an absent microvascular (MV) pattern.
2. S: The WOS had a regular mazelike pattern with symmetrical distribution and regular arrangement. Thus, this lesion was graded as a regular MS pattern using WOS as a marker for the MS pattern. The VS classification of this lesion was an absent MV pattern plus a regular MS pattern (WOS+) with a DL. Therefore, the M-NBI diagnosis was nonmalignant (Figs. e and f). Histopathological findings of the biopsy specimen showed low-grade tubular adenoma (Figs. g and h).

Tips

Duodenal polyposis affects nearly all patients with familial adenomatous polyposis, and adenomas of the papilla occur in the majority [1]. Uchiyama et al. [2] classified the surface structures of the ampullary lesions using M-NBI into three groups: I, oval-shaped villi; II, pinecone/leaf-shaped villi; or III, irregular/nonstructured. All cases of adenoma and adenocarcinoma had type II/III surface structures [2]. WOS has been demonstrated in the superficial parts of many duodenal tumors, as in this case. In cases where WOS is present, magnified endoscopic evaluation of the V component is impossible. Therefore, it is important to distinguish between benign and malignant lesions in order to assess the WOS morphology as a marker of the S component.

References

1. Bhatt A, Mankaney G, Burke CA. Approach to resection of adenomas of the papilla: should familial adenomatous polyposis change the approach? Gastrointest Endosc. 2020;92:331–3.
2. Uchiyama Y, Imazu H, Kakutani H, et al. New approach to diagnosing ampullary tumors by magnifying endoscopy combined with a narrow-band imaging system. J Gastroenterol. 2006;41:483–90.

Early Duodenal Cancer (Papillary Region)

Shigetsugu Tsuji and Hisashi Doyama

Content

Y. Sano et al. (eds.), *Atlas of Advanced Endoscopy*, https://doi.org/10.1007/978-981-97-2732-2_51

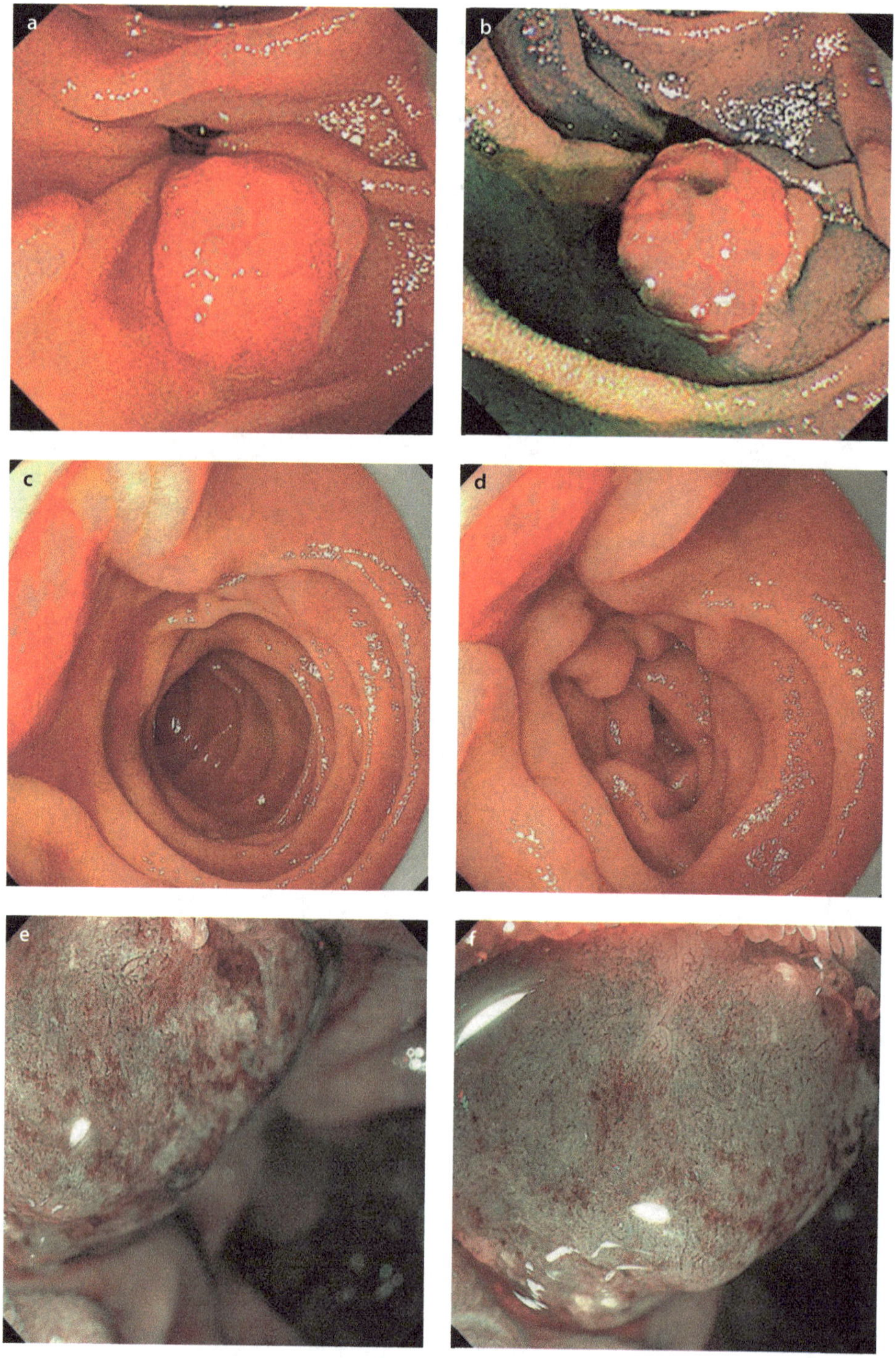
a
b
c
d
e
f

Pathological Diagnosis

Histopathological examination of the resected specimen showed poorly differentiated adenocarcinoma.

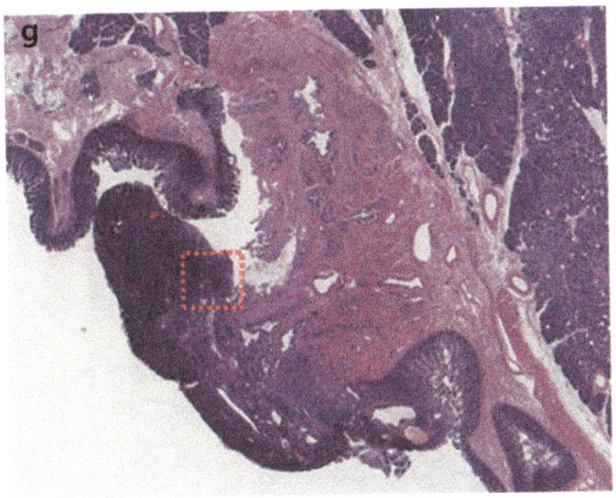

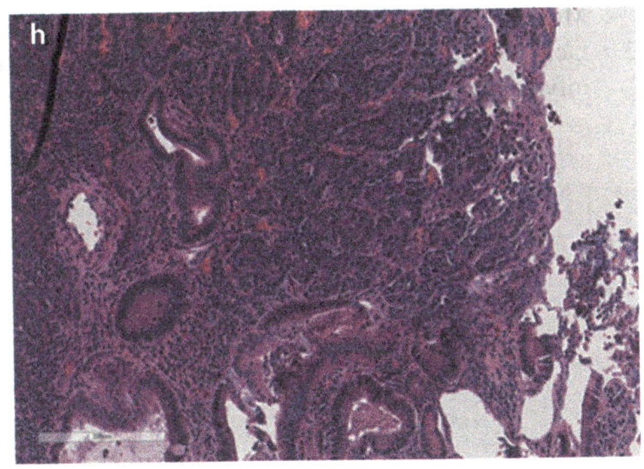

Explanation

Endoscopic examination using conventional endoscopy with side-viewing endoscopy (JF 260, Figs. a and b) and forward-viewing (X1 system, Figs. c and d) and white light imaging revealed a readily bleeding, reddish lesion (10 mm in diameter) at the ampulla of Vater. Magnifying endoscopy with narrowband imaging (M-NBI) showed a clear demarcation line (DL) due to the difference in the vessel plus surface (VS) component between the cancerous and noncancerous mucosa. V: The proliferation of microvessels with variable sizes, asymmetrical distribution, and irregular arrangement made this an irregular microvascular pattern. S: There were areas in which the marginal crypt epithelium could not be visualized. This lesion was assessed as an absent microsurface pattern. The VS classification of this lesion was an irregular microvascular pattern plus absent microsurface pattern with a DL. Therefore, the M-NBI diagnosis was cancer (Figs. e and f). The final histological diagnosis was poorly differentiated adenocarcinoma (Figs. g and h).

Tips

The presence of malignancy is likely when surface nodularity, increased friability, and spontaneous bleeding are present during endoscopic examination. These findings help to distinguish benign adenoma from cancer in typical cases. However, the macroscopic appearance of ampullary malignancies varies widely; thus, standard endoscopic assessment alone is not entirely reliable for diagnosis [1]. Uchiyama et al. [2] developed a new approach to the differential diagnosis of ampullary tumors using M-NBI. In their study, tortuous, dilated, and network-like vessels were defined as abnormal vessels, seen only in adenocarcinomas and never in adenomas [2]. In the present case, M-NBI showed proliferation of microvessels with variable sizes, asymmetrical distribution, and irregular arrangement. M-NBI may therefore help to distinguish adenoma from cancer even in the papillary region. M-NBI has the ability and potential to predict histological characteristics of ampullary lesions.

References

1. Yamao T, Isomoto H, Yamaguchi N, et al. Magnified endoscopic observation using narrow-band imaging of periampullary adenoma in a patient with familial adenomatous polyposis. Med Sci Monit. 2009;15:CS169-173.
2. Uchiyama Y, Imazu H, Kakutani H, et al. New approach to diagnosing ampullary tumors by magnifying endoscopy combined with a narrow-band imaging system. J Gastroenterol. 2006;41:483–90.

Duodenal Adenoma (Gastric-Type, Non-papillary Region)

Kenichi Goda and Kazuyuki Ishida

Y. Sano et al. (eds.), *Atlas of Advanced Endoscopy*, https://doi.org/10.1007/978-981-97-2732-2_52

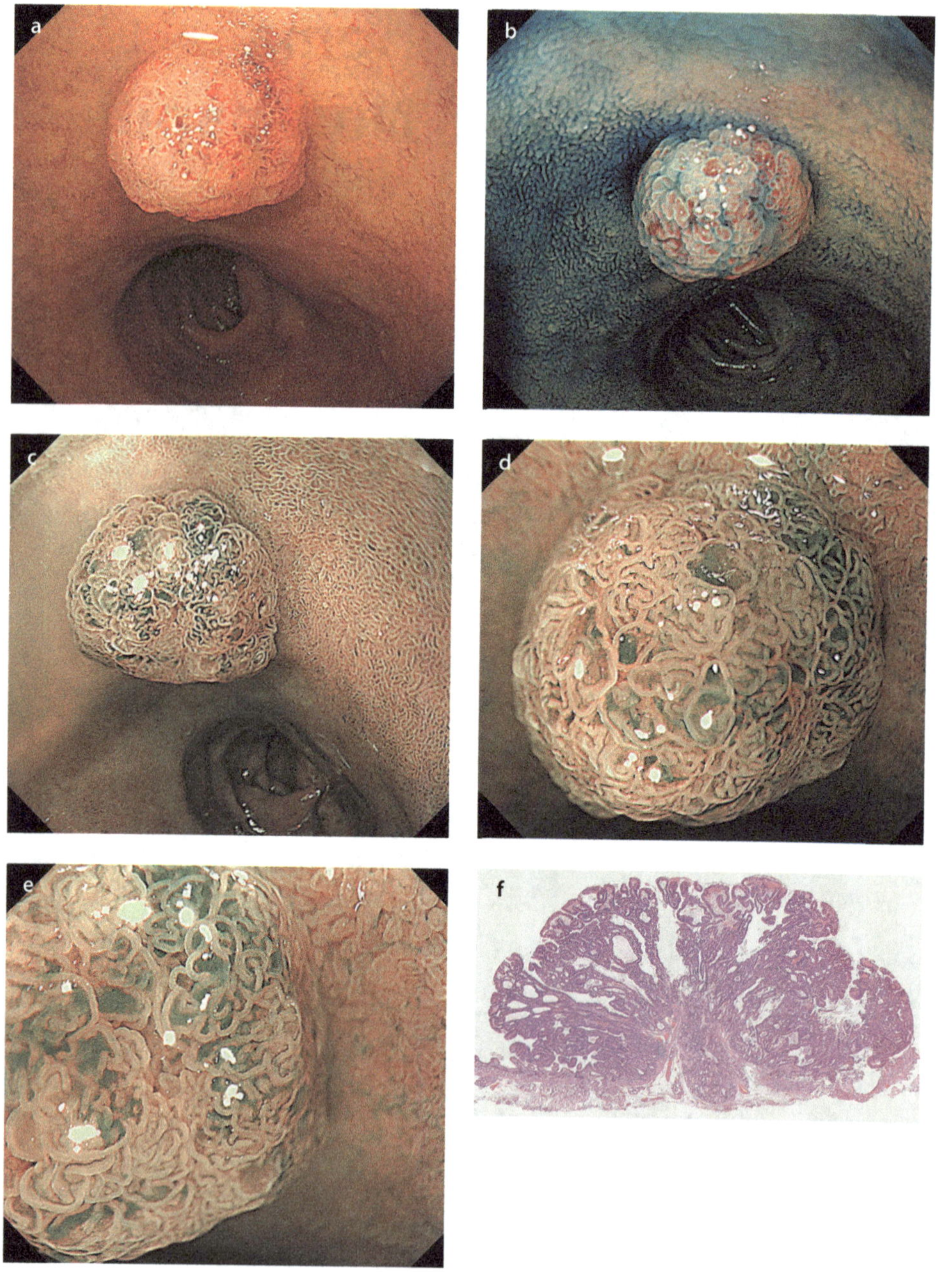
a
b
c
d
e
f

▪ Pathological Diagnosis

Non-papillary duodenal adenoma, gastric phenotype.

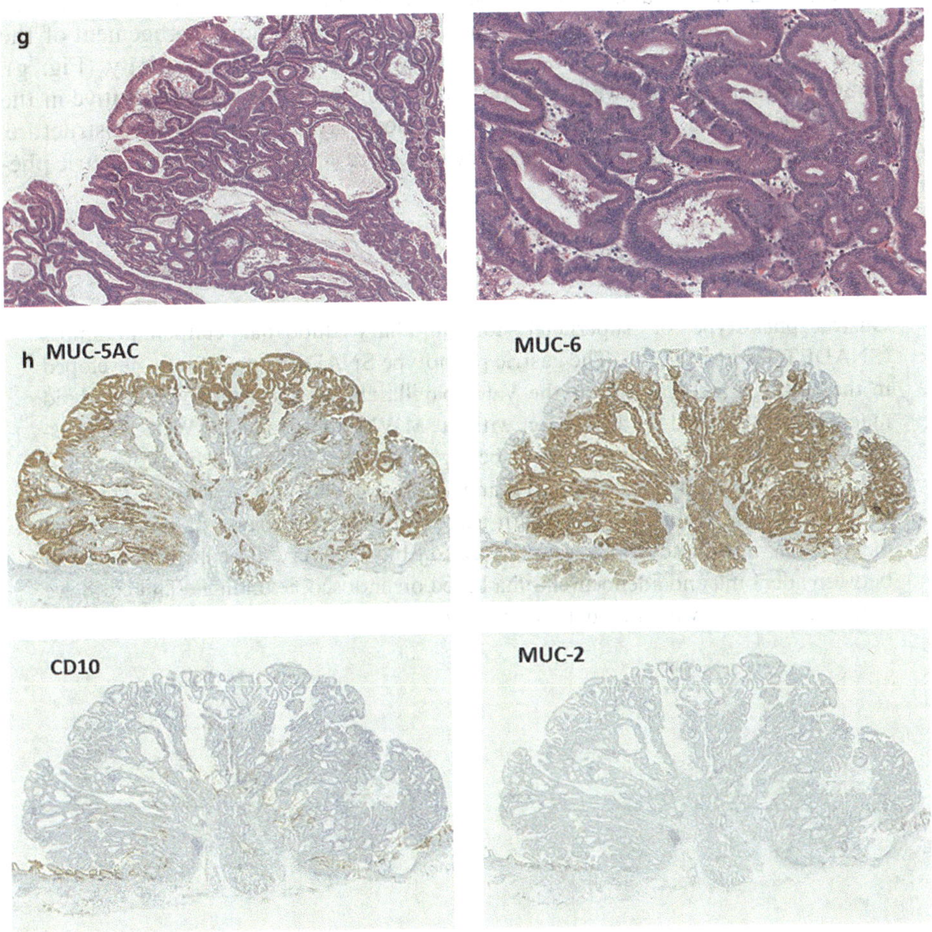

▪ Explanation

High-definition white-light endoscopy shows a polypoid lesion with patchy erythema at the superior wall in the bulb (Fig. a). The polypoid lesion corresponds to the macroscopic type of 0-Is (protruded/polypoid) with 6 mm in diameter. No milk-white mucosa (MWM) is seen in the lesion on white-light imaging (Fig. a). Indigo carmine chromoendoscopy shows a clear border between the polypoid lesion and the surrounding duodenal mucosa (Fig. b). The chromoendoscopy and NBI endoscopy visualize many fingerlike mucosal structures corresponding to the normal duodenal villi around the lesion, whereas NBI endoscopy showed many enlarged structures (larger than fingerlike normal duodenal villi by more than threefolds) without MWM appearance or white opaque substance (WOS) in the polypoid lesion (Figs. c and d). NBI magnifying endoscopy shows the polypoid lesion consisted of enlarged villous structures like round leaves and dilated or congested vessels (Figs. d and e). ESD-resected specimens showed a polypoid lesion with atypical glands with tubulopapillary structures and multiple glands with significant dilation (Figs. f and g). Atypical glands had mild structural atypia and consisted of tall columnar cells with enlarged nuclei, inconspicuous nucleoli, and eosinophilic cytoplasm. The arrangement of the enlarged nuclei showed low-grade irregularity and no loss of polarity (Fig. g). Immunohistochemistry showed that the tumor cells were MUC5AC-positive in the superficial layers and MUC6-positive in the deep layer, with a layered structure. CD10 and MUC2 were negative. These findings were consistent with the gastric phenotype (Fig. h).

Tips

Gastric phenotype of superficial non-ampullary duodenal epithelial tumors (SNADETs) is uncommon. The gastric phenotype SNADETs are usually developed in the bulbs or the oral side of the Vater papilla and to show protruded/polypoid macroscopic type and reddish lesion without MWM appearance or WOS. The gastric-type SNADETs are more likely to be higher grade of dysplastic histology (high-grade adenoma or adenocarcinoma) than intestinal phenotype SNADETs. This case can be predicted for a SNADET with gastric phenotype because of its color and macroscopic morphology and lesion's location but was very difficult to distinguish between adenoma and adenocarcinoma based on endoscopic findings. Therefore, we removed the lesion with adequate safety margin by ESD.

Duodenal Adenoma (Gastrointestinal-Type, Non-papillary Region)

Kenichi Goda and Kazuyuki Ishida

Y. Sano et al. (eds.), *Atlas of Advanced Endoscopy*, https://doi.org/10.1007/978-981-97-2732-2_53

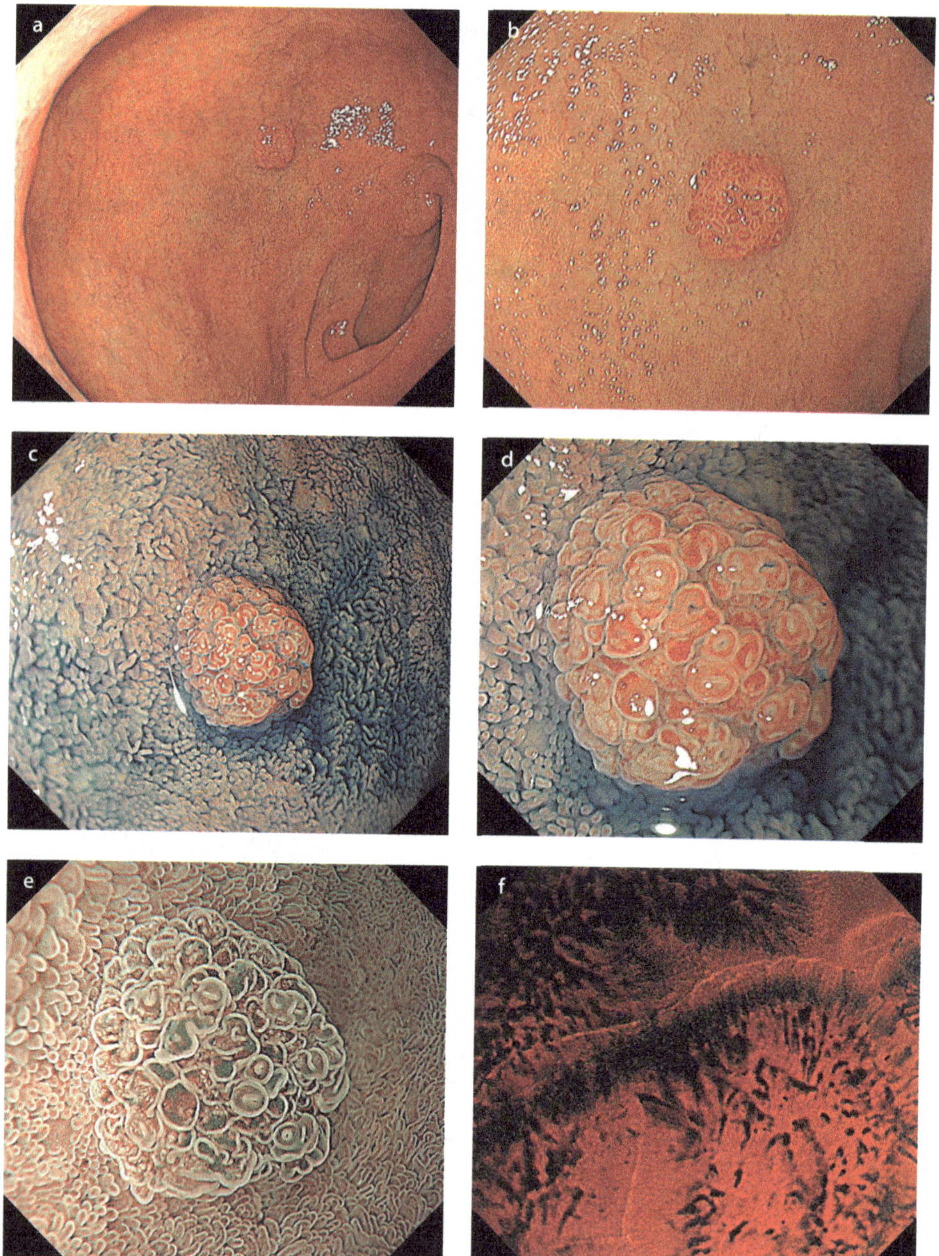

Pathological Diagnosis

Non-papillary duodenal adenoma, gastrointestinal phenotype.

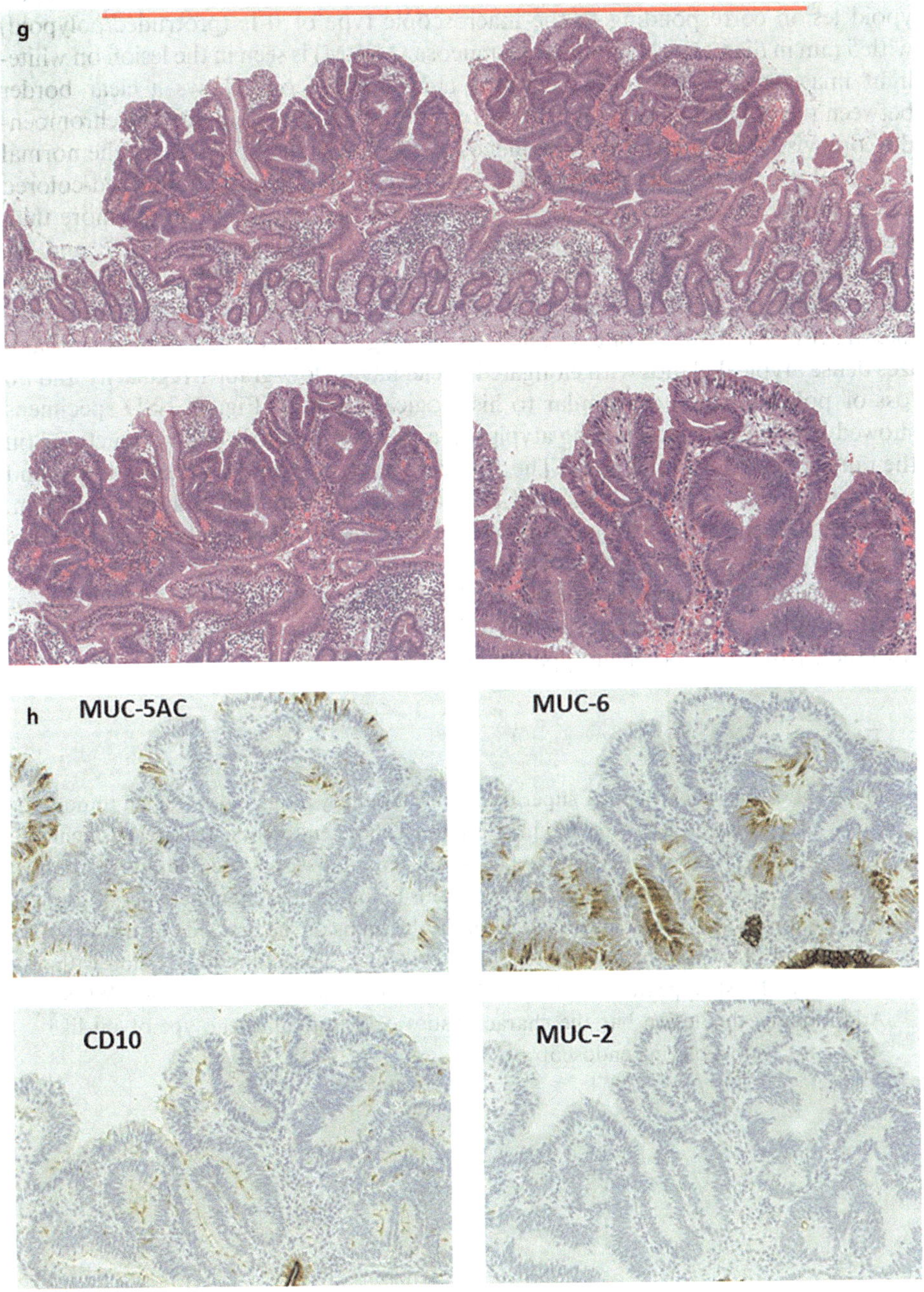

■ Explanation

High-definition white-light endoscopy shows a polypoid lesion at the opposite side of the supraduodenal angle in the bulb (Fig. a). The lesion shows a reddish and polypoid lesion corresponding to the macroscopic type of 0-Is (protruded/polypoid) with 5 mm in diameter. No milk-white mucosa (MWM) is seen in the lesion on white-light imaging (Fig. b). Indigo carmine chromoendoscopy shows a clear border between the polypoid lesion and the surrounding duodenal mucosa. The chromoendoscopy visualizes many fingerlike mucosal structures corresponding to the normal duodenal villi around the lesion, whereas the chromoendoscopy showed red-colored and enlarged structures (larger than fingerlike normal duodenal villi by more than threefolds) without MWM appearance in the polypoid lesion (Figs. c and d). Magnifying endoscopy with indigo carmine spraying and NBI shows the polypoid lesion consisted of enlarged villous structures like round leaves often including pit-like structures and dilated or congested vessels (Figs. d and e). Endocytoscopy visualizes dense atypical glands with elongated nuclei having low-grade irregularity and no loss of polarity which are similar to histological findings (Fig. f). ESD specimens showed a polypoid lesion having atypical glands with tubulopapillary structures (in the range of a red line of Fig. g). The atypical glands had mild structural atypia and consisted of tall columnar cells with elongated nuclei and eosinophilic cytoplasm. The arrangement of the elongated nuclei showed low-grade irregularity and no loss of polarity (Fig. g). Immunohistochemistry for phenotypes showed multifocal positivity for MUC5AC, partial positivity for MUC6 (Fig. h), diffuse weak positivity for CD10 (Fig. h), and negativity for MUC2 (Fig. h). These findings were consistent with the gastrointestinal phenotype.

Tips

Gastrointestinal phenotype of superficial non-ampullary duodenal epithelial tumors (SNADETs) may be very rare, and little is known about endoscopic features of them. The gastrointestinal phenotype SNADETs may tend to have endoscopic characteristics both gastric and intestinal phenotypes. This lesion has the characteristics of gastric phenotype as follows: lesion's location in the bulbs or the oral side of the Vater papilla and showing protruded/polypoid macroscopic-type and reddish lesion without milky whitish appearance named as MWM and white opaque substance (WOS). Additionally, this lesion has the characteristics of intestinal phenotype of pit-like structures on magnifying endoscopy.

Duodenal Adenoma (Intestinal-Type, Non-papillary Region)

Kenichi Goda and Kazuyuki Ishida

Y. Sano et al. (eds.), *Atlas of Advanced Endoscopy*, https://doi.org/10.1007/978-981-97-2732-2_54

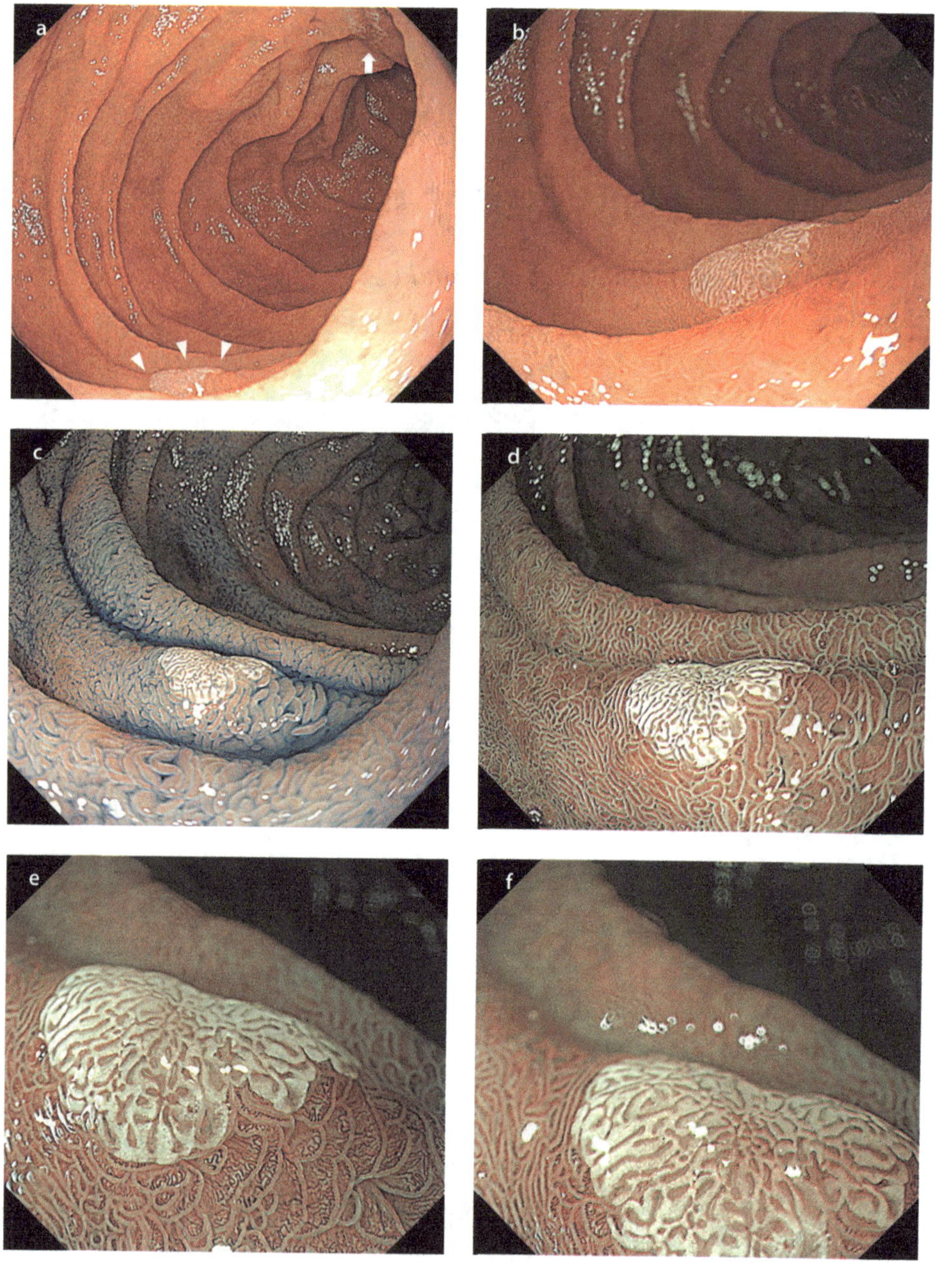
a
b
c
d
e
f

■ Pathological Diagnosis

Non-papillary duodenal adenoma, intestinal phenotype.

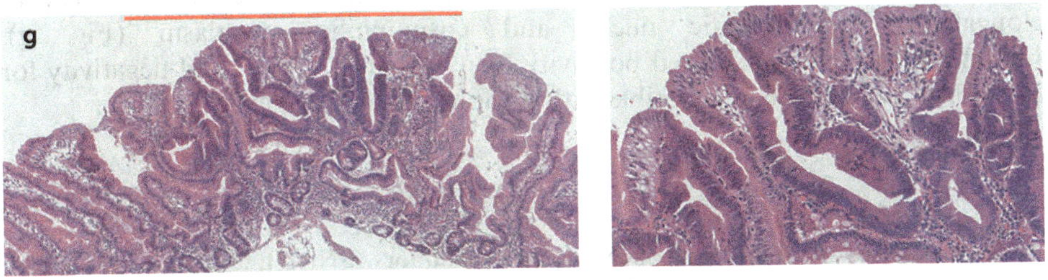

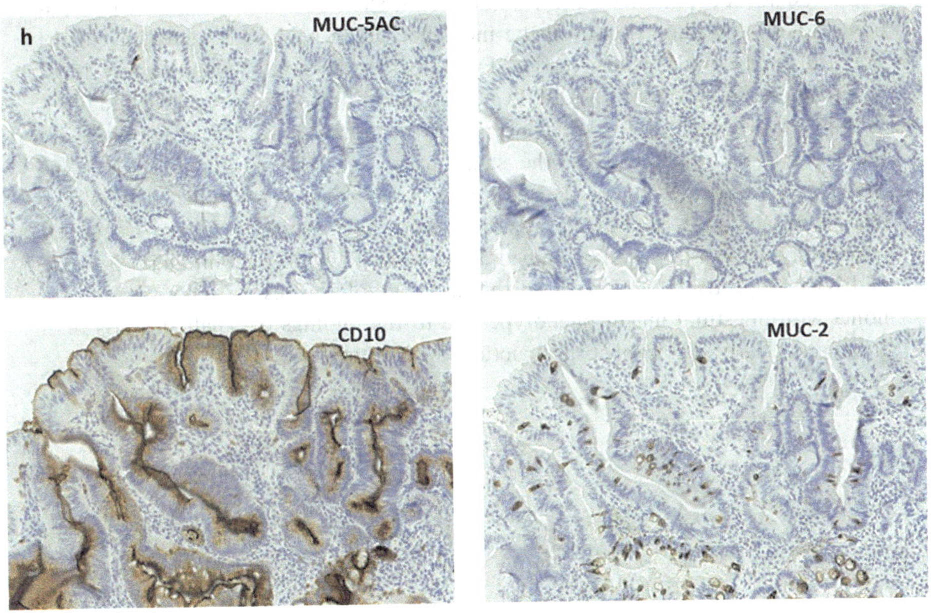

■ Explanation

High-definition white-light endoscopy showed a small whitish lesion at the opposite/oral side of the Vater papilla (a white arrow, Fig. a) in the second portion of the duodenum (white arrowheads, Fig. a). The whitish lesion showed the macroscopic type of 0-IIa (slightly elevated) with 4 mm in diameter (Fig. b). Indigo carmine chromoendoscopy showed many fingerlike mucosal structures corresponding to the normal duodenal villi around the lesion, whereas the chromoendoscopy small circular or tubular mucosal structures with milky white edging which was named "milk-white mucosa: MWM (entire type)" in the slightly elevated lesion (Fig. c). The mucosal structures with milky white edging are called as "white opaque substance (WOS)" on narrowband imaging (NBI) which will have the same histopathological backgrounds as MWM on white-light imaging (Fig. d). As with indigo carmine chromoendoscopy, NBI endoscopy showed a well-demarcated whitish lesion with slight elevation. Mucosal structures in the lesion are completely different from those in surrounding sites showing normal duodenal villi. NBI magnifying endoscopy (NBI-ME) shows mucosal structures consisted of circular, tubular, and ridged patterns which are clearly visible (Figs. e and f). We resected the lesion by cold snare polypectomy. The resected specimens showed atypical glands (in the range of a red line, Fig. g). The atypical glands had mild structural atypia and consisted of tall columnar cells with elongated, hyperchromatic nuclei and eosinophilic cytoplasm (Fig. g). Immunohistochemistry showed positivity for CD10 and MUC2 and negativity for MUC5AC and MUC6, compatible with the intestinal phenotype (Fig. h).

Tips

Superficial non-ampullary duodenal epithelial tumors (SNADETs) including adenoma and adenocarcinoma which confined to the submucosa have been rare in previous studies. Nowadays, however, the incidence of SNADETs is increasing in the practical endoscopy as widespread of EGD and progression of aging society. Most recent studies showed that SNADETs are divided into two groups, gastric and intestinal mucin phenotypes. Intestinal phenotype adenomas and adenocarcinomas are the most common subtype in the small intestine. Most of the intestinal-type SNADETs are located in the descending (second) portion or deeper parts of the duodenum and show slightly elevated/depressed macroscopic types and frequently have milky whitish appearance named as MWM and WOS on white-light imaging and NBI, respectively. We predicted the histology of this lesion as low-grade adenoma with intestinal mucin phenotype due to the findings of small size less than 5 mm and clearly visible mucosal structures with entire-type MWM.

Early Duodenal Cancer (Gastric-Type, Non-papillary Region)

Kenichi Goda and Kazuyuki Ishida

Y. Sano et al. (eds.), *Atlas of Advanced Endoscopy*, https://doi.org/10.1007/978-981-97-2732-2_55

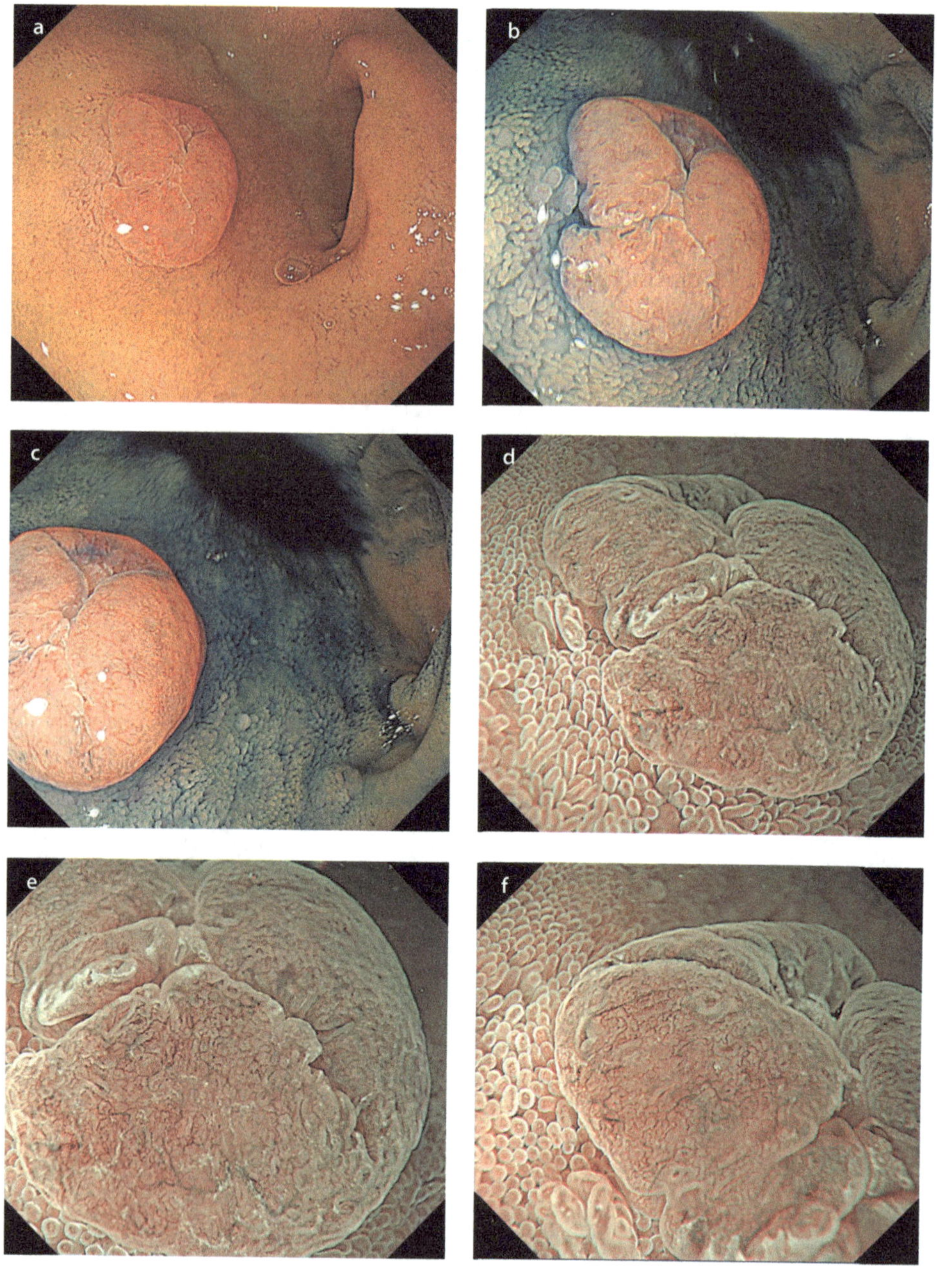
a
b
c
d
e
f

■ Pathological Diagnosis

Non-papillary duodenal adenocarcinoma (intramucosal carcinoma), gastric phenotype.

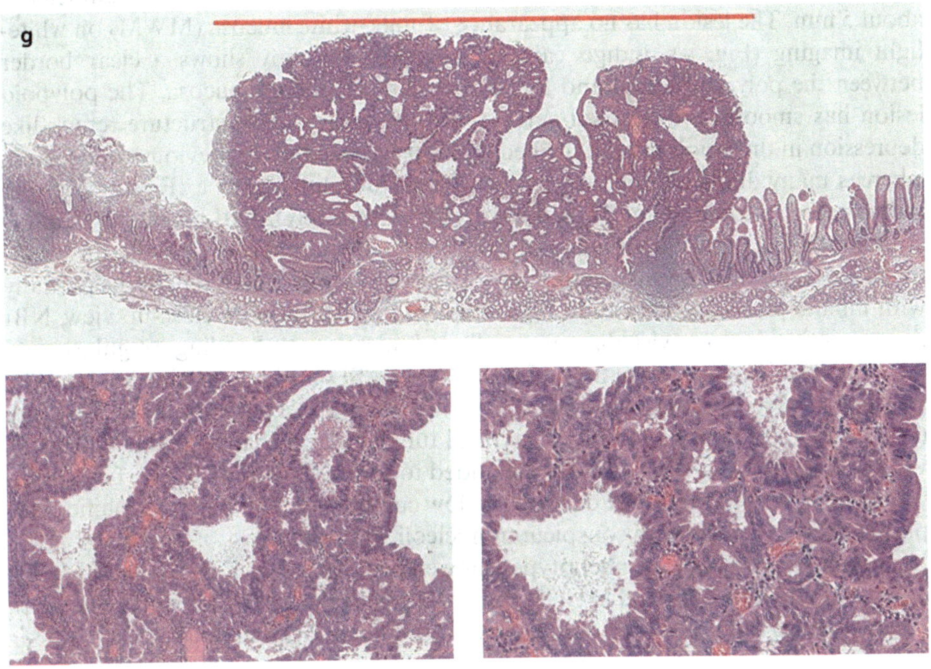

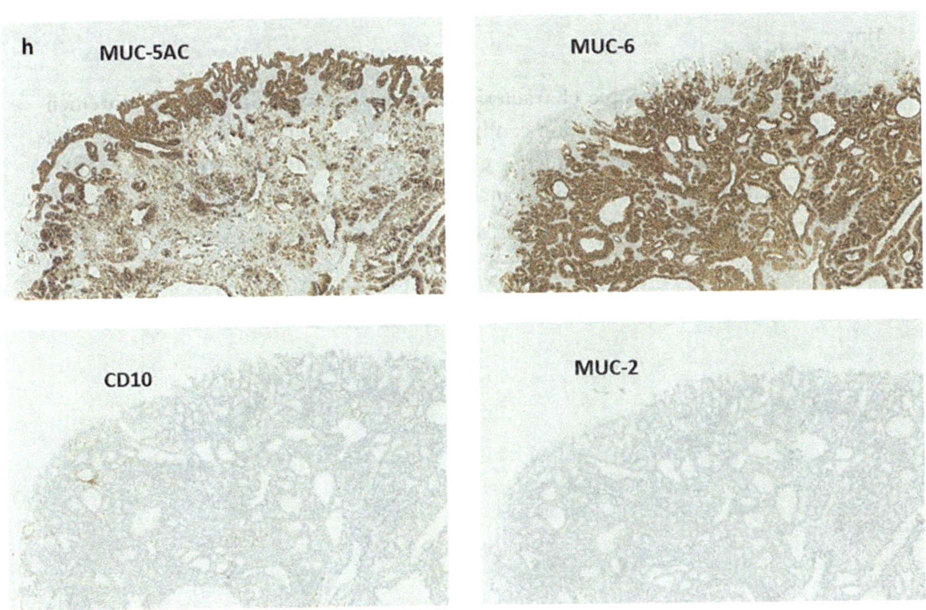

▪ Explanation

High-definition white-light endoscopy shows a polypoid lesion with reddish color at the opposite side of the supraduodenal angle in the bulbs (Fig. a). Macroscopic-type of this lesion is corresponding to 0-Is (protruded/polypoid), and the diameter is about 5 mm. The lesion has no appearance of milk-white mucosa (MWM) on white-light imaging (Fig. a). Indigo carmine chromoendoscopy shows a clear border between the polypoid lesion and the surrounding duodenal mucosa. The polypoid lesion has smooth and lustered surface with no clear visible structure (crack-like depression in the lesion may be formed by biopsy procedure in previous endoscopy), whereas many duodenal villi are seen in the surrounding mucosa (Figs. b and c in indigo carmine staining and Figs. d–f in NBI, respectively). NBI magnifying endoscopy shows fingerlike villous mucosal structures with light blue crest in the surrounding mucosa, whereas it visualizes lesion surface as obvious mucosal patterns partially with minute pit-like structures in high density (Fig. d). In more close-up view, NBI magnifying endoscopy visualizes many dilated microvessels bending irregularly and partially forming network pattern (Figs. e and f). ESD-resected specimens showed a steeply protruded lesion consisting of atypical glands with a packed tubular structure (in the range of a red line, Fig. g). Atypical tubular glands had irregular lumens of various sizes and shapes, which corresponded to severe structural atypia. Each atypical gland consisted mainly of cuboidal or low columnar cells with eosinophilic cytoplasm, enlarged nuclei with conspicuous nucleoli, and severe loss of polarity (Fig. g). Immunohistochemistry for phenotypes showed positivity for MUC5AC and MUC6 and negativity for CD10 and MUC2, compatible with the gastric phenotype (Fig. h). These findings suggested adenocarcinoma (intramucosal carcinoma).

Tips

According to the endoscopic characteristics of gastric-type SNADETS aforementioned, the histology of this lesion can be predicted as high-grade adenoma or adenocarcinoma with gastric mucin phenotype due to the location, color, macroscopic-type, and an obvious (unclear or very minute) mucosal pattern with irregular macrovessels.

Early Duodenal Cancer (Intestinal-Type, Non-papillary Region)

Kenichi Goda and Kazuyuki Ishida

Y. Sano et al. (eds.), *Atlas of Advanced Endoscopy*, https://doi.org/10.1007/978-981-97-2732-2_56

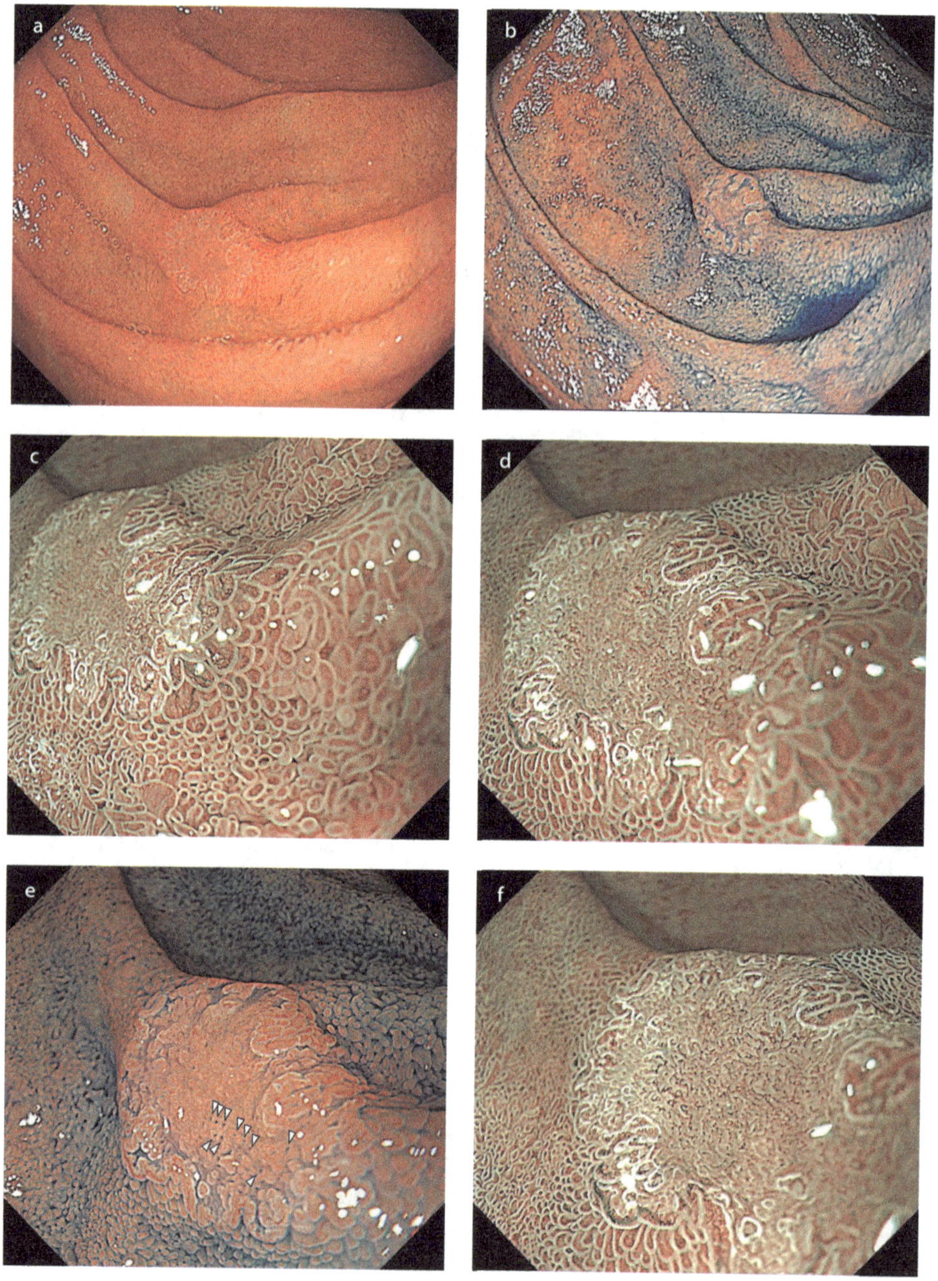
a
b
c
d
e
f

▪ Pathological Diagnosis

Non-papillary duodenal adenocarcinoma, intestinal phenotype.

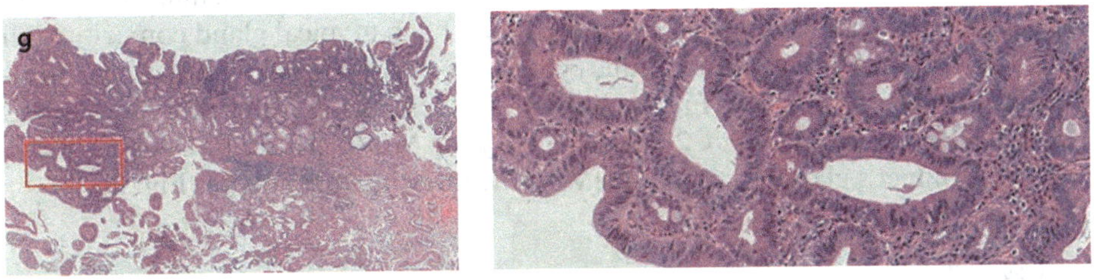

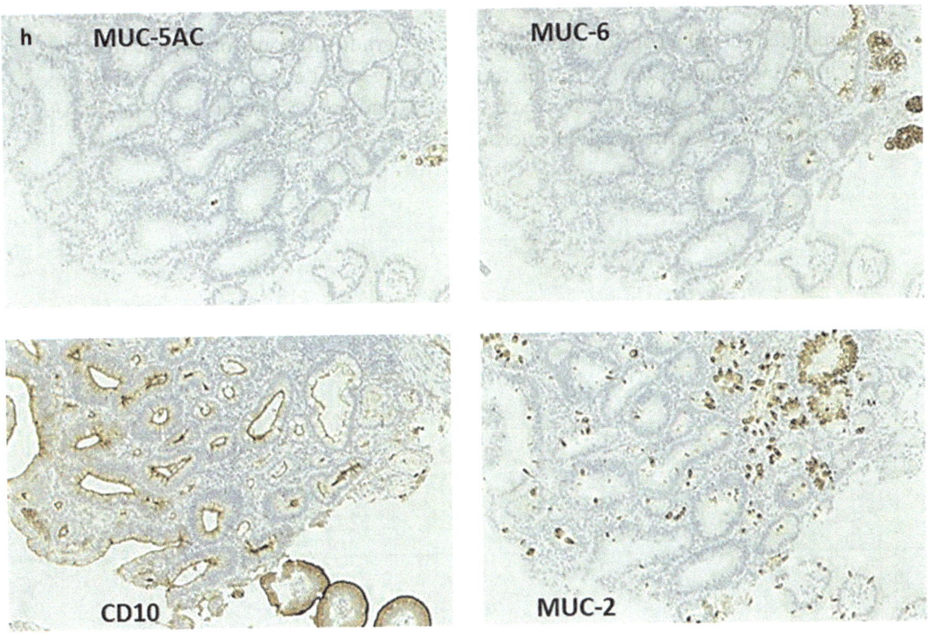

▪ Explanation

High-definition white-light endoscopy showed a small lesion on the Kerckring's fold at the lower duodenal angle in the second portion. The small lesion is edged with white lines or spots suggesting of milk-white mucosa (MWM: non-entire type) (Fig. a). Indigo carmine chromoendoscopy visualizes a clear demarcation line and shows a slightly depressed lesion with irregular margin (Fig. b). NBI endoscopy with low magnification shows many round leaf-like mucosal structures corresponding to the non-tumorous duodenal villi around the lesion, whereas the NBI endoscopy very small pit-like and various shaped (heterogeneous) mucosal structures were seen in the depressed lesion (Fig. c). NBI magnifying endoscopy shows very small or obvious structures in the depressed lesion. Very small pit-like structures scattering in the depressed lesion (Fig. d) will correspond to the micro-spots of indigo carmine solution (arrowheads, Fig. e). Indigo carmine and NBI magnifying endoscopy demonstrates multifocally located MWM and WOS in the marginal portion (Figs. d, e, and f), respectively. NBI magnifying endoscopy shows very small or obvious mucosal structures and many microvessels irregularly bending with caliber change (Fig. f). The EMR specimens had packed atypical glands. The atypical glands consisted of a tubular structure with lumens of various sizes and irregular branching that corresponded to moderate to severe structural atypia. Each atypical gland comprised tall to low columnar cells with eosinophilic cytoplasm, enlarged nuclei, and nuclear stratification and loss of polarity (Fig. g). Immunohistochemistry for phenotypes showed positivity for CD10 and MUC2 and negativity for MUC5AC and MUC6 (positive only for Brunner's glands), compatible with the intestinal phenotype (Fig. h).

Tips

We predicted the histology of this lesion as high-grade adenoma or adenocarcinoma with intestinal mucin phenotype due to the findings of the location, non-entire type of MWM, and obvious (unclear or very minute) and heterogeneous (various shapes) mucosal structures with irregular microvessels.

Colon

Contents

Hyperplastic Polyp (HP)

Naresh Bhat and Mineo Iwatate

Supplementary Information The online version contains supplementary material available at https://doi.org/10.1007/978-981-97-2732-2_57. The videos can be accessed individually by clicking the DOI link in the accompanying figure caption or by scanning this link with the SN More Media App.

Y. Sano et al. (eds.), *Atlas of Advanced Endoscopy*, https://doi.org/10.1007/978-981-97-2732-2_57

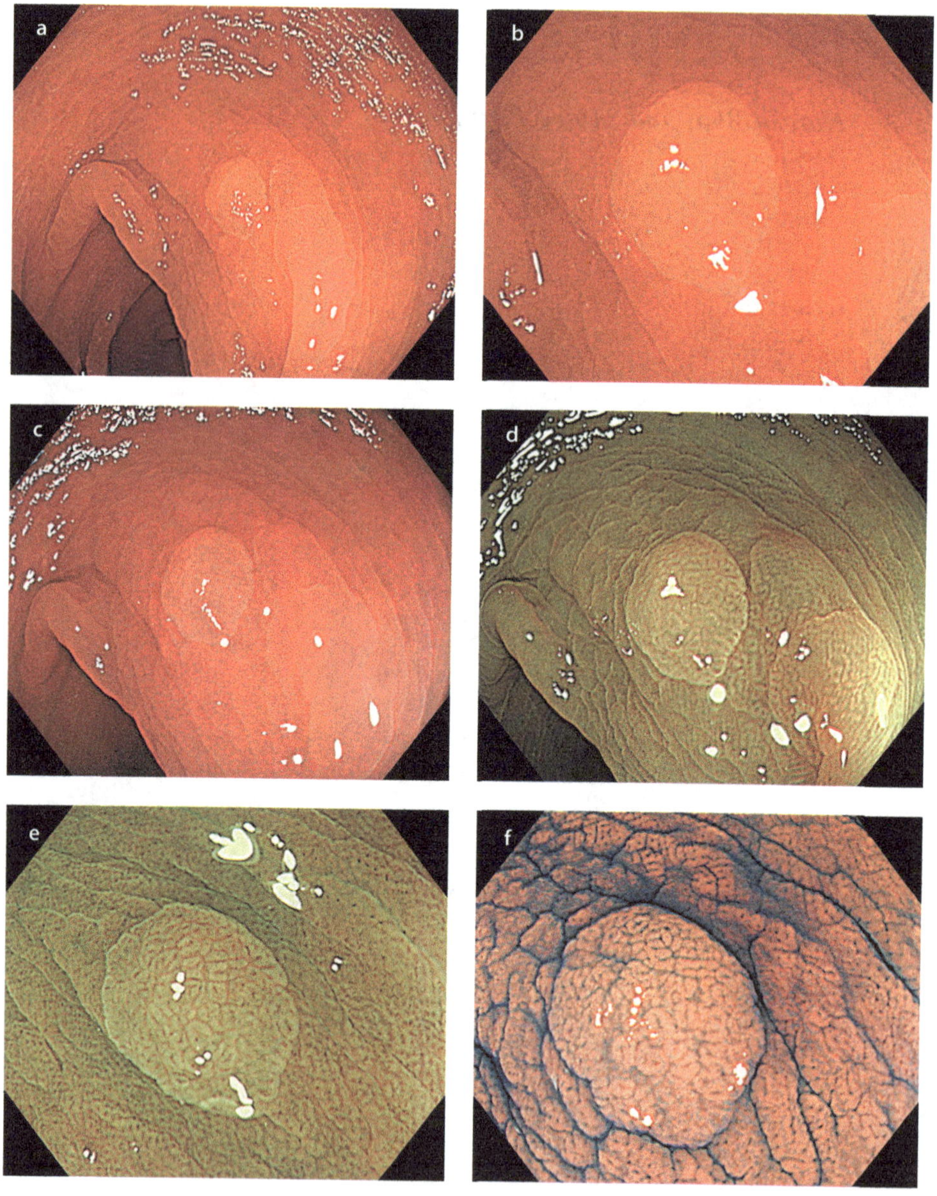
a
b
c
d
e
f

Pathological Diagnosis

Hyperplastic polyp.

g

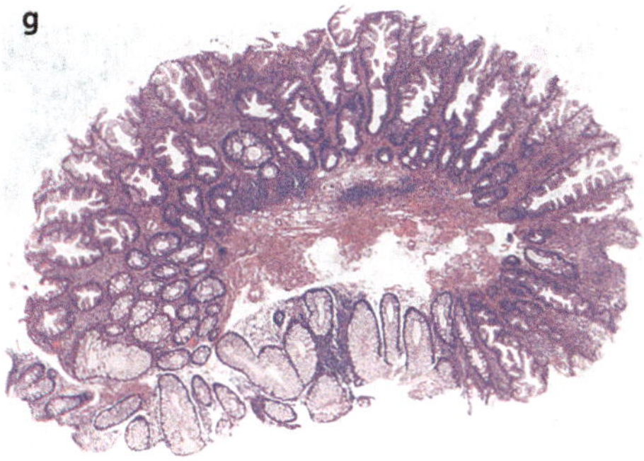

h

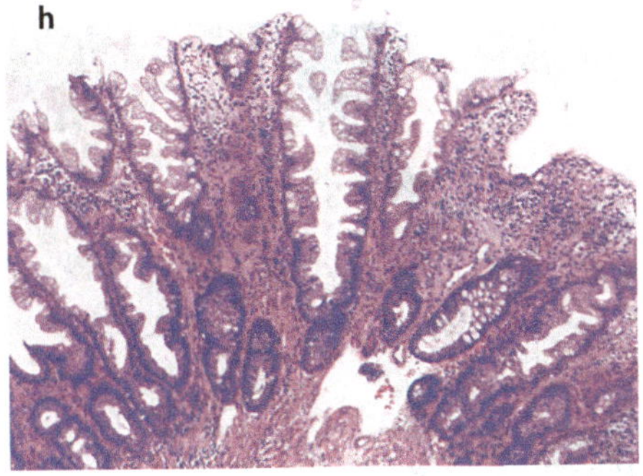

Explanation

This is a typical hyperplastic polyp found in the rectosigmoid on surveillance colonoscopy. The polyp measures 4 mm in size and has a Paris 0-IIa morphology (Fig. a). It has a pale and smooth surface merging with the background mucosa of the colon (Fig. b). TXI provides the contrast to highlight the lesion and margins (Fig. c). With non-magnifying NBI, the vessels are barely visible (Fig. d). Even with magnified NBI, vessels are inconspicuous, and brown or white spots are seen with regular shapes (Fig. e). The surface crypts, which are starlike shapes (Kudo's type II pit), are well demonstrated on chromoendoscopy (Fig. f).

The polyp was resected by cold snare polypectomy. Histopathology reveals no branching or dilatation of the basal crypts without dysplasia (Figs. g and h). Some serration can be appreciated only in the upper half of crypts, unlike sessile serrated lesion (SSL) where it also involves deeper parts of crypts.

Tips

Hyperplastic polyps are characterized as nonneoplastic with extremely low potential for advancing cancer unless they are part of a polyposis syndrome. Typical hyperplastic polyps have regular black or white dots on the surface without visible vessels by NBI and starlike pits (type II pits) by chromoendoscopy. The hyperplastic polyps need to be accurately differentiated from SSL with higher malignant potential, which may look similar in being flat and pale during colonoscopy. We should suspect SSL for the lesion with type II open pits and consider traditional serrated adenoma for the lesion with IIIH or IVH pits.

■ Video

A short video is available as ◘ Fig. 1.

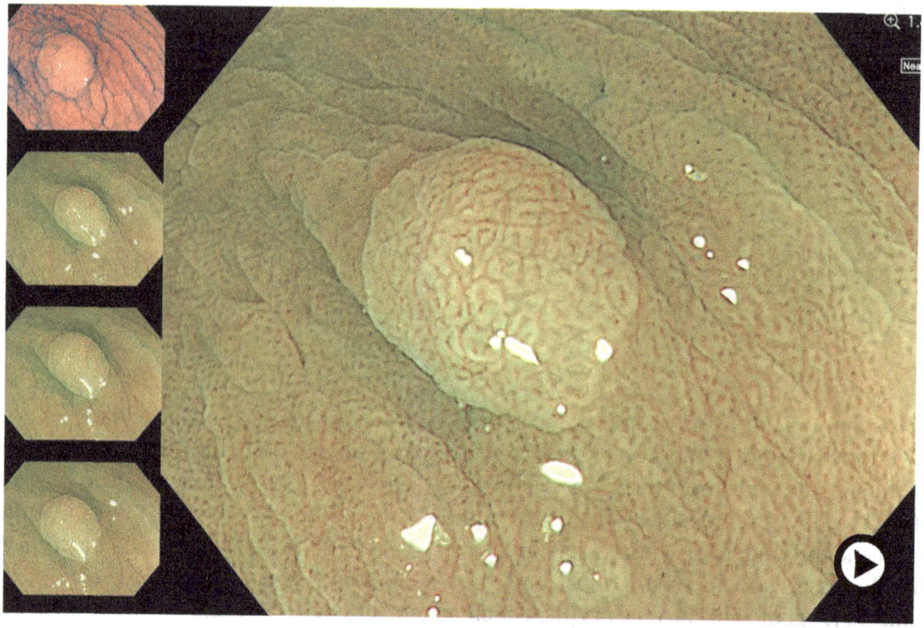

◘ **Fig. 1** Practical observation and diagnosis of a hyperplastic polyp with WLI, TXI, NBI, and chromoendoscopy (▶ https://doi.org/10.1007/000-c8n)

Sessile Serrated Lesion (SSL)

Daizen Hirata and Wataru Sano

Supplementary Information The online version contains supplementary material available at https://doi.org/10.1007/978-981-97-2732-2_58. The videos can be accessed individually by clicking the DOI link in the accompanying figure caption or by scanning this link with the SN More Media App.

Y. Sano et al. (eds.), *Atlas of Advanced Endoscopy*, https://doi.org/10.1007/978-981-97-2732-2_58

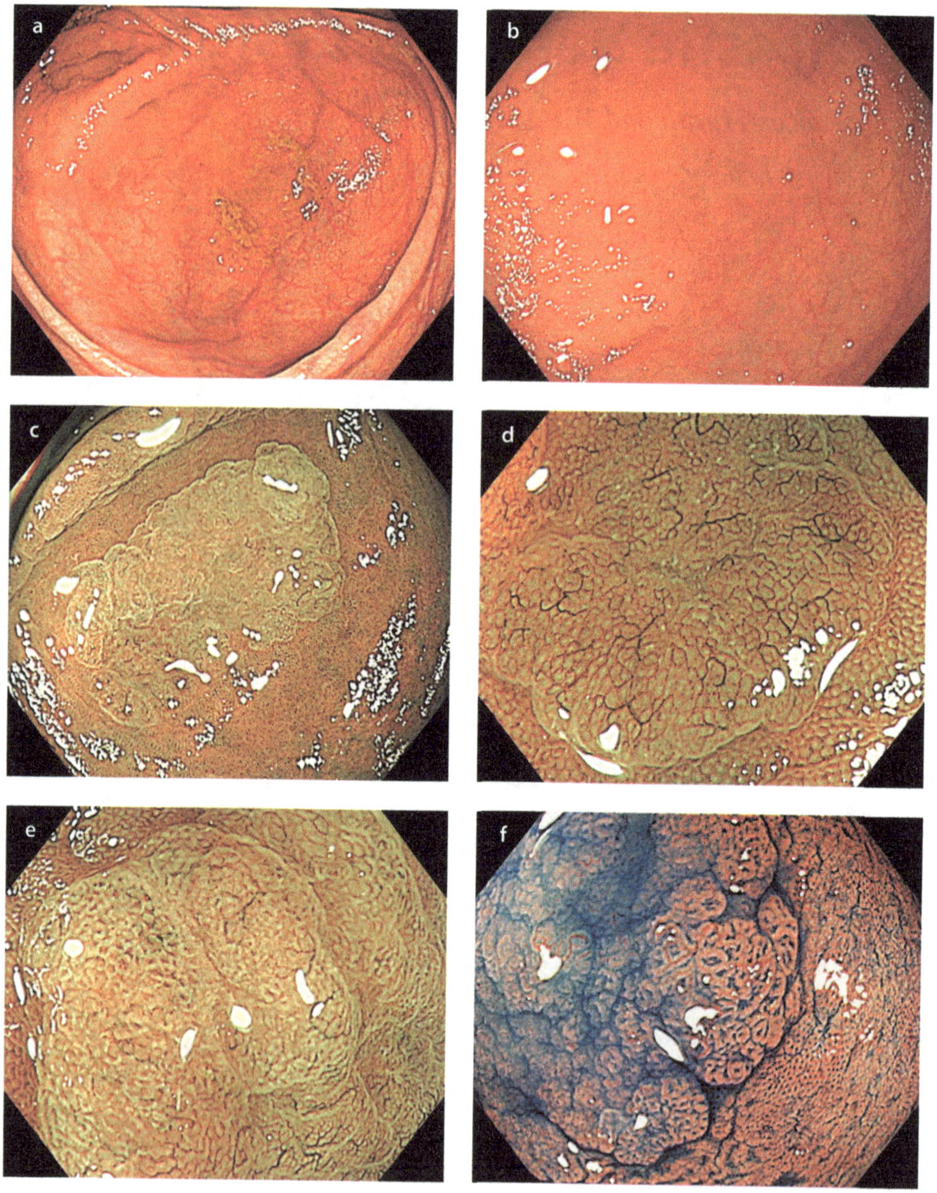
a
b
c
d
e
f

Pathological Diagnosis

Sessile serrated lesion, pHM (−).

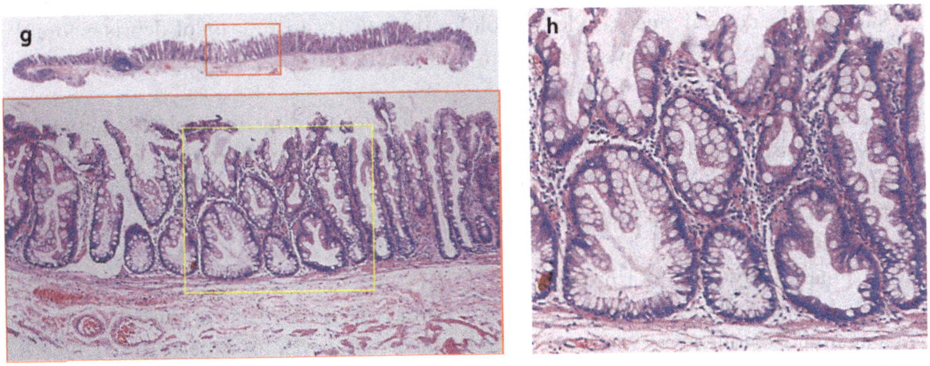

Explanation

This lesion was a flat lesion, 15 mm in size, and first detected with mucus cap and rim of debris in the cecum (Fig. a). The lesion had an irregular shape, and its borders were indistinctive in WLI observation when washed (Fig. b).

In non-magnifying NBI, the lesion exhibited a subtle paler hue compared to the surrounding mucosa. The surface of the lesion was cloudlike, and its borders were more distinct than those in WLI (Fig. c). Switching to the magnifying NBI, dilated branched vessels (Fig. d) and expanded crypt opening (Fig. e) were observable throughout the lesion. This lesion was diagnosed as JNET type 1 and showed multiple characteristic findings of sessile serrated lesions: irregular shape, indistinctive border, cloudlike surface, mucus cap, rim of debris, dilated vessels, and dilated crypts. Chromoendoscopy with indigo carmine dye provided better definition of the boundaries and showed type II-Open pit pattern (Fig. f).

Based on these findings, this lesion was diagnosed as a sessile serrated lesion and resected en bloc by EMR. Unequivocal distorted crypts showing inverted T and L shapes were observable in the specimen. According to the fifth edition of the *World Health Organization Classification of Tumors of the Digestive System*, this lesion was diagnosed as a sessile serrated lesion (Figs. g and h).

Tips

First, SSL is difficult to detect because of its flat morphology and similar color to the surrounding normal mucosa. Please look for the mucus cap and rim of debris before washing the colon. Intensive stool/mucus attachment and no vessel visibility are considered useful for detecting SSLs.

Second, the boundaries of SSL are often indistinct, causing incomplete resection. Indigo spraying is effective, but sometimes it is still difficult to determine the boundary. In such cases, try acetic acid spraying (1.7%). Within seconds, the aceto-whitening reaction will occur, and only the lesion will turn white, dramatically clarifying the boundary.

Finally, 90% of SSLs present with JNET type 1, while JNET type 1 lesions include both SSLs and HPs. Accurate differential diagnosis of SSLs and HPs using magnifying colonoscopy was reported to be challenging even for experts. JNET type 1 lesions ≥6 mm are recommended to be resected because selective endoscopic resection has a disadvantage of leaving approximately 20% of SSLs on site.

Video

A short video is available as Fig. 1.

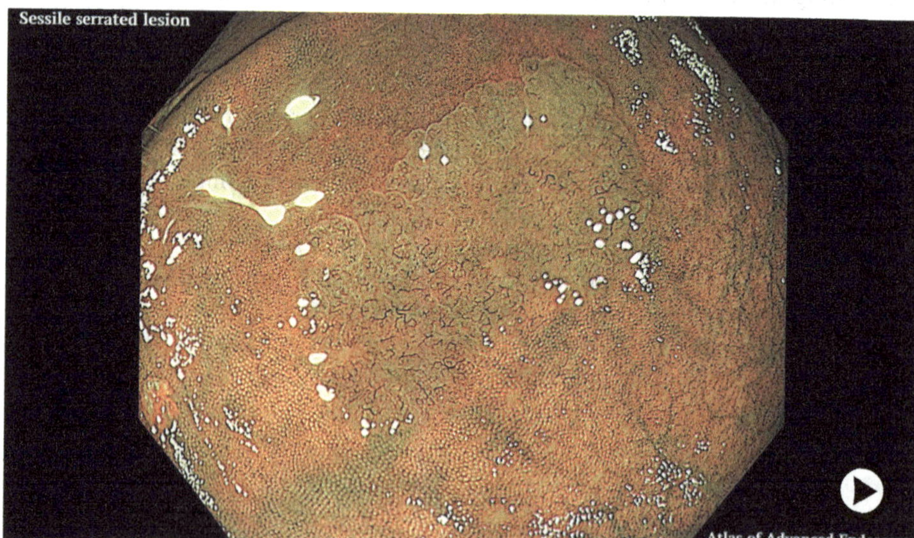

Fig. 1 A 15-mm flat lesion in the cecum (sessile serrated lesion) (▶ https://doi.org/10.1007/000-c8p)

Sessile Serrated Lesion with Dysplasia (SSLD)

Yoji Takeuchi

Supplementary Information The online version contains supplementary material available at https://doi.org/10.1007/978-981-97-2732-2_59. The videos can be accessed individually by clicking the DOI link in the accompanying figure caption or by scanning this link with the SN More Media App.

Y. Sano et al. (eds.), *Atlas of Advanced Endoscopy*, https://doi.org/10.1007/978-981-97-2732-2_59

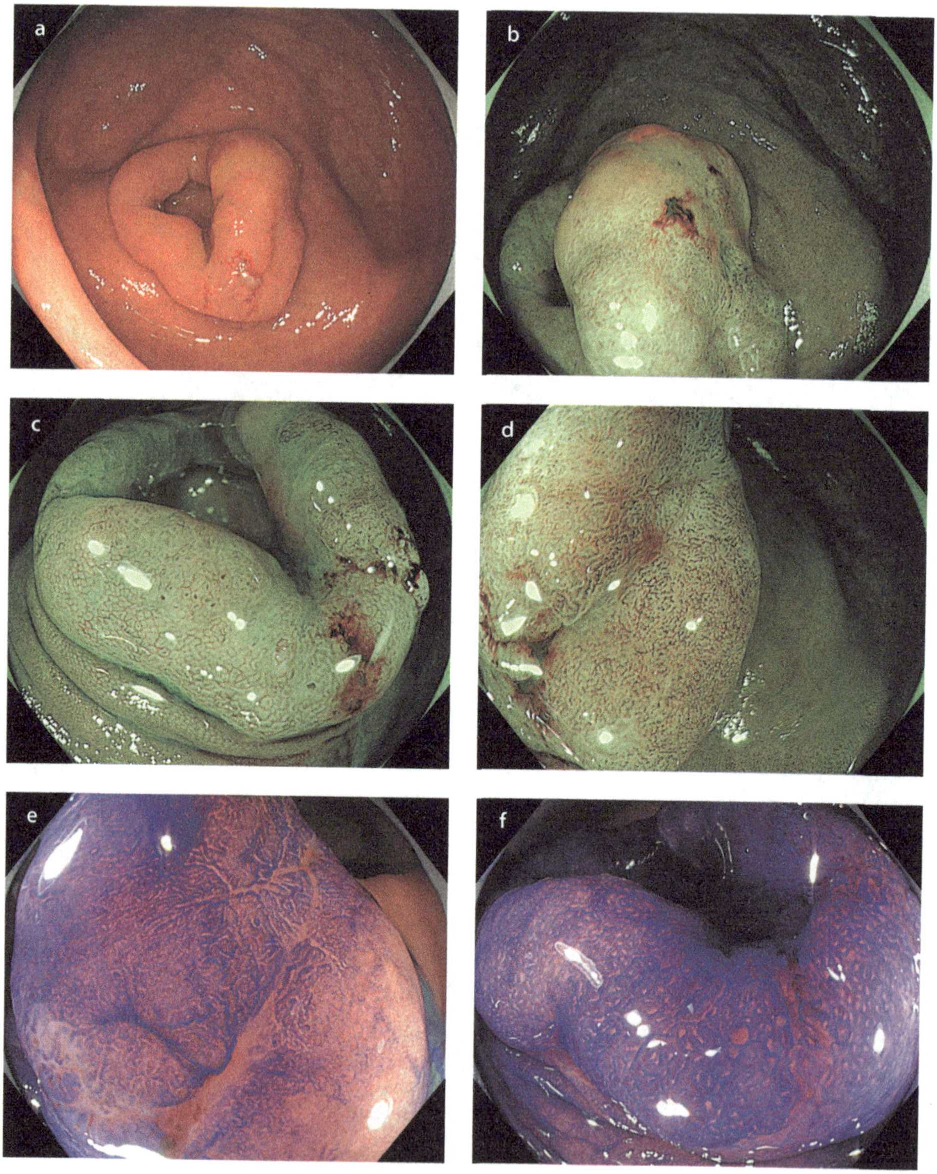
a
b
c
d
e
f

▪ Pathological Diagnosis

Sessile serrated lesion (SSL) with dysplasia (high-grade dysplasia in SSL).

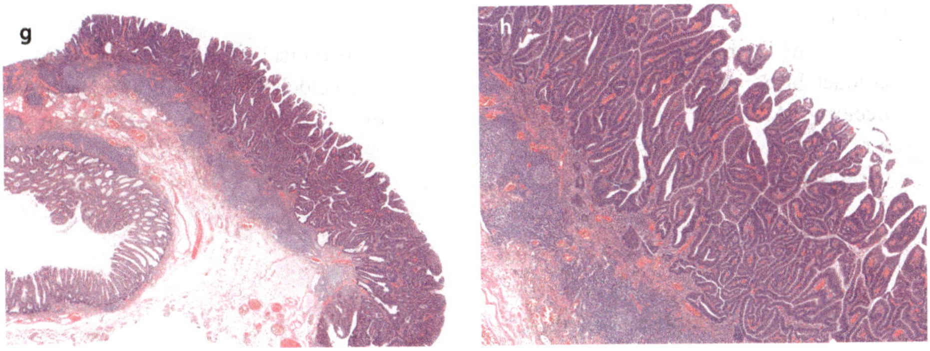

▪ Explanation

A protruding lesion measuring approximately 20 mm has been detected in the cecum. A doughnut-like-shaped protrusion surrounds the appendiceal orifice, and its surface is covered with thick mucus. The color of the lesion is almost similar or somewhat whitish compared to the surrounding mucosa when observed with a non-magnifying white light image (WLI) (Fig. a). The doughnut-like protrusion is asymmetry, and the right half of the lesion shows a thick embankment with an uneven surface (Fig. a). Narrowband imaging (NBI) offers the right half of the lesion as brownish and darker, while the left half is whitish and brighter (Figs. b and c). Our close observation of the right-side brownish area under non-magnifying NBI revealed its surface pattern and brownish vessels which is classified as NICE type 2 (Fig. b). On the contrary, we can recognize lacy vessels with dark brownish spots on the left-sided lighter whitish area, suggesting a sessile serrated lesion (SSL) (Fig. c). With magnifying NBI, varied calibers, meandering dilated microvessels, distributed irregularly, indicating JNET type 2B on the right side (Fig. d). Tubular but branching and irregular pits (Kudo's VI low-grade) on the right side (Fig. e) and starlike open pits (type II-O) on the left side (Fig. f) were shown with magnifying chromoendoscopy after crystal violet dye spraying. The lesion was diagnosed as SSL with high-grade dysplasia and was removed en bloc using the ESD technique.

Histopathological examination revealed an adenocarcinoma in SSL, invading muscularis mucosa (Fig. g and h).

Tips

SSL is a noninvasive, benign lesion that can be treated with piecemeal cold snare resection. It is also known that SSL is occasionally accompanied by dysplasia, which can be invasive. Since the dissection plane of cold snare resection is superficial and can leave muscularis mucosa in situ, it cannot be recommended for the lesions suspected to be high-grade dysplasia, which has a risk of invading to muscularis mucosa. Therefore, meticulous observation before deciding on a treatment strategy is required. In this case, we could identify the dysplastic area during the pretreatment assessment and choose to remove the lesion en bloc with ESD. Actually, the lesion invaded to muscularis mucosa; en bloc resection with ESD would be an optimal strategy for this lesion.

During the mucus on the mucosal surface, which is typical in SSL, was removed, contact bleeding of the mucosal surface at the six o'clock position of the lesion occurred. Although the contact bleeding, in this case, did not influence the treatment strategy decision because it happened outside the diagnostically important area, careful manipulation to avoid damaging the lesion is desirable.

▪ Video

A short video is available as Fig. 1.

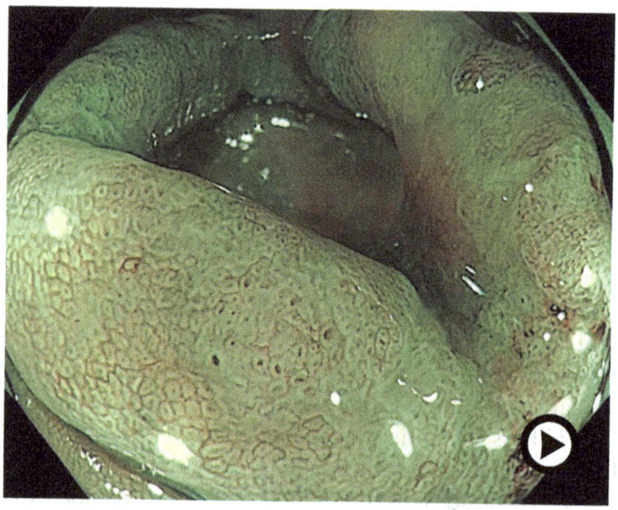

Fig. 1 Practical observation and diagnosis of sessile serrated lesion accompanied with intramucosal cancer (▶ https://doi.org/10.1007/000-c8q)

Traditional Serrated Adenoma (TSA)

Supakij Khomvilai and Yasushi Sano

Y. Sano et al. (eds.), *Atlas of Advanced Endoscopy*, https://doi.org/10.1007/978-981-97-2732-2_60

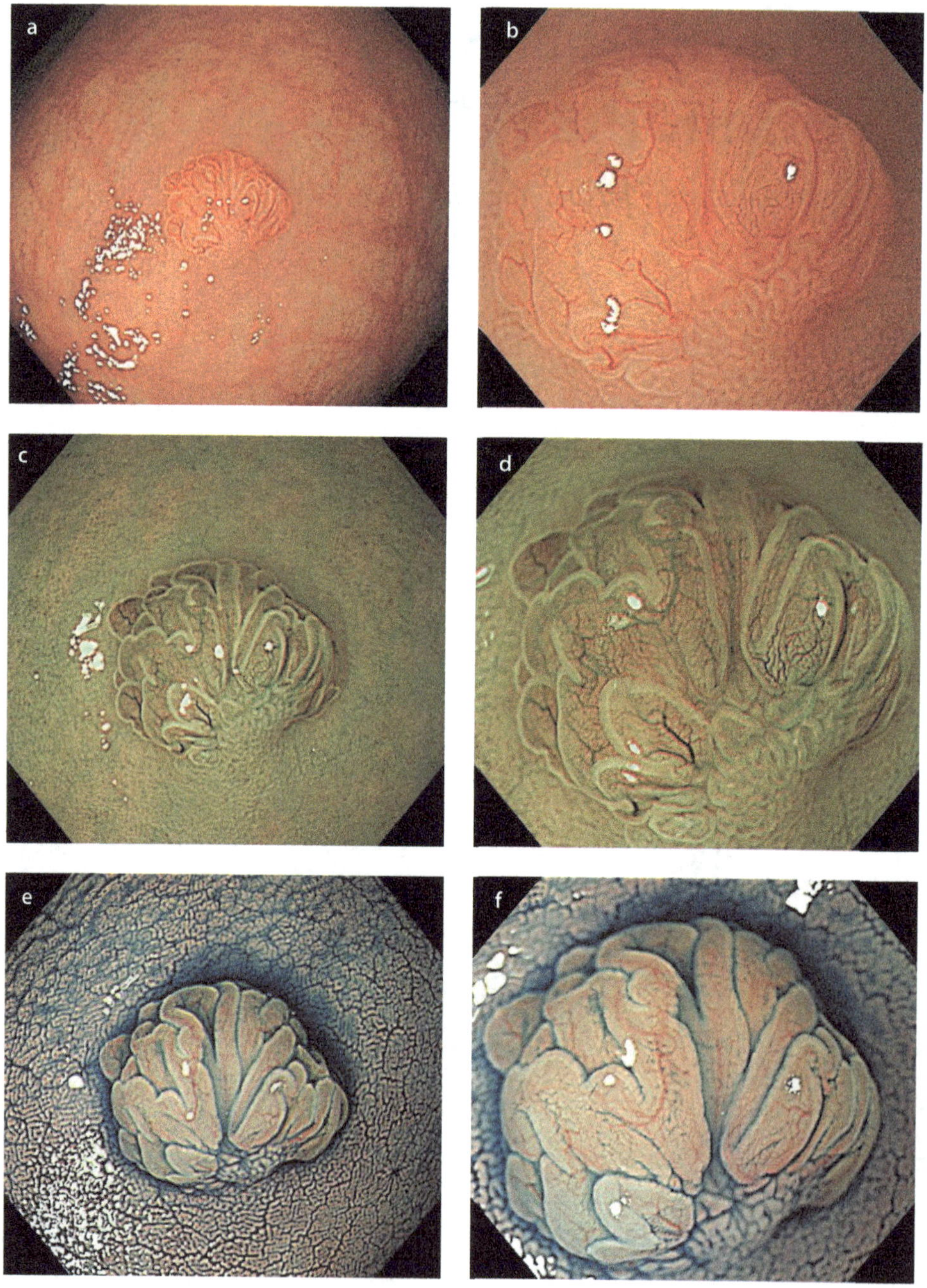

■ Pathological Diagnosis

Traditional serrated adenoma.

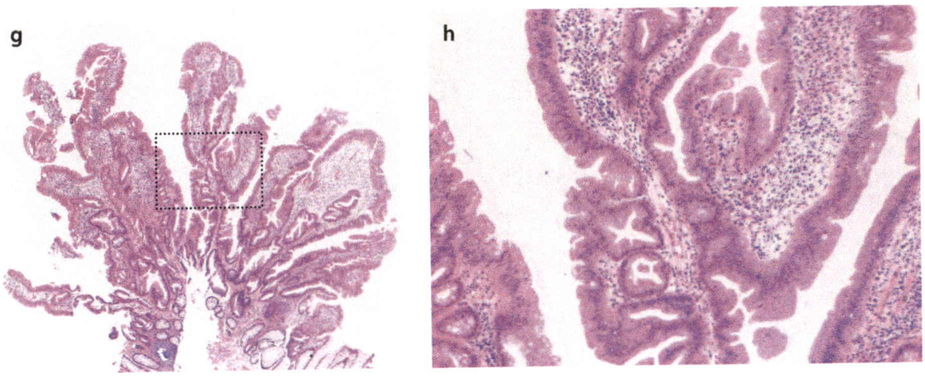

■ Explanation

A protruded lesion measuring approximately 5 mm in size has been detected in the sigmoid colon. The lesion presents a pinecone-like appearance with a red coloration when observed with non-magnifying WLI (Figs. a and b). Switching to NBI, the surface pattern demonstrates swelling villous forms (Fig. c). Spiral or cicada wing-like vessels can be observed on the villous surface by magnifying NBI (Fig. d). Chromoendoscopy with indigo carmine dye reveals serrated features on the edge of the villous pits, indicating Kudo's IV$_{\text{H}}$ pit (Figs. e and f). Based on these findings, this lesion is diagnosed as a traditional serrated adenoma (Figs. e and f).

The histopathological images demonstrate the protuberant villiform growth pattern with slit-like serrated glands (Fig. g). The lesional cells in TSA exhibit elongated, penicillate nuclei with eosinophilic cytoplasm (Fig. h). Some glands focally demonstrate a key feature of ectopic crypt foci which are abnormally positioned crypts with bases not seated at the muscularis mucosae. The final histopathological diagnosis was traditional serrated adenoma.

Tips

Traditional serrated adenoma (TSA) is primarily located in the rectum and sigmoid colon. The endoscopic diagnosis of TSA is relatively simple due to the characteristic villous appearance, which resembles a pinecone or branch coral and is often accompanied by a deep red coloration. The most common macroscopic types of TSA are protruded or pedunculated although the flat type may also be observed. The typical pit patterns of TSA are identified as Kudo's III$_{\text{H}}$ (III$_{\text{L}}$ with serrations or IV$_{\text{H}}$ with serrations) pits by chromoendoscopy. Additionally, spiral or cicada wing-like vessels can be observed in the large villus by NBI. When the surface villus is uneven in size and undergoes atrophy, it suggests that TSA has progressed to cancer.

Superficially Serrated Adenoma (SuSA)

Yasuhiko Mizuguchi and Yutaka Saito

Content

Y. Sano et al. (eds.), *Atlas of Advanced Endoscopy*, https://doi.org/10.1007/978-981-97-2732-2_61

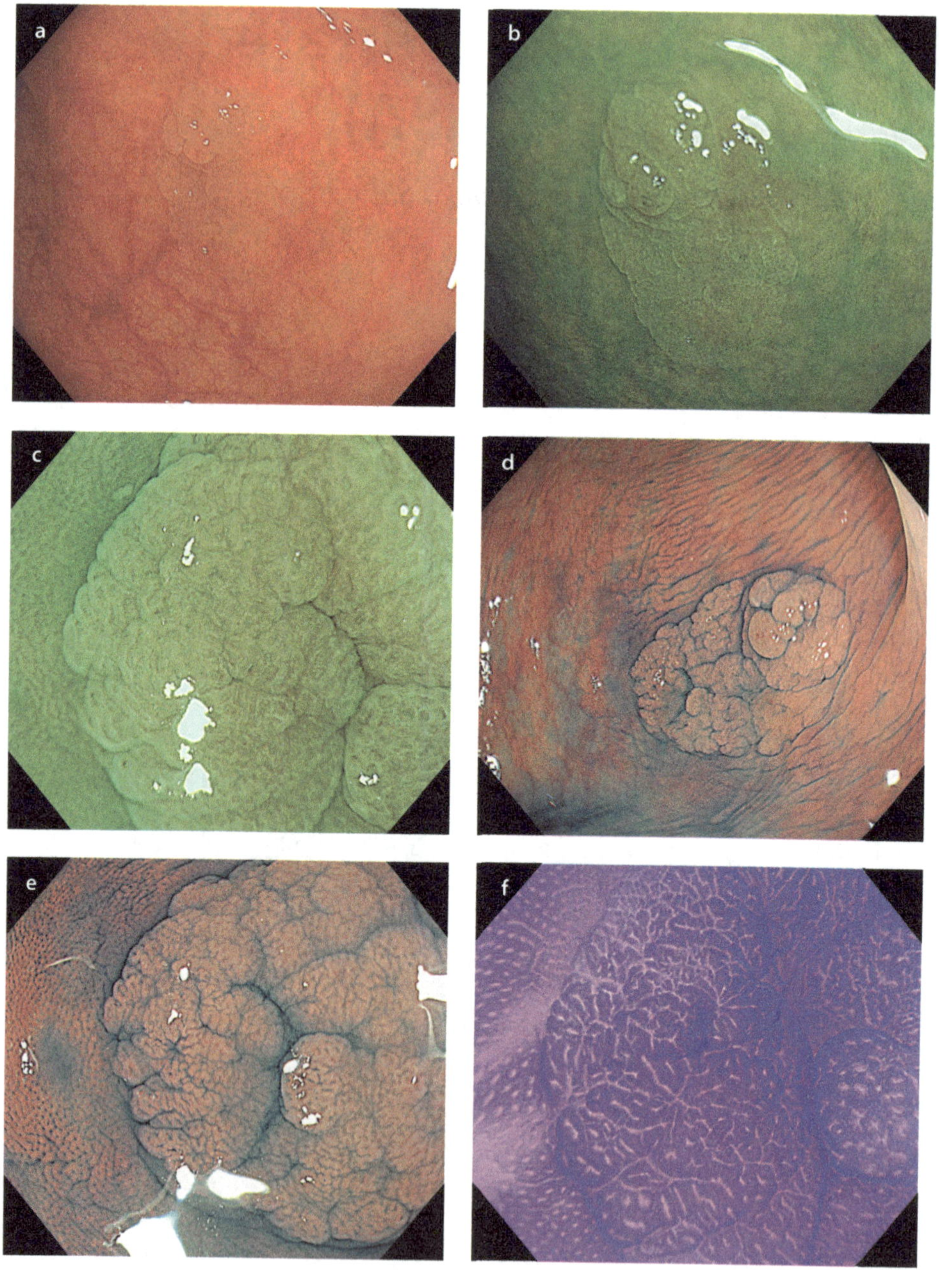
a
b
c
d
e
f

Pathological Diagnosis

Superficially serrated adenoma.

g

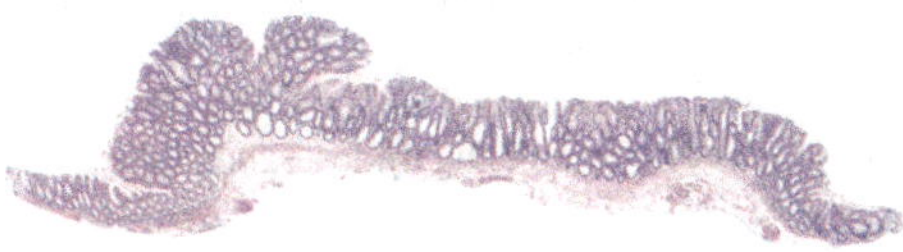

h

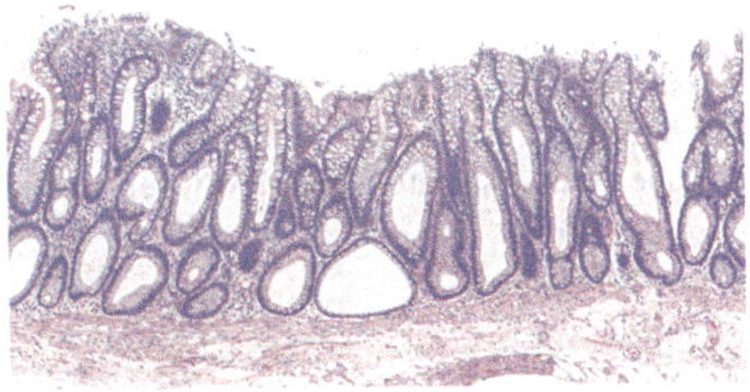

Explanation

A superficial elevated lesion, measuring 20 mm in diameter, is located in the rectum. The color is whitish compared to the surrounding mucosa (Fig. a). Magnifying NBI revealed an absence of visible vessels, consistent with the vessel pattern of JNET type 1 (Figs. b and c). Chromoendoscopy with indigo carmine dye showed more sharply defined borders, exhibiting convex and concave contours rather than a circular shape. Furthermore, the lobulated surface was more pronounced (Fig. d). Magnifying chromoendoscopy using indigo carmine or crystal violet unveiled a characteristic fernlike appearance. Among the pits, some displayed a stellar shape (Kudo's II), while others were fused or branched (fernlike, type IIIH pit) (Figs. e and f). The lesion resembles an adenoma more than a serrated lesion under WLI or non-magnifying chromoendoscopy; however, it also showed findings similar to serrated lesions when observed through magnifying NBI or magnifying chromoendoscopy. Based on the combination of characteristic findings of macroscopically adenoma-like but microstructurally serrated lesions, the patient was endoscopically diagnosed with SuSA. The lesion was subsequently resected en bloc using the EMR technique. Histologically, the lesion comprised adenomatous glands with superficial serration, consistent with superficially serrated adenoma (Figs. g and h).

Tips

SuSA is a recently described subtype of serrated lesion characterized by adenomatous glands with serration confined to the superficial layer [1]. These lesions are mostly found in the rectosigmoid colon. The current WHO classification does not recognize SuSA as a distinct subtype; consequently, they are classified as unclassified serrated adenomas. In current clinical practice, many SuSAs are histologically diagnosed as tubular adenomas due to their minimal serration. However, the surface serrated architecture of SuSAs can be readily recognized through magnifying endoscopy [2]. The majority of SuSA harbor a *KRAS* mutation and *RSPO* fusion/overexpression, indicating that they possess distinct molecular characteristics from conventional adenomas. SuSA is considered a precursor to traditional serrated adenoma, and approximately 10% of TSAs are associated with a SuSA component.

Acknowledgment We thank Shigeki Sekine for his advice and pathology images.

References

1. Hashimoto T, Tanaka Y, Ogawa R, et al. Superficially serrated adenoma: a proposal for a novel subtype of colorectal serrated lesion. Mod Pathol. 2018;31:1588–98.
2. Mizuguchi Y, Tanaka Y, Cho H, et al. Endoscopic features of isolated and traditional serrated adenoma-associated superficially serrated adenomas of the colorectum. Dig Endosc. 2022;34:153–62.

Tubular Adenoma (0-IIa)

Jonard T. Co and Patricia Anne I. Cabral-Prodigalidad

Contents

Supplementary Information The online version contains supplementary material available at https://doi.org/10.1007/978-981-97-2732-2_62. The videos can be accessed individually by clicking the DOI link in the accompanying figure caption or by scanning this link with the SN More Media App.

Y. Sano et al. (eds.), *Atlas of Advanced Endoscopy*, https://doi.org/10.1007/978-981-97-2732-2_62

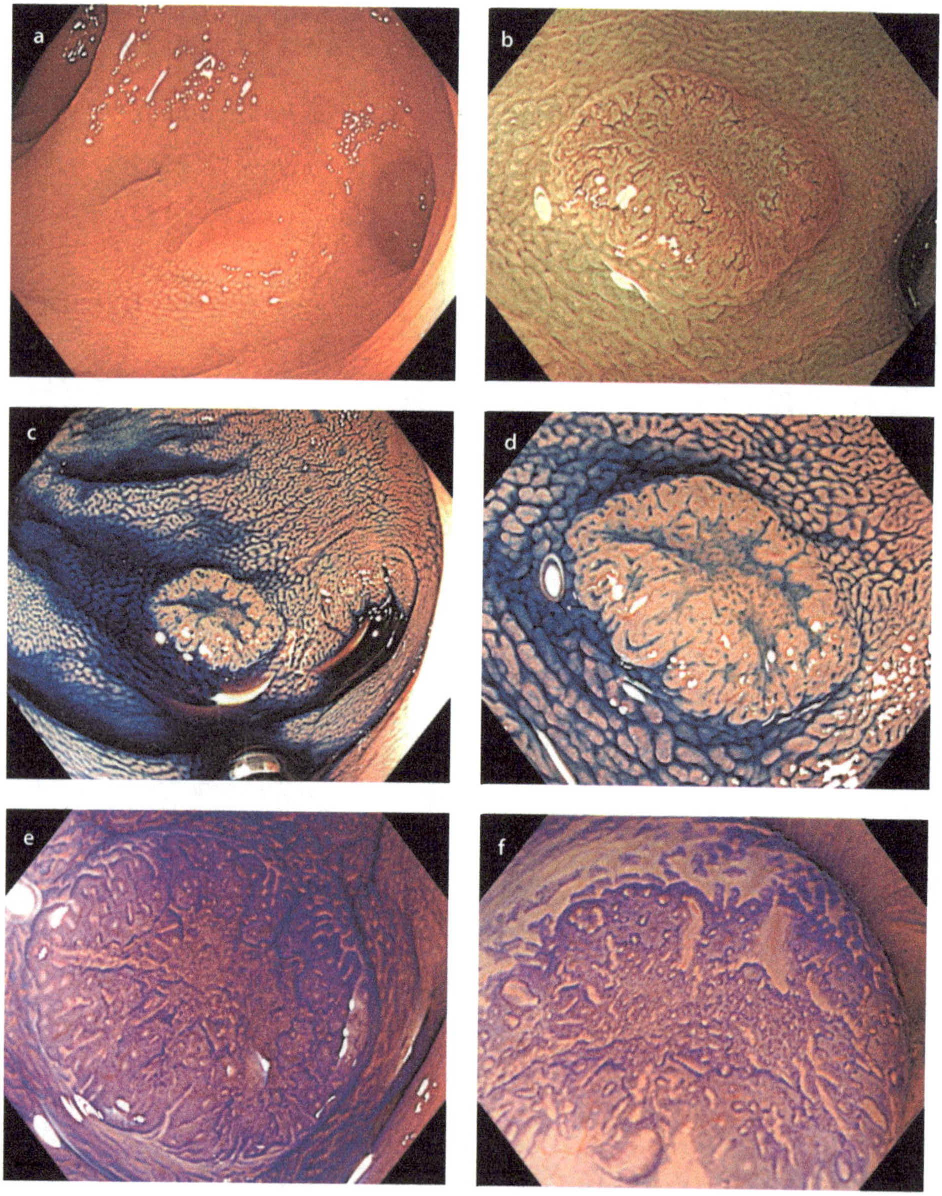
a
b
c
d
e
f

▪ Pathological Diagnosis

Tubular adenoma with low-grade dysplasia.

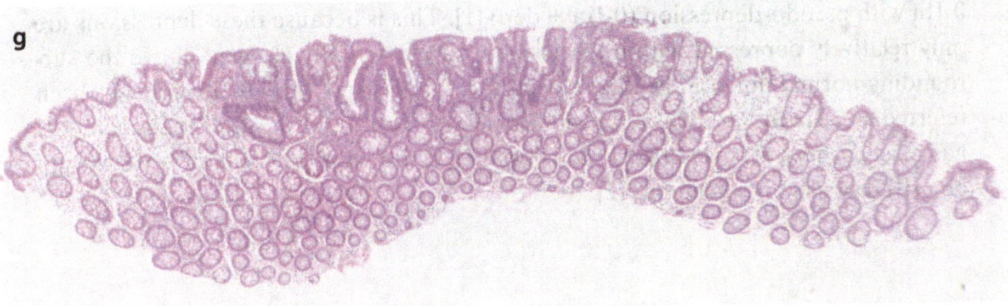

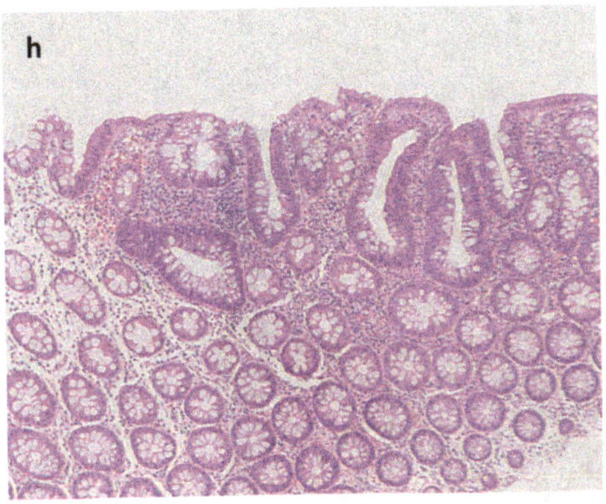

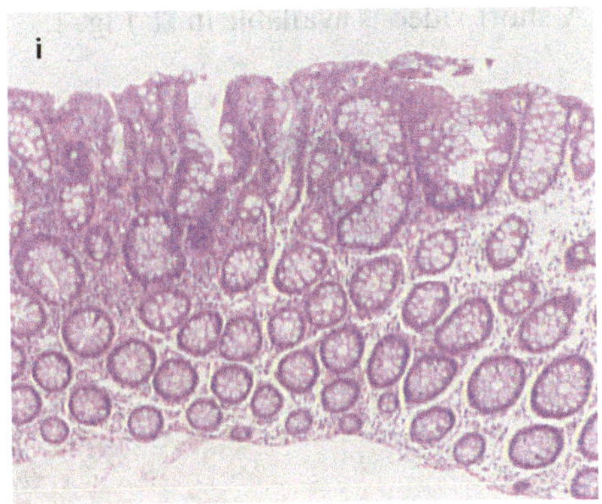

▪ Explanation

A sessile polyp measuring approximately 4 mm was seen in the sigmoid colon, Paris 0-IIa. On white light imaging, the color of the lesion was slightly reddish and appeared to contain a depression in the center, so-called as 0-IIa with pseudo-depression (0-IIa + dep) or valley sign (Fig. a) [1, 2]. On NBI, the lesion exhibited vascular pattern with high density of capillary vessels in a uniform and regular, relatively well-ordered reticular pattern, classified as JNET 2A (Fig. b). Chromoendoscopy with indigo carmine spraying provided better definition of the surface and boundaries (Figs. c and d). With crystal violet dying, the pit patterns showed elongated pits (Kudo Type IIIL) at the margins but were not clearly visible in the central area (Fig. e). When switching to high magnification with under-water condition, small round pits (Kudo Type IIIs) can be appreciated in the central area (Fig. f). Based on these findings, the lesion was diagnosed as a tubular adenoma and resected *en bloc* through cold snare polypectomy. The histopathologic examination revealed a tubular adenoma (Fig. g). A closer look at the specimen showed that in the central area, the ductal apertures were smaller, and the ducts were nearly perpendicular to the mucosa (Fig. h). In contrast, at the margins, the ductal apertures were larger, and the structure was more oblique (Fig. i).

Tips

Tubular adenomas like lesion have been reported in the early 2000s and referred to as 0-IIa with pseudo-depression (0-IIa + dep) [1]. This is because these depressions are only relatively depressed within the lesion and are actually elevated above the surrounding normal mucosa. In this respect, they are distinct from 0-IIc and have been referred as 0-IIa in Paris classification because of their lower grade of malignancy. It has recently been reported that lesions smaller than 5 mm, which show a valley sign on NBI, are usually adenomas [2].

■ Video

A short video is available in Fig. 1.

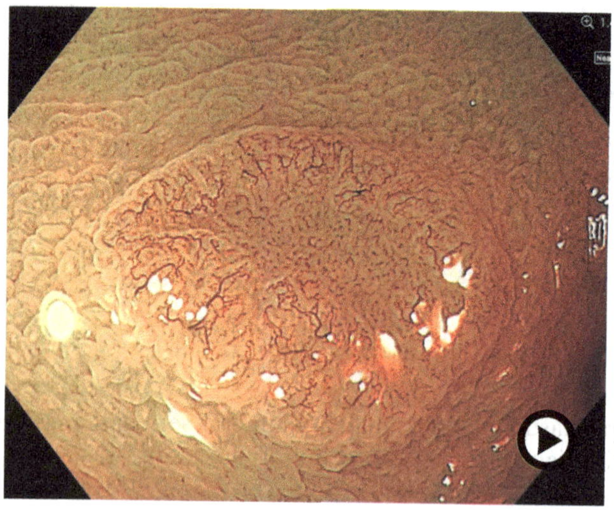

Fig. 1 Practical observation and diagnosis of a tubular adenoma by WLI, NBI, TXI, and chromoendoscopy (▶ https://doi.org/10.1007/000-c8r)

Acknowledgments Special thanks to the following: Dr. Enrik John Aguila and Dr. Maxine Andrea Garcia.

References

1. Sano Y, Tanaka S, Teixeira CR, Aoyama N. Endoscopic detection and diagnosis of 0-IIc neoplastic colorectal lesions. Endoscopy. 2005;37(3):261–7.
2. Rex DK, Ponugoti P, Kahi C. The "valley sign" in small and diminutive adenomas: prevalence, interobserver agreement, and validation as an adenoma marker. Gastrointest Endosc. 2017;85(3):614–21.

Tubulovillous Adenoma

Louis H. S. Lau and Yunbi Ni

Y. Sano et al. (eds.), *Atlas of Advanced Endoscopy*, https://doi.org/10.1007/978-981-97-2732-2_63

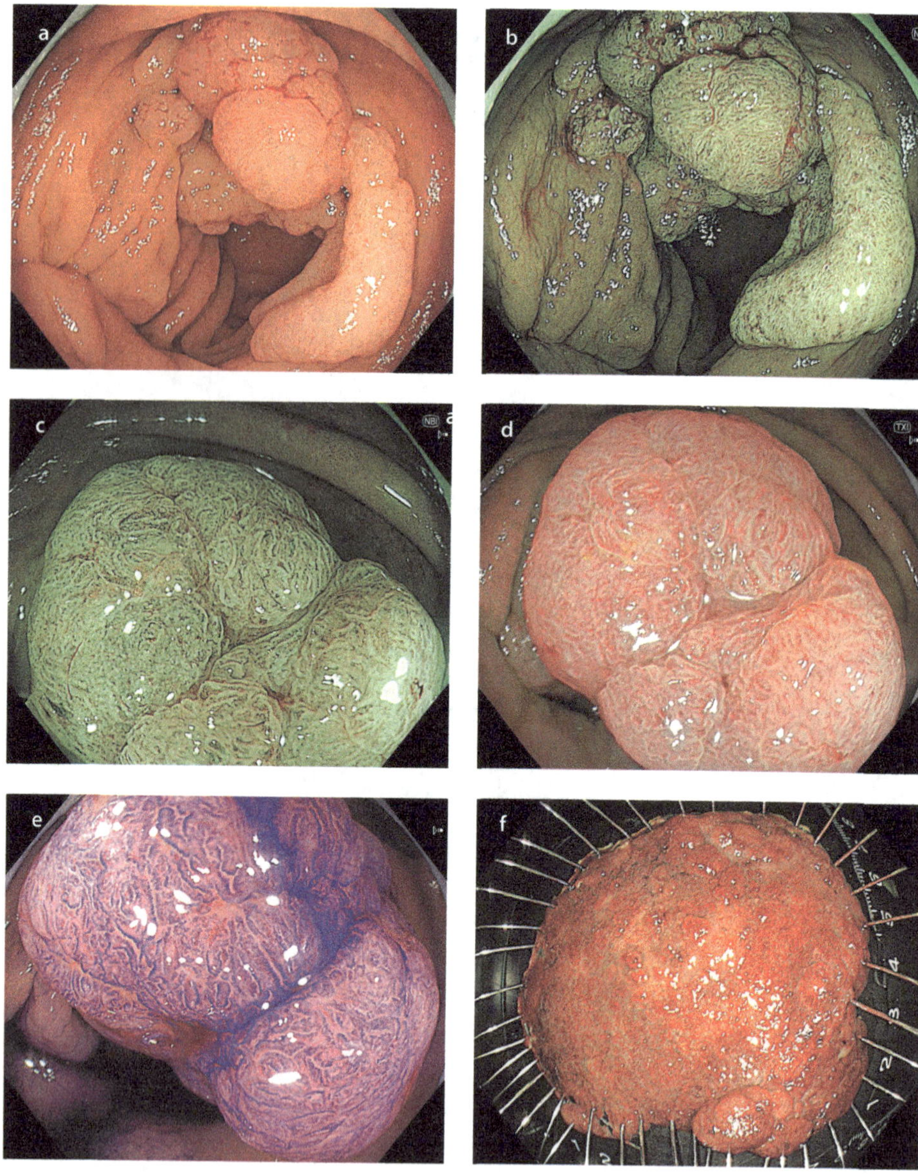
a
b
NBI
c
NBI
d
TXI
e
f

Pathological Diagnosis

Tubulovillous adenoma, with low-grade and focal high-grade dysplasia

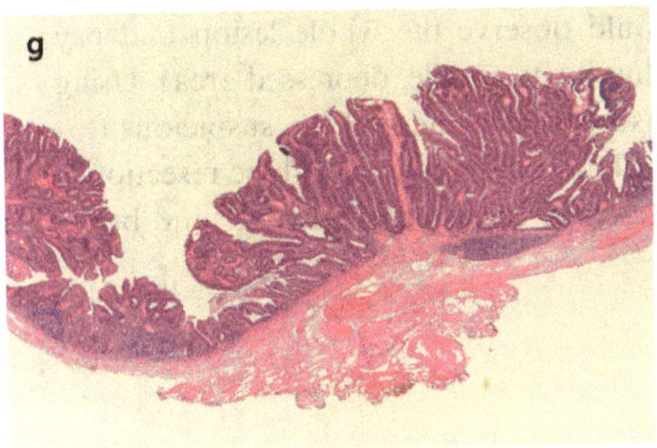

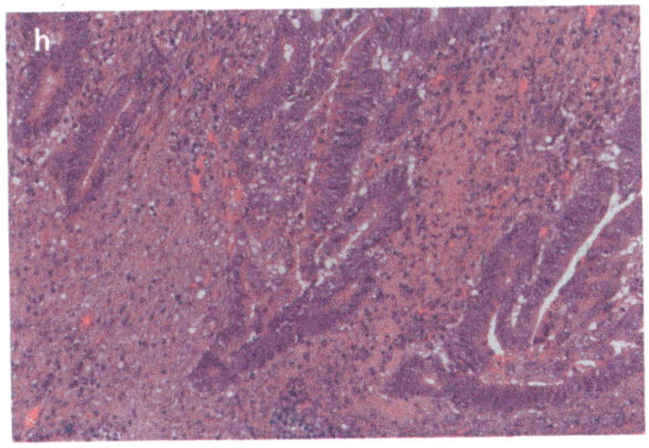

Endoscopic Findings

A lateral spreading tumor was detected in the cecum, which did not involve the ileocecal valve or the appendiceal opening. During observation with non-magnifying white-light image, the lesion occupied half of the lumen circumference and consisted of two major components—a broad-based sessile part and a dominant nodule part (Paris classification 0-Is) (Fig. a). Under non-magnifying narrowband image (NBI), the lesion was brown in color with tubular-shape surface patterns (NICE classification type 2) (Fig. b). Upon high-power magnification NBI, there was extensive villi formation at the dominant nodule, with tortuous and thickened meshed capillary vessels (Sano classification IIIA) (Fig. c). On TXI mode, the visibility of lesion was enhanced, which highlighted a regular tubular surface pattern and an irregular meshed vessel pattern (JNET classification type 2B) (Fig. d). Using chromoendoscopy with 0.05% crystal violet, the lesion demonstrated a dendritic-like pit pattern (Kudo classification type IV) (Fig. e). Based on the above observations, the overall impression of this lesion was suggestive of a tubulovillous adenoma with focal high-grade dysplasia. It was resected en bloc by endoscopic submucosal dissection technique. The final specimen size was 70 mm × 65 mm (Fig. f).

Histopathology Findings

A complete section of specimen showed colonic mucosa and submucosa with adenomatous proliferation exhibiting tubulovillous pattern (Fig. g). The glandular lining cells mainly showed nuclear pseudostratification and hyperchromasia. Focally, the glands were complex and lined by tumor cells with round nuclei, loss of nuclear polarity, and moderate nuclear pleomorphism (Fig. h). The findings confirmed the diagnosis of tubulovillous adenoma, with low-grade and focal high-grade dysplasia.

Tips

Any large colorectal polyp is known to be at risk of harboring occult high-grade dysplasia or malignancy. Endoscopists should observe the whole lesion and pay attention to areas of special interest (e.g., dominant nodule, depressed area). Using high-power magnification and chromoendoscopy, we could pick up suspicious features like irregular vessel or pit patterns. For high-grade lesions, en bloc resection is preferred than piecemeal resection, due to a lower local recurrence rate and better histopathological examination.

Adenoma (LST-NG)

Kenichiro Imai and Yasushi Sano

Contents

Supplementary Information The online version contains supplementary material available at https://doi.org/10.1007/978-981-97-2732-2_64. The videos can be accessed individually by clicking the DOI link in the accompanying figure caption or by scanning this link with the SN More Media App.

Y. Sano et al. (eds.), *Atlas of Advanced Endoscopy*, https://doi.org/10.1007/978-981-97-2732-2_64

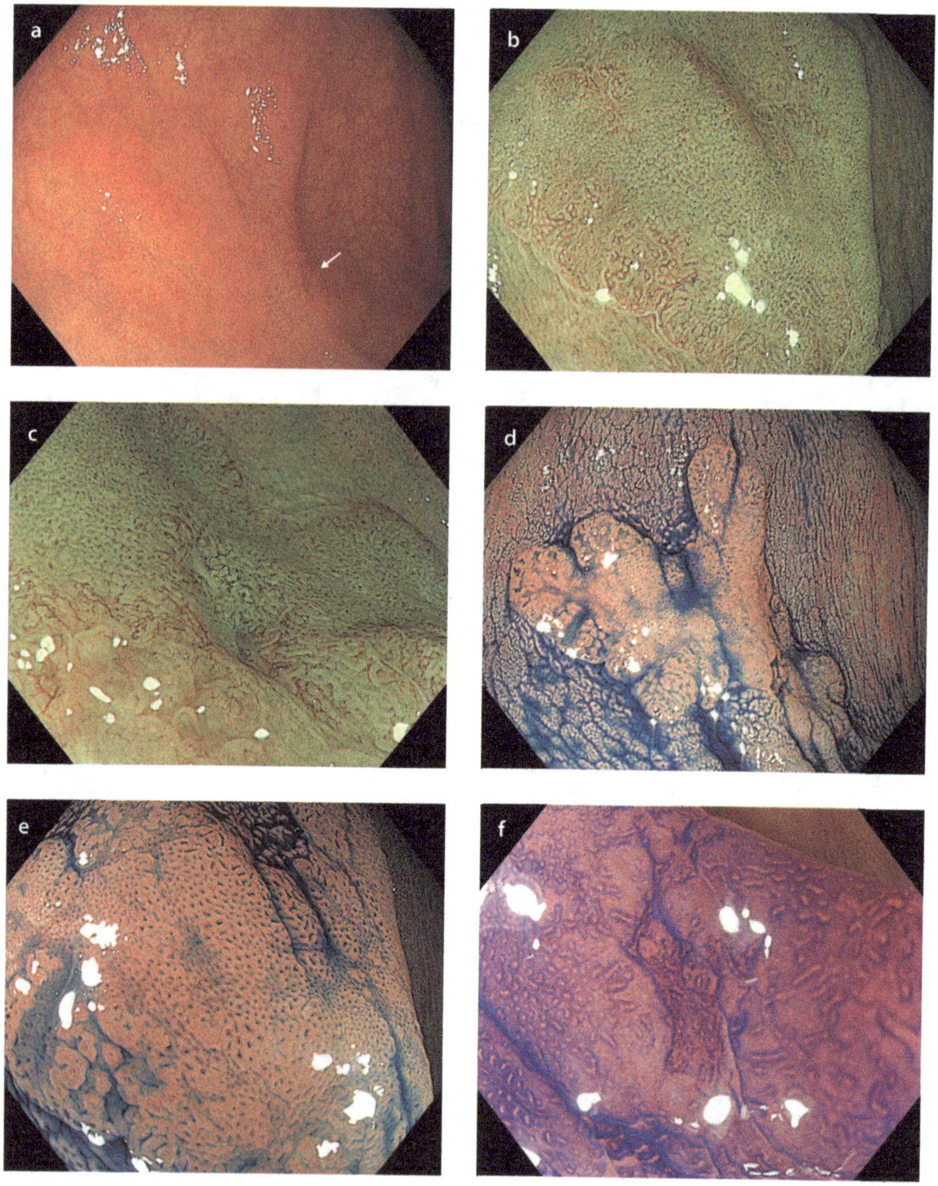
a
b
c
d
e
f

Pathological Diagnosis

Tubulovillous adenoma, low-grade dysplasia, including small foci of high-grade dysplasia.

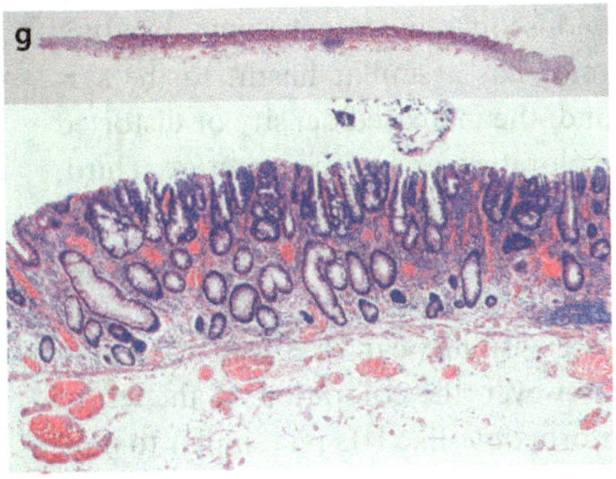

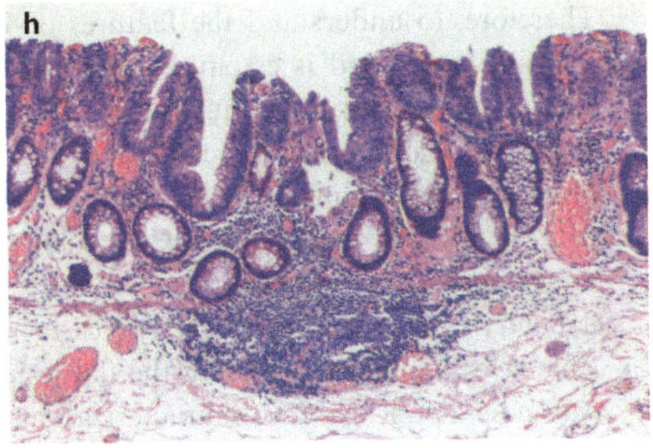

Explanation

A flat lesion 25 mm in size with a small depressed area (arrow) has been detected in the transverse colon (Fig. a). NBI with near-focus magnification shows small tubular or oval structures and regular brown meshed vessels, indicating JNET type 2A (Fig. b). In comparison, the depressed area has irregular meshed vessels with indistinct surface structures that could be categorized as JNET type 2B (Fig. c). Chromoendoscopy with indigo carmine reveals the octopus leg-like wavy margin of the flat lesion and gentle sloping (pseudo-depression) at the edge of the surface (Fig. d). In the center of the lesion, small round pits are observed and diagnosed as Kudo's IIIs pit pattern (Fig. e). Chromoendoscopy with crystal violet staining shows a small irregular pit of uneven size in the depressed area, indicating Kudo's Vɪ low-grade pit pattern (Fig. f).

These findings suggest that the lesion is a laterally spreading tumor (LST) pseudo-depressed nongranular type and high-grade adenoma. The lesion was resected *en bloc* with the ESD technique. The histopathological diagnosis was tubulovillous adenoma, low-grade dysplasia for non-depressed areas (Fig. g), and high-grade dysplasia with a lymphoid follicular formation for the depressed area (Fig. h).

Tips

The LST non-granular type is known to have a subtle appearance, which causes an overlooking during colonoscopic withdrawal more frequently than expected [1]. Therefore, to understand the features of its endoscopic appearance with a pathological background is essential. First, this tumor has a similar height to the surrounding non-neoplastic mucosa margin. Second, the increased density or distorted shape of tumor glands may cause a distinctive coloration or surface structure. Third, fibrosis in the submucosal layer may cause deformation of the folds. Based on this pathological knowledge, the FIND clues (*F*old deformation, *I*ntensive stool/mucus attachment, *N*o vessel visibility, *D*emarcated reddish area) will help detect high-risk flat and depressed lesions [2]. In this case, no vessel visibility can be considered a possible key finding for detecting the flat lesion; however, the coloration of the lesion was indistinct. Regarding characterization, uniform dots-like IIIs pits similar to normal pits of the surrounding normal mucosa may suggest a low-grade dysplasia, and their uneven changes indicate a more malignant nature.

Video

A short video is available in Fig. 1.

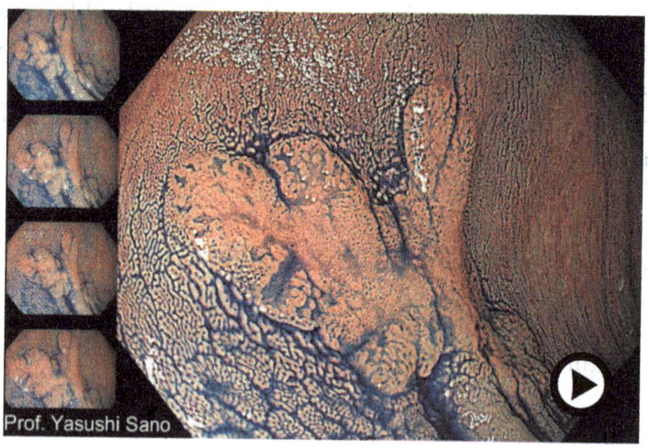

Fig. 1 Practical observation of LST pseudo-depressed non-granular type (▶ https://doi.org/10.1007/000-c8s)

References

1. Matsuda T, Fujii T, Sano Y, et al. Randomised comparison of postpolypectomy surveillance intervals following a two-round baseline colonoscopy: the Japan Polyp Study Workgroup. Gut. 2021;70:1469–78.
2. Iwatate M, Hirata D, Francisco CPD, et al. Efficacy of international web-based educational intervention in the detection of high-risk flat and depressed colorectal lesions higher (CATCH project) with a video: randomized trial. Dig Endosc. 2022;34:1166–75.

Adenoma (LST-G, Homogeneous Type)

Takuji Kawamura

Supplementary Information The online version contains supplementary material available at https://doi.org/10.1007/978-981-97-2732-2_65. The videos can be accessed individually by clicking the DOI link in the accompanying figure caption or by scanning this link with the SN More Media App.

Y. Sano et al. (eds.), *Atlas of Advanced Endoscopy*, https://doi.org/10.1007/978-981-97-2732-2_65

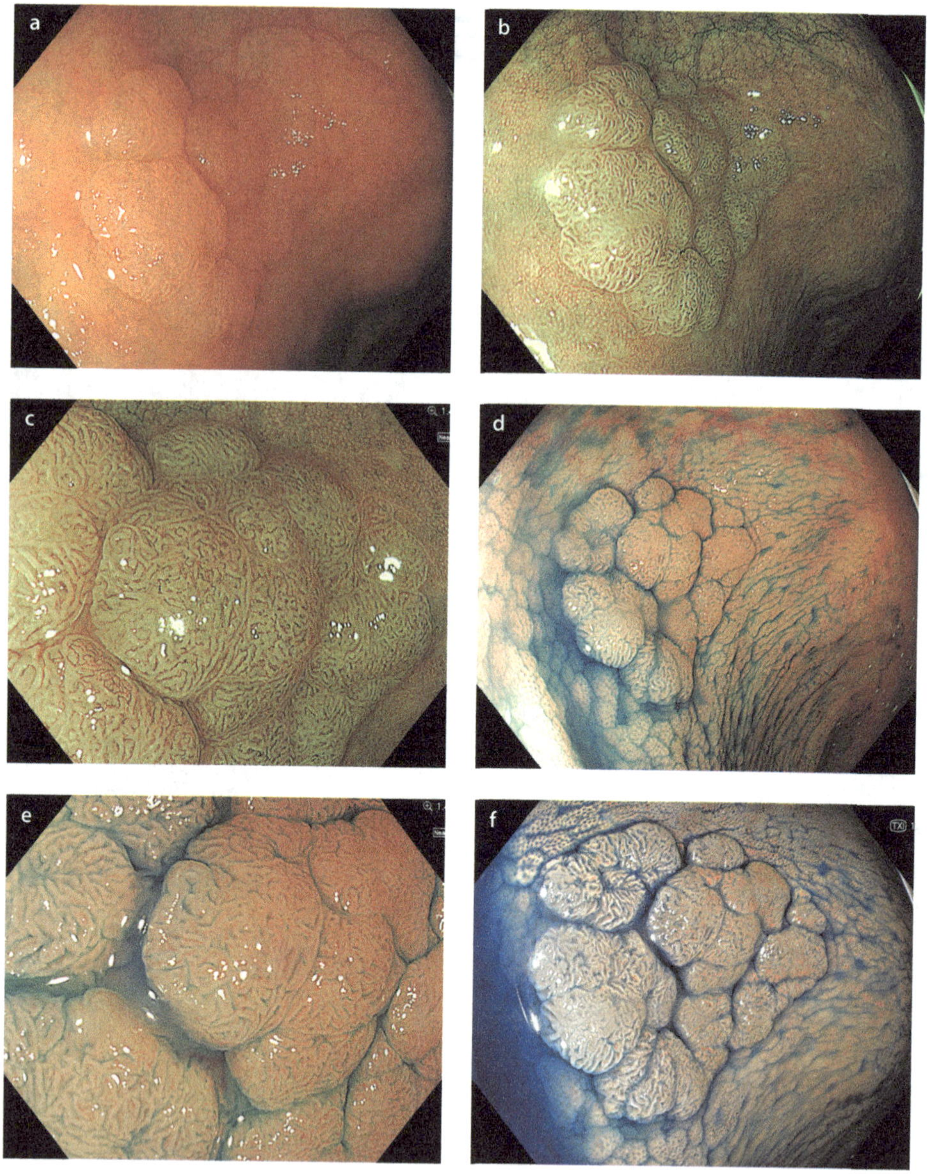
a
b
c
d
e
f

Pathological Diagnosis

High-grade tubular adenoma, margin (-).

g

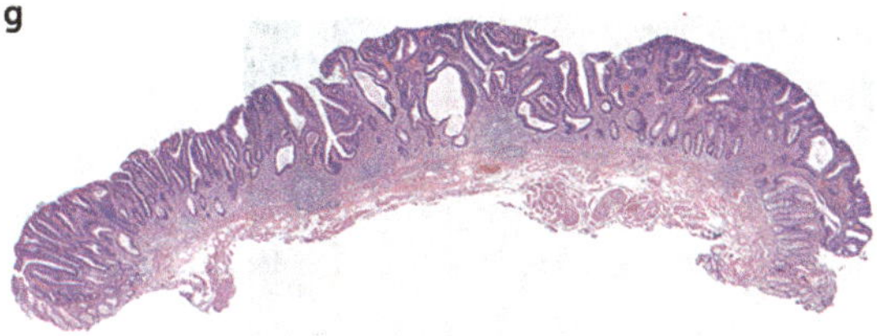

h

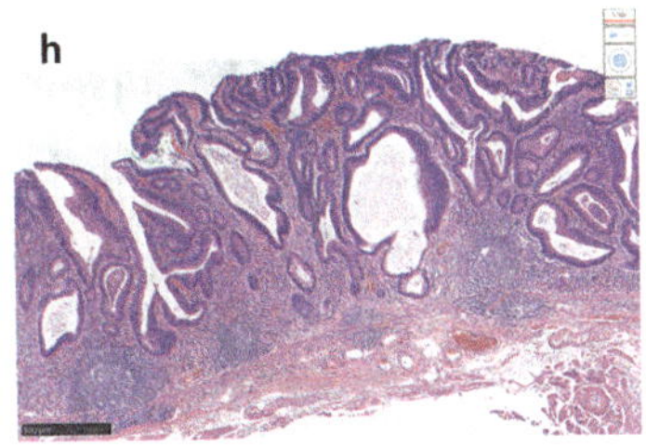

Explanation

We present a 12-mm flat lesion observed in the cecum (Fig. a). Narrowband imaging (NBI) revealed the surface structure of the lesion, providing a proper pathologic evaluation of the lesion. Using the NBI International Colorectal Endoscopic (NICE) classification, the slightly brown lesion with tubular and branched ductal structures as surface patterns and brown vessels around white structures as vessel patterns were diagnosed as adenomatous lesions (NICE type 2, Fig. b). Regular tubular surface and regular caliber vascular patterns were observed using magnification with NBI, and the lesion was diagnosed as Japan NBI Expert Team classification type 2A (Fig. c).

Flat tumors larger than 10 mm are called laterally spreading tumors (LST), which are differentiated using indigo carmine spray and classified into granular (LST-G) and non-granular (LST-NG) types. In the present case, the indigo carmine spray revealed an almost homogeneous nodule-like structure, diagnosed as a homogenous type of LST-G (Fig. d). Regular tubular structures were observed using magnification with indigo carmine spray, and pit pattern classification Kudo type IIIL was diagnosed (Fig. e). Contrast is clearer using the recently developed texture and color enhancement imaging (TXI) mode in combination with indigo carmine (Fig. f).

This lesion was endoscopically estimated as a low-grade adenoma, underwent underwater endoscopic mucosal resection (EMR), and achieved en bloc resection. Pathologically, the final diagnosis was a high-grade tubular adenoma (Figs. g and h).

Tips

When LST-G homogenous type is diagnosed, piecemeal resection is acceptable because of the low probability of carcinoma. However, considering the risk of recurrence after piecemeal resection, en bloc resection is recommended unless technically challenging. For adenomatous lesions larger than 10 mm, cold snare polypectomy (CSP) is not recommended by the Japanese guidelines, and EMR is preferable. Recently, underwater EMR has been reported to be useful for en bloc resection of lesions ranging from 10 to 20 mm.

Video

A short video is available as Fig. 1.

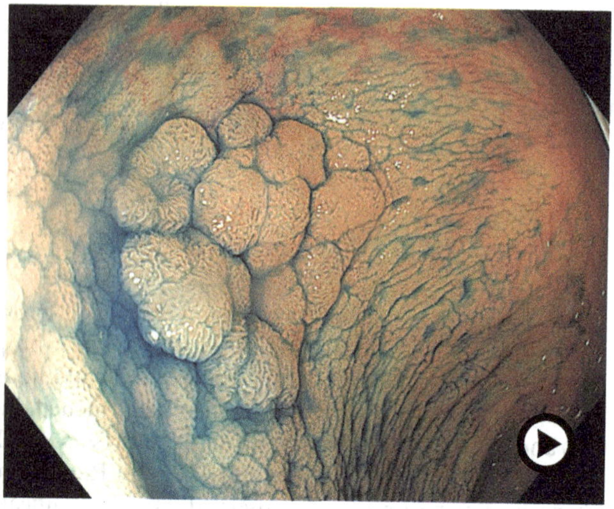

Fig. 1 A 12-mm flat lesion observed in the cecum (a granular laterally spreading tumor, homogenous type) (▶ https://doi.org/10.1007/000-c8t)

Adenoma (LST-G, Mixed Type)

Naoki Sugimura and Mikio Fujita

Supplementary Information The online version contains supplementary material available at https://doi.org/10.1007/978-981-97-2732-2_66. The videos can be accessed individually by clicking the DOI link in the accompanying figure caption or by scanning this link with the SN More Media App.

Y. Sano et al. (eds.), *Atlas of Advanced Endoscopy*, https://doi.org/10.1007/978-981-97-2732-2_66

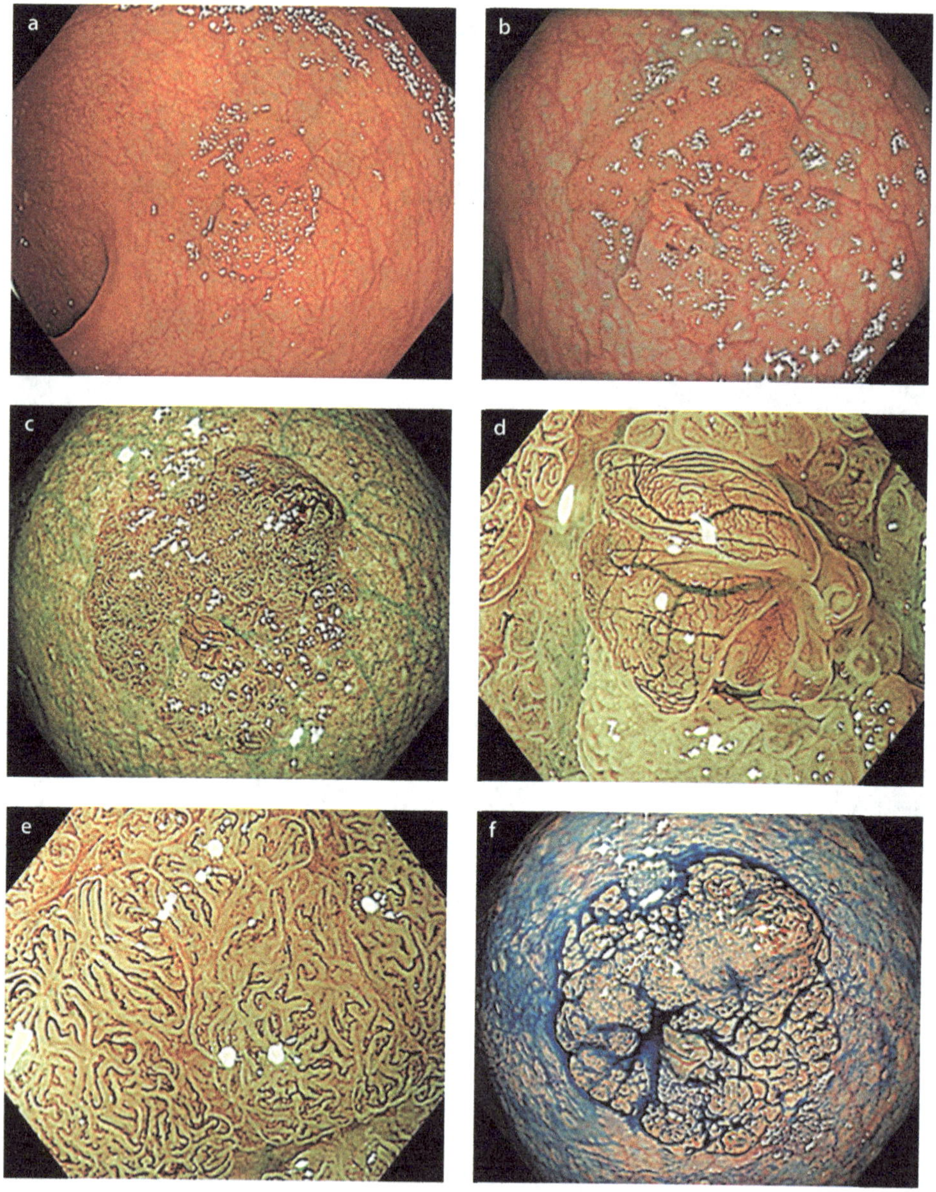

a
b
c
d
e
f

Pathological Diagnosis

Tubulovillous adenoma, including small foci of severe atypia, and TSA-like appearance.

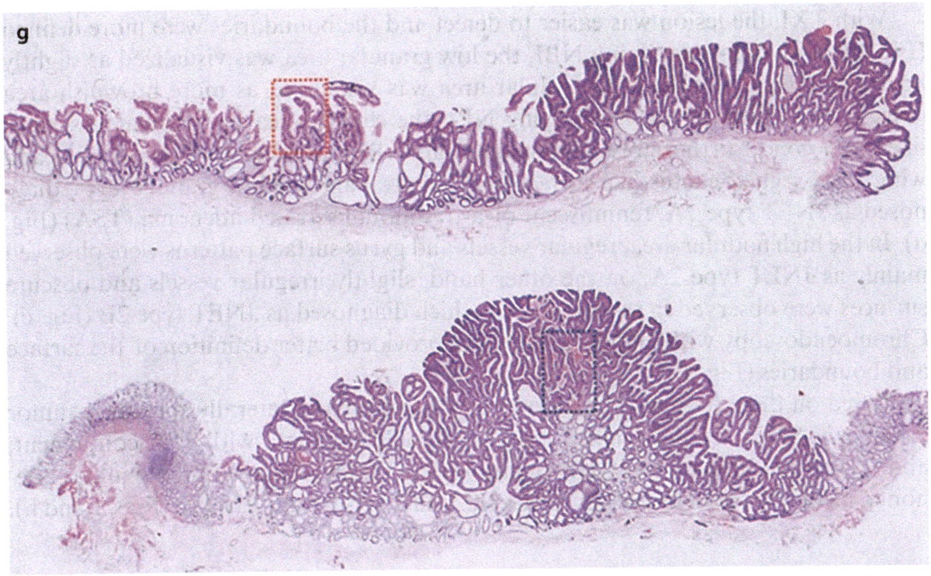

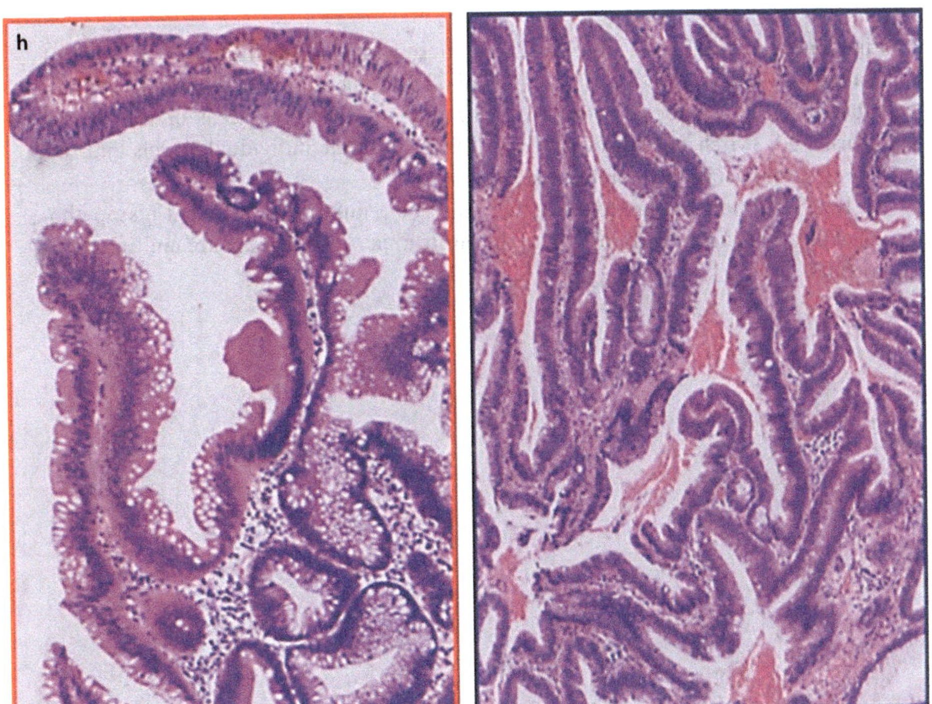

▪ Explanation

This lesion was detected approximately 20 mm in size in the rectum. Non-magnifying WLI observation revealed a mixture of low granular and high nodular areas within the lesion (Fig. a).

With TXI, the lesion was easier to detect and the boundaries were more defined (Fig. b). With non-magnifying NBI, the low granular area was visualized as slightly brownish area, while the high nodular area was visualized as more brownish area (Fig. c). Switching to low-magnifying NBI, the central area had lobulated surface like a pine cone. Furthermore, high-magnifying NBI showed lacy vessels like a cicada wing and irregular cactus-like surface in the central and the low granular area, diagnosed as JNET type 2A, reminiscent of a traditional serrated adenoma (TSA) (Fig. d). In the high nodular area, regular vessels and gyrus surface patterns were observed mainly as JNET type 2A; on the other hand, slightly irregular vessels and obscure surfaces were observed in the small area which diagnosed as JNET type 2B (Fig. e). Chromoendoscopy with indigo carmine dye provided better definition of the surface and boundaries (Fig. f).

Based on these findings, the lesion was diagnosed as a laterally spreading tumor granular (LST-G) nodular mixed type, high-grade adenoma with TSA component, and resected en bloc by ESD. Histopathological findings showed tubulovillous adenoma, including small foci of severe atypia, and TSA-like appearance (Figs. g and h).

Tips

First, check the change in vessel/surface pattern with low-magnifying NBI. Areas of change are often suggestive of severe atypia. Focusing on the vessel density and distribution is the tip for detecting irregular changes. Once the areas of interest are determined, switch to high-magnifying NBI to evaluate individual vessels and surface structure using the JNET classification.

In LST-G nodular mixed type, there may be a mixture of different histopathological features, so be sure to look at the entire area to see if there are multiple components.

■ Video

A short video is available as ◘ Fig. 1.

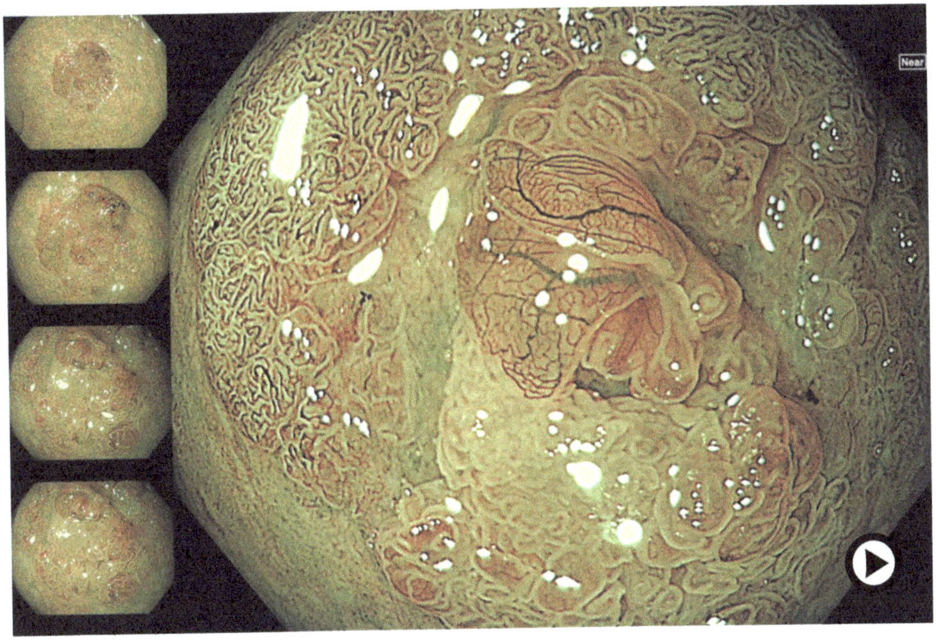

◘ **Fig. 1** Practical observation and diagnosis of LST-G with magnifying TXI, NBI, and chromoendoscopy (▶ https://doi.org/10.1007/000-c8v)

Early Colorectal Cancer (LST-G, Mixed Type)

Hiroshi Kashida

Y. Sano et al. (eds.), *Atlas of Advanced Endoscopy*, https://doi.org/10.1007/978-981-97-2732-2_67

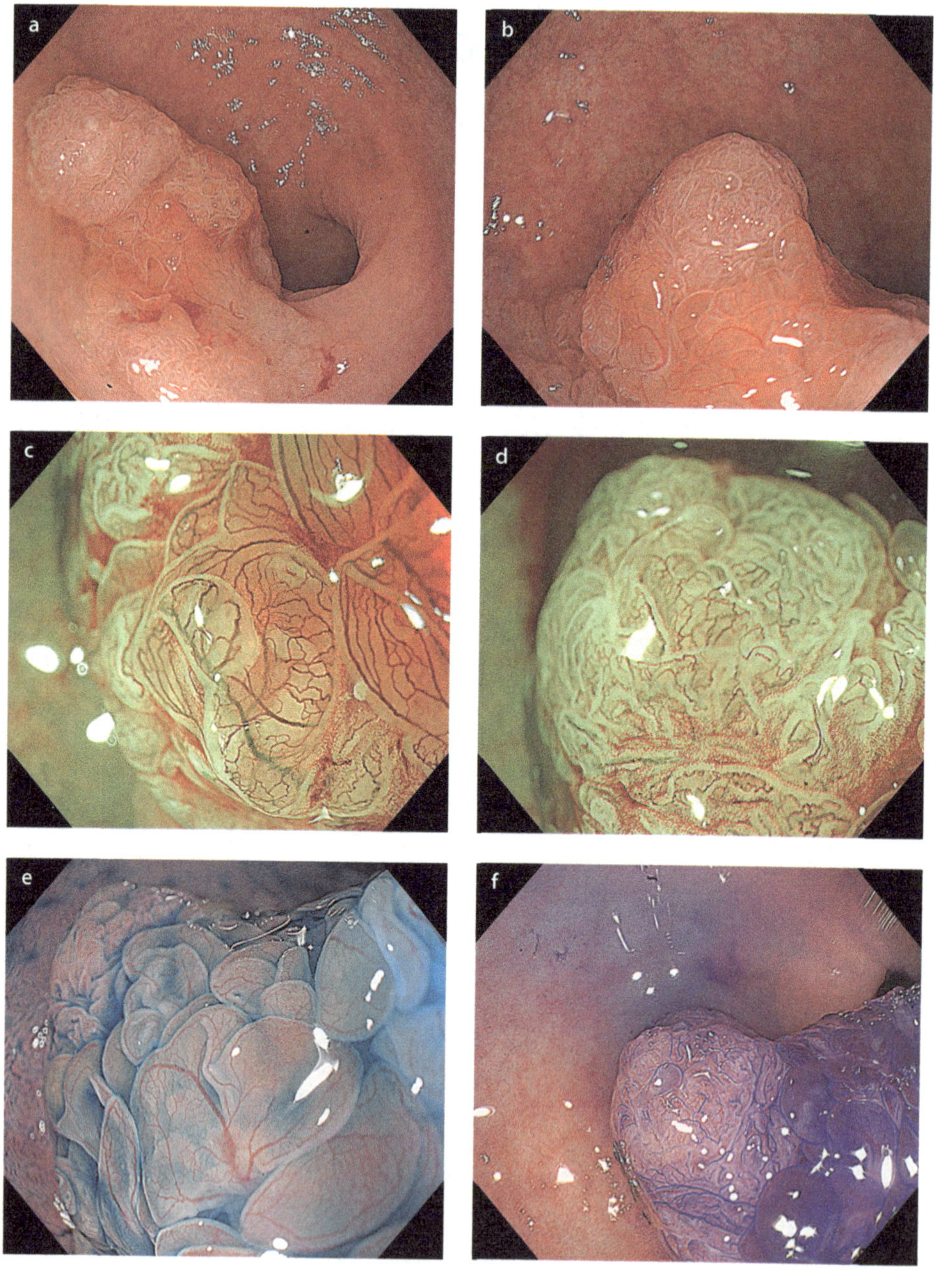
a
b
c
d
e
f

Pathological Diagnosis

Tubular adenocarcinoma (tub1 > 2, pTis, Ly0, V0) in traditional serrated adenoma.

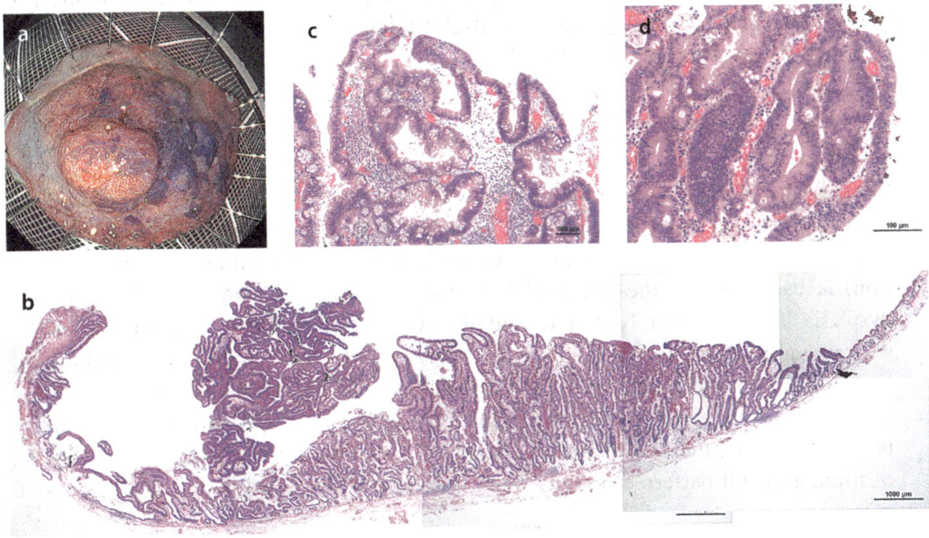

Explanation

The patient underwent colonoscopy due to positive fecal occult blood. Non-magnifying white light view (scope: Olympus CF-XZ 1200 I) revealed approximately 30 mm elevated lesion in the rectosigmoid. There was a nodular protrusion within the lesion; thus, the configuration was judged as laterally spreading tumor granular (LST-G) nodular mixed-type LST-G (Mix) (Fig. a). At a closer look, the nodular area looked paler than the adjacent part, and the surface structure looked different between the two parts (Fig. b). With magnifying NBI, the central part of the lesion showed characteristic structure and vessel pattern (Fig. c). We named such image as "pinecone appearance" and "leaf vein pattern" and have been reporting it as a typical view of a traditional serrated adenoma (TSA). In the protruded part of the lesion, NBI revealed slightly irregular surface and vessel patterns, judged to be JNET type 2B (Fig. d). Indigo carmine spraying and magnification of the central part of the lesion enhanced the characteristic lobular structure of the lesion (Fig. e). Crystal violet staining and magnification clearly demonstrated that the central protruding part was associated with irregular pit pattern (VI low-grade), different from the surrounding area (Fig. f).

Based on these findings, the lesion was predicted to be a TSA harboring a highly dysplastic area which might be slightly invasive. We determined the lesion should be resected en bloc and performed ESD considering its size (Fig. g). The low-power microscopic view of the pathological specimen showed a large elevated lesion with a protruding nodular segment (Fig. h). The high-power magnification view showed mildly serrated structure and ectopic crypt formation consisting of cells with eosinophilic cytoplasm and pencillike nucleus and thus diagnosed as TSA (Fig. i). The high-power microscopic view of the protruded part of the lesion showed compactly arranged crypts with complicated structure. The nuclei showed higher N/C ratio and were arranged stratification and loss of polarity (Fig. j). There was no sign of submucosal invasion. The final diagnosis was tubular adenocarcinoma (tub1 > 2, pTis, Ly0, V0) in traditional serrated adenoma.

Tips

First, wash off the stool residue or mucus overlying the lesion. Observe the whole lesion with non-magnifying white light view, and look for the structural or color difference within the lesion. Then take a closer look (preferably with slight magnification) at each part of the lesion to look for any focus suspicious of high-grade dysplasia. Then switch to NBI mode and observe the surface and vessel pattern. Start with non-magnifying view and gradually magnify. Indigo carmine spraying without magnification is useful for clarifying the configuration and boundary of the lesion and looking for a depressed or protruded focus and can be used to observe the pit pattern with magnification as well. Crystal violet staining is useful for precisely demonstrating the pit pattern especially when the lesion is suspected to be partially cancerous.

Early Colorectal Cancer (0-IIc) (1)

Kenjiro Shigita and Shinji Nagata

Y. Sano et al. (eds.), *Atlas of Advanced Endoscopy*, https://doi.org/10.1007/978-981-97-2732-2_68

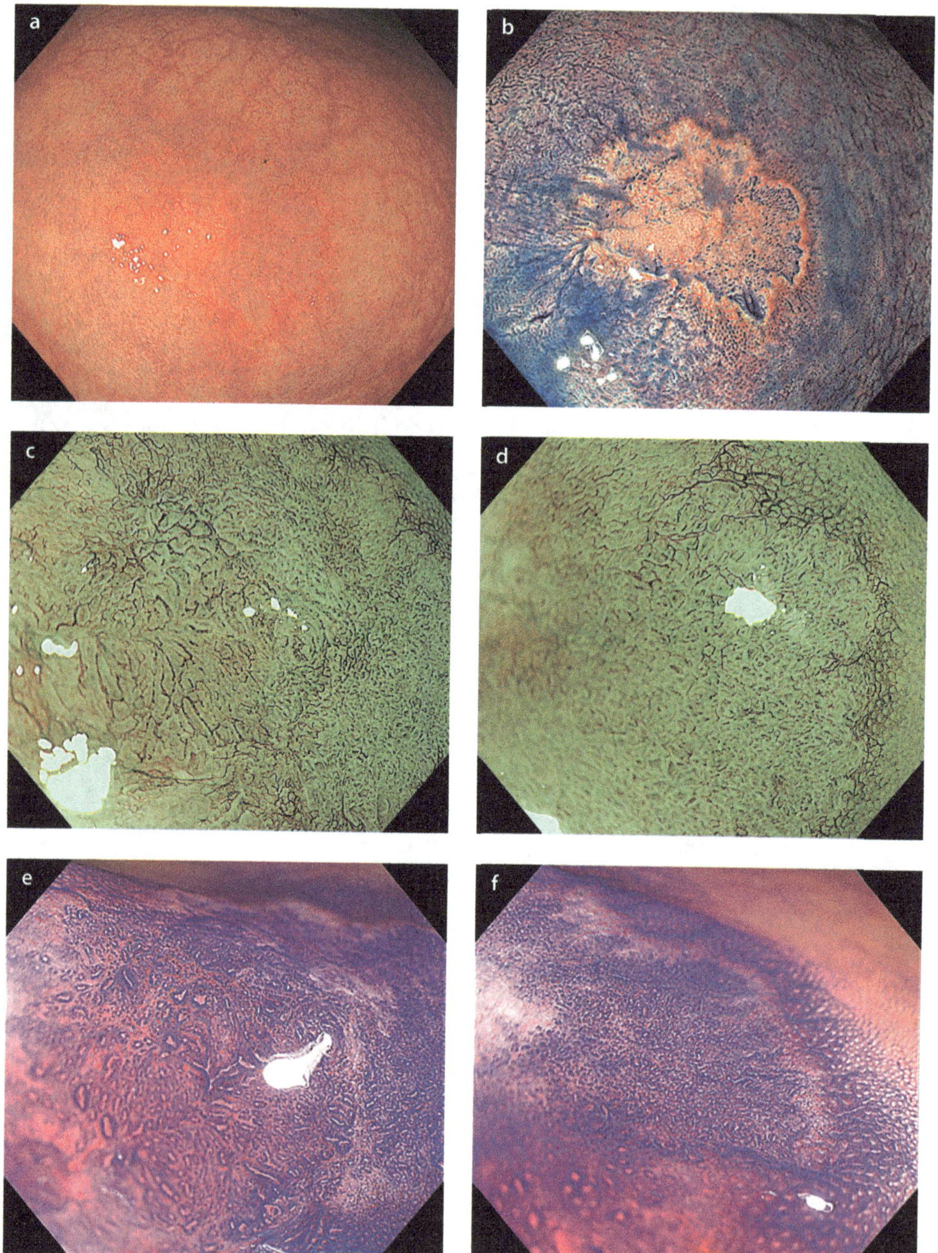
a
b
c
d
e
f

▪ Pathological Diagnosis

Adenocarcinoma (tub1 > tub2), intramucosal carcinoma, Ly0(D2-40), V0(EVG), pHM0, pVM0.

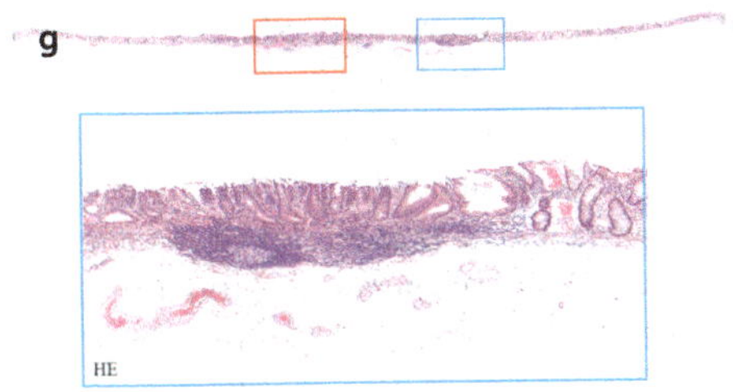

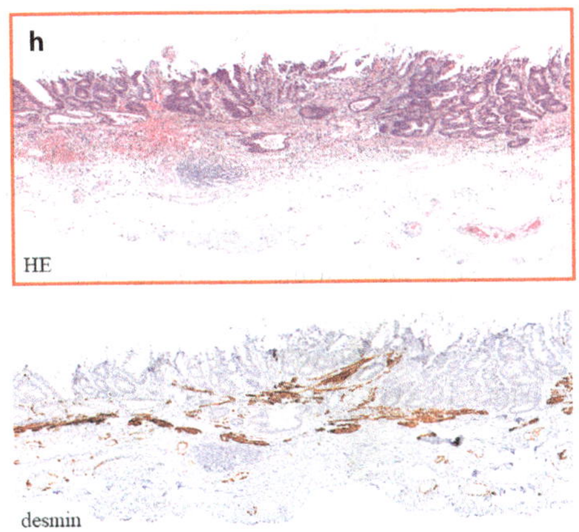

▪ Explanation

A depressed lesion measuring approximately 10 mm in size was identified in the sigmoid colon. Non-magnified WLI observation revealed the lesion as a depressed area with a reddish border and a loss of vascular translucency. Within the depression, distinct regions of white to normal color and a pale reddish area were observed (Fig. a). Chromoendoscopy using indigo carmine dye under TXI (mode 2) revealed clear delineation of the margin of the depression, with the pale reddish area in the WLI view appearing slightly elevated within the depression (Fig. b). Upon magnified NBI, the lesion exhibited an irregular surface pattern and heterogeneous vessel diameter and distribution within the depression, leading to the diagnosis of JNET type 2B. The slightly elevated area within the depression (Fig. c) displayed a more irregular surface pattern and greater disparities in vessel diameter and distribution compared to the depressed area (Fig. d). On magnification observation with crystal violet staining, the slightly elevated areas exhibited pit patterns of varying sizes, classified as VI low-grade (Fig. e), while the depressed area exhibited a type IIIs pit pattern (Fig. f). Based on these findings, we diagnosed early-stage colorectal cancer of type 0-IIc, with a potential for submucosal invasion, and we decided to perform ESD for total excisional biopsy. In the histopathological examination, the depressed area was identified as an intramucosal carcinoma (Fig. g), and there was infiltration into the mucosal muscularis propria in the elevated area within the depression (Fig. h).

Tips

Depressed tumors are often associated with a more aggressive course and a higher malignant potential. To detect these lesions, we should pay attention to endoscopic findings such as mucosal surface erythema, depressions, and areas of reduced vascular permeability. During NBI examination, the "O-ring sign," which appears as a ring-shaped-enhanced image of dilated vessels surrounding the reactive ridge encircling the depression, is helpful. The use of TXI can be beneficial for detecting erythema and depressions because of its enhanced contrast. The presence of prominent elevations within the depression is a critical indicator for suspicion of submucosal invasion and should be further investigated with dye endoscopy. The presence of irregular surface or vessel patterns within the depression on NBI magnification or V-shaped pit patterns on crystal violet staining increases the likelihood of SM invasion. In such cases, en bloc resection rather than segmental resection is recommended.

Early Colorectal Cancer (0-IIc) (2)

Masashi Misawa and Shin-ei Kudo

Content

Supplementary Information The online version contains supplementary material available at https://doi.org/10.1007/978-981-97-2732-2_69. The videos can be accessed individually by clicking the DOI link in the accompanying figure caption or by scanning this link with the SN More Media App.

Y. Sano et al. (eds.), *Atlas of Advanced Endoscopy*, https://doi.org/10.1007/978-981-97-2732-2_69

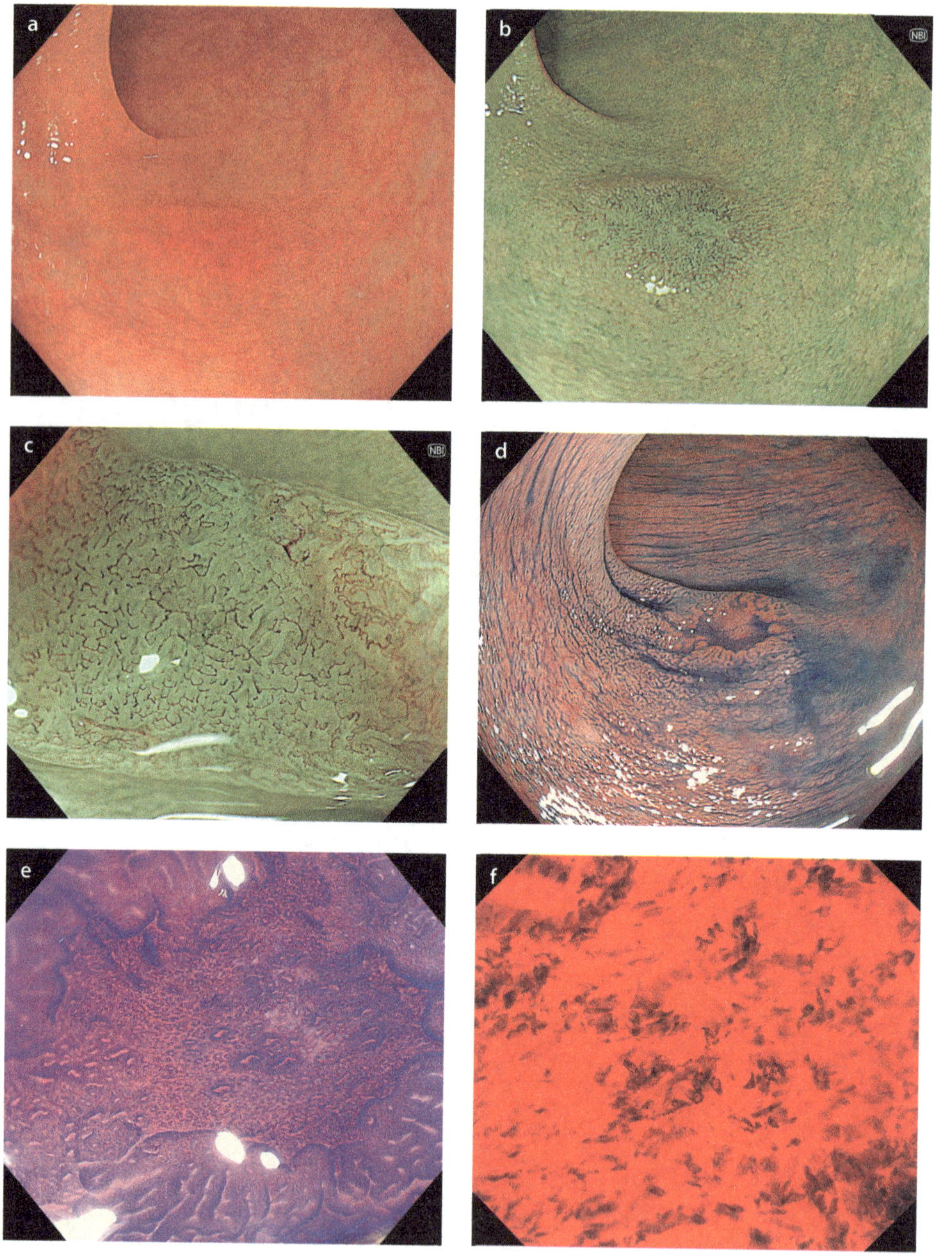
a
b
NBI
c
NBI
d
e
f

▪ Pathological Diagnosis

Well-differentiated adenocarcinoma T1a with 51 μm of submucosal invasion.

g

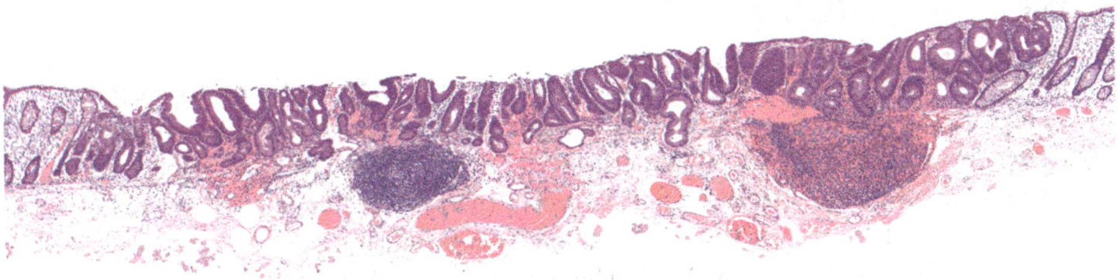

h

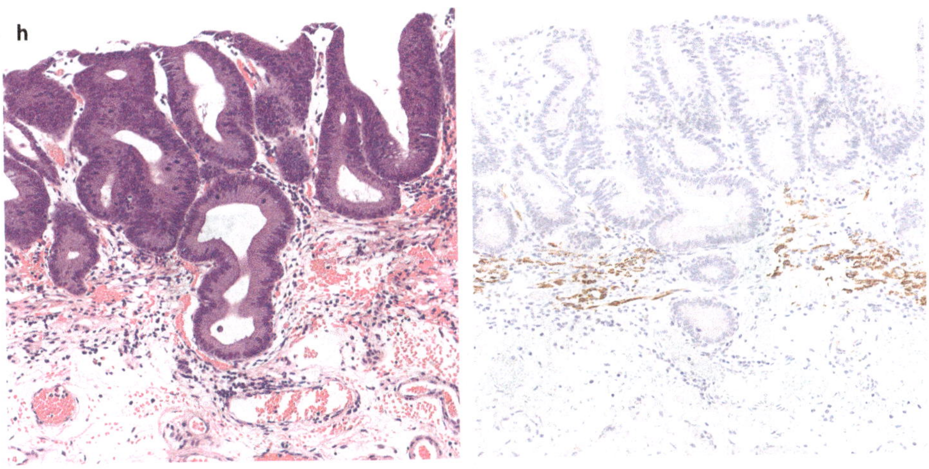

▪ Explanation

During the surveillance colonoscopy, a slightly reddish lesion was detected in the descending colon. The lesion, approximately 6 mm in diameter, exhibited a shallow depression, even under white-light endoscopy (Fig. a). Under non-magnified NBI, the vessels within the depression were observed to be slightly dilated when compared to the adjacent normal mucosa (Fig. b). The use of magnifying NBI revealed slightly thickened vessels of variable caliber, resulting in a diagnosis of JNET type 2B (Fig. c). Indigo carmine dye spraying showed clear depression of the lesion, categorizing its macroscopic type as type 0-IIc (Fig. d). While the pit pattern observed in the peripheral area of depression was small, roundish pits (Kudo's type IIIs pit pattern), the central area of the lesion demonstrated irregularly shaped pit pattern (Kudo's type V_I low-grade pit pattern) (Fig. e). Endocytoscopy with methylene blue staining revealed swollen and roundish nuclei, indicative of an EC3a diagnosis (Fig. f). The lesion was diagnosed as intramucosal or slightly invasive submucosal cancer, and ESD was performed. The pathological specimen revealed a well-differentiated adenocarcinoma (Fig. g). Despite its size of 6 mm, the cancerous tissue invaded the submucosal layer to a depth of 51 μm. Invasive areas were indicated in Fig. h (left, HE staining; right, desmin staining). No lymphovascular invasion was observed.

Tips

Depressed-type colorectal cancers are also known to develop via the de novo carcinogenic pathway. Typically, depressed-type cancers originate from 0-IIc lesions characterized by small, roundish pit pattern (Kudo's type IIIs). Once the cancer tissue invades the submucosal layer, it tends to grow expansively, changing the macroscopic type to 0-IIa + IIc or 0-Is + IIc. The most significant clinical feature of depressed-type cancers is their malignant potential. The incidence of submucosal invasive carcinomas in lesions ≤5 mm is higher in the depressed type (8.4%) compared to the protruded and flat types (0–0.03%) [1]. Therefore, it is essential to detect depressed cancers while they are still of diminutive size. As demonstrated in this case, depressed-type lesions exhibit a slightly reddish color. Thus, in addition to focusing on polypoid-type lesions, attention should be paid to slight color changes, as evidenced in this case.

▪ Video

A short video is available as Fig. 1.

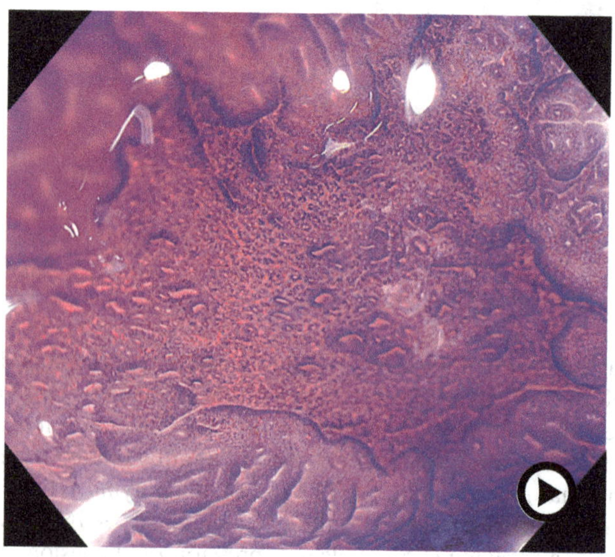

Fig. 1 Practical observation and diagnosis of early colorectal cancer (0-IIc) (▶ https://doi.org/10.1007/000-c8w)

Acknowledgment We would like to thank Professor Tetsuo Nemoto for his contribution to pathological diagnosis. Some images of this case that are not presented in this paper have already been published as part of a case report [2].

References

1. Kudo S, Lambert R, Allen JI, et al. Nonpolypoid neoplastic lesions of the colorectal mucosa. Gastrointest Endosc. 2008;68:S3–S47.
2. Kudo SE, Misawa M, Mori Y, et al. Identification of a small, depressed type of colorectal invasive cancer by an artificial intelligence-assisted detection system. Endoscopy. 2022;54(10):E592–3.

Early Colorectal Cancer (0-Is + IIc)

Hidenori Tanaka and Shiro Oka

Y. Sano et al. (eds.), *Atlas of Advanced Endoscopy*, https://doi.org/10.1007/978-981-97-2732-2_70

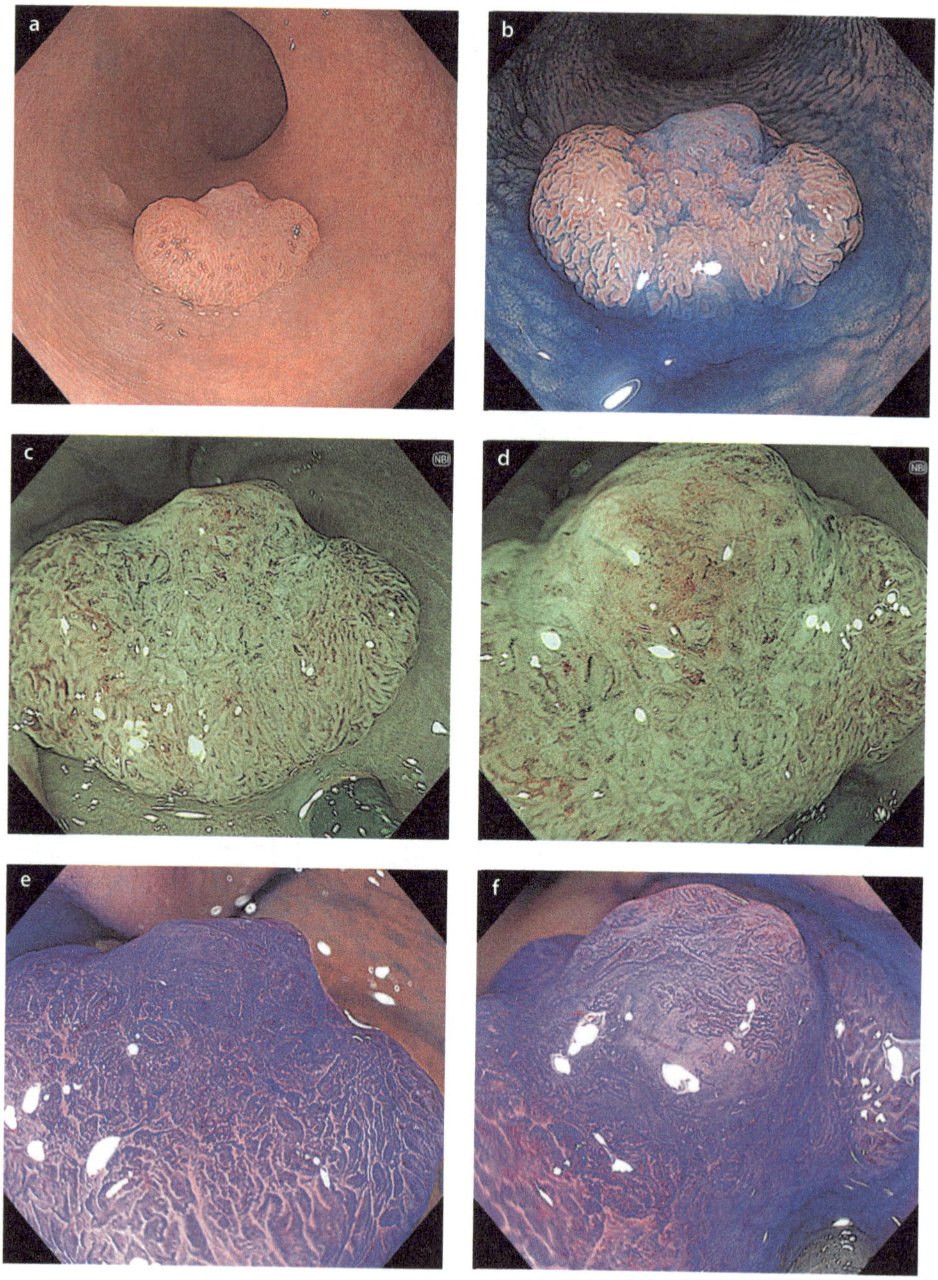
a
b
c
NBI
d
NBI
e
f

Pathological Diagnosis

Adenocarcinoma with adenoma, tub1 > tub2 ≫ por2, pT1b (SM 2799 μm), INFb, BD2, Ly0, V0.

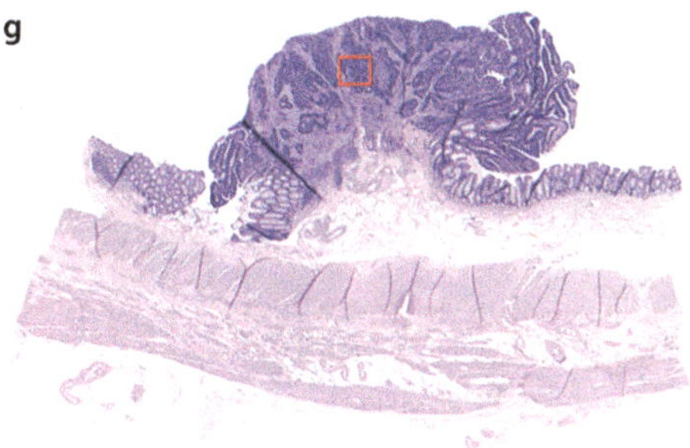

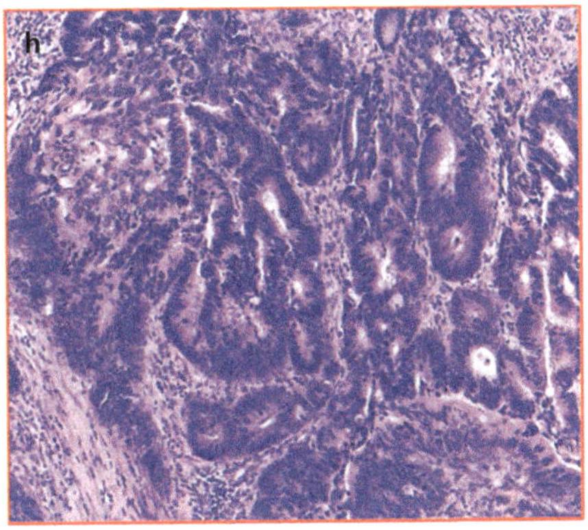

Explanation

A small polypoid lesion 10–15 mm in diameter has been detected in the upper rectum by non-magnifying WLI observation. The surface of the top of the lesion is whitish, rough, and depressed (Fig. a). Depressed area is clearly observed at the top of the lesion by non-magnifying chromoendoscopy with indigo carmine dye. Elevation is observed at the oral side of the depressed area (Fig. b). NBI magnification at the center depressed area reveals irregular surface patterns and irregular microvessels with variable caliber and irregular distributions (JNET type 2B) (Fig. c). Amorphous and loose vessel areas are observed at the elevation in the depressed area (JNET type 3) (Fig. d). Magnifying chromoendoscopy with crystal violet staining at the center depressed area reveals highly irregular pit pattern with irregular, rough, and unclear pit margins (Kudo's type VI high-grade pit pattern) (Fig. e). Highly irregular pit pattern is also observed, and nonstructured area is not observed at the elevation in the depressed area although there are some parts where pit pattern is invisible due to mucous adhesion on surface (Fig. f). Based on these findings, the tumor was diagnosed as deep submucosal invasive carcinoma; therefore, surgical resection was performed. Histopathological examination reveals adenocarcinoma with adenoma: tub1 > tub2 ≫ por2, pT1b (SM 2799 μm), INFb, BD2, Ly0, V0 (Figs. g and h).

Tips

The presence of depression, as well as an elevation in the depression, suggests submucosal invasive carcinoma. It is important to find a depression by WLI observation by paying attention to the difference of color tone and surface texture. The depression is clearly observed by chromoendoscopy. If the depression is identified, magnification using NBI and chromoendoscopy should be performed intensively. Although JNET classification is useful for diagnosis of colorectal tumor, type 2B contains lesions with a variety of histology, ranging from low-grade dysplasia to submucosal invasive carcinoma; therefore, pit pattern diagnosis by magnified chromoendoscopy is required for accurate diagnosis.

Early Colorectal Cancer (0-IIa + IIc)

Hiroaki Ikematsu and Kensuke Shinmura

Supplementary Information The online version contains supplementary material available at https://doi.org/10.1007/978-981-97-2732-2_71. The videos can be accessed individually by clicking the DOI link in the accompanying figure caption or by scanning this link with the SN More Media App.

Y. Sano et al. (eds.), *Atlas of Advanced Endoscopy*, https://doi.org/10.1007/978-981-97-2732-2_71

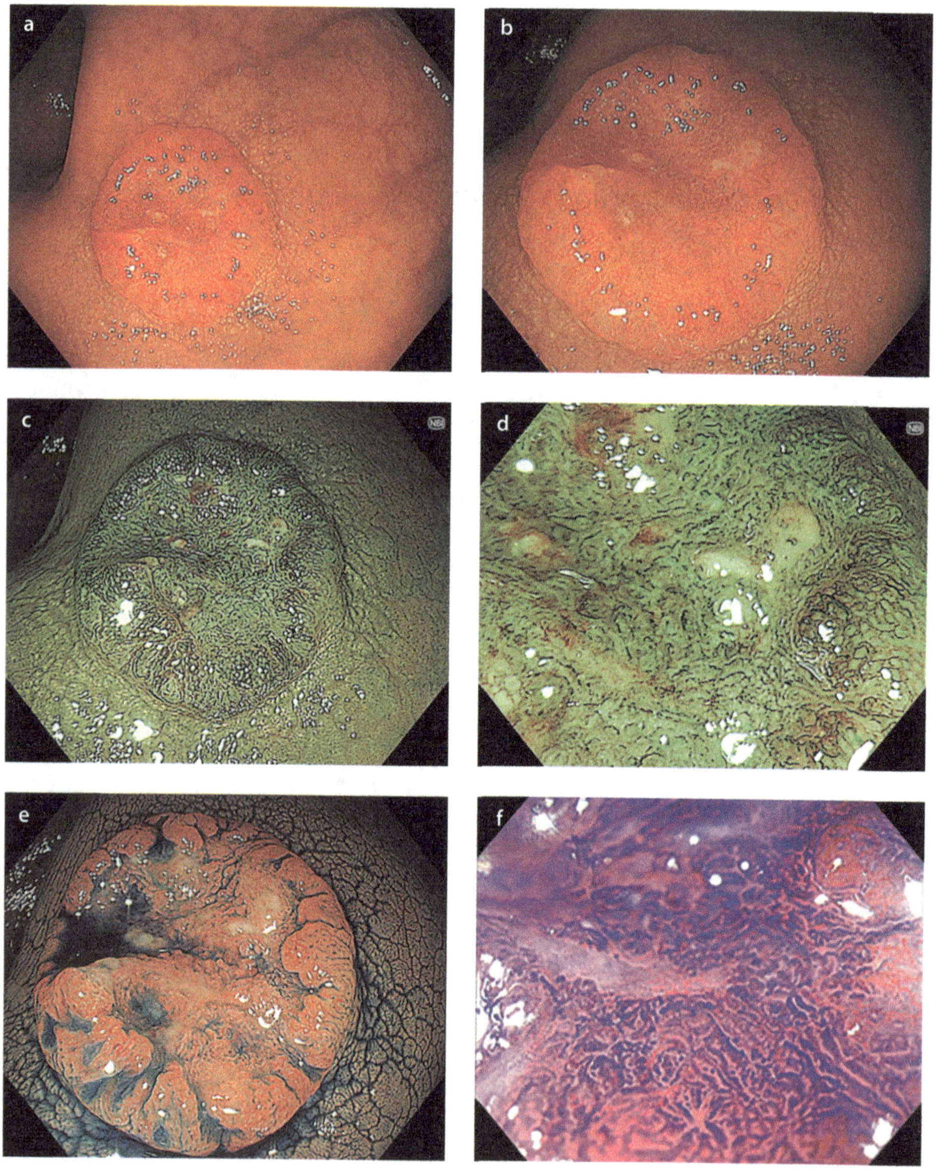
a
b
c
d
e
f

▪ Pathological Diagnosis

Adenocarcinoma, tub2, pT1b (SM: 2200 μm), Ly0, V1a, Pn0, pPM(-), pDM(-), pRM(-), pN0 (total 0/11).

g

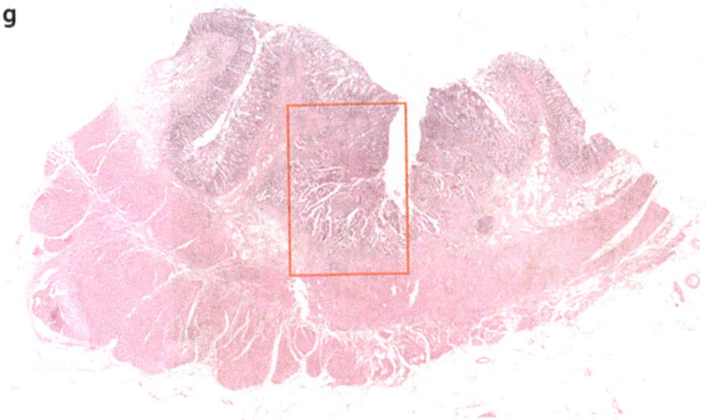

h

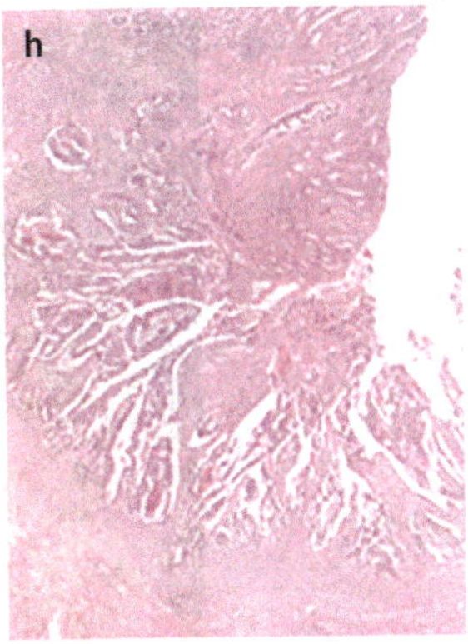

▪ Explanation

A flat lesion with a central depression measuring approximately 10 mm has been detected in the lower rectum (Figs. a and b). The surface of the depression area showed irregularities. The chicken skin mucosa was present around the lesion. Non-magnifying NBI showed different vessel pattern within the depressed and the surrounding flat area (Fig. c). Magnifying endoscopy with NBI showed that the lesion had irregular surface and vessel patterns and was diagnosed as JNET type 2B (Fig. d). Chromoendoscopy using indigo carmine dye enhances the margin of the depressed area (Fig. e). In addition, several small depressed areas suspected of retrograde infiltration were observed in the surrounding flat area of reactively raised normal mucosa. Magnifying endoscopy with 0.05% crystal violet staining in the depressed area showed Kudo's type Vɪ high-grade pit pattern (Fig. f). Based on these findings, the lesion was diagnosed as a high-grade adenoma with deep submucosal invasive carcinoma, and surgically resected by low anterior resection. Histopathological examination revealed a deep submucosal invasive carcinoma that invaded just below the muscularis propria with vascular invasion (Figs. g and h). No lymph node metastasis was observed.

Tips

Lesions less than 10 mm have a low cancer rate. However, we sometimes encounter deep submucosal invasive cancer like this case. Many small lesions of deep submucosal invasive carcinoma have depressed area. Therefore, when a lesion is detected, it is important to first observe whether or not the lesion has a depressed area. Next, observe the vessel findings in the depressed area with NBI magnification, and if JNET type 2B is observed, it must be suspected to be a cancerous lesion. In addition, it is necessary to add a pit pattern observation with magnifying endoscopy with crystal violet staining. If the pit pattern shows highly irregular Vɪ or Vɴ, it can be diagnosed as deep submucosal invasive carcinoma. Small lesions may be preceded by endoscopic resection for the purpose of total biopsy; however, in this case, the endoscopic resection may have resulted in vertical margin positive because the lesion was just above the intrinsic muscularis propria.

▪ Video

A short video is available as ◘ Fig. 1.

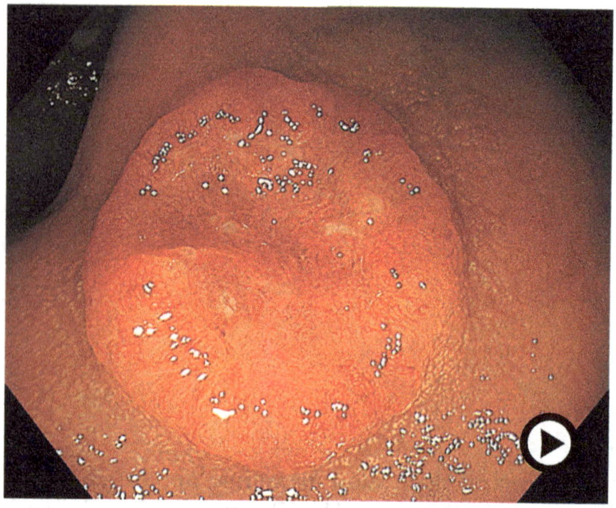

◘ **Fig. 1** Practical observation and diagnosis of early colorectal cancer with magnifying NBI and chromoendoscopy. A flat lesion with a central depressed area measuring approximately 10 mm in the lower rectum (▶ https://doi.org/10.1007/000-c8x)

Adenoma with Pagetoid Spread

Hiroyuki Takamaru and Nozomu Kobayashi

Y. Sano et al. (eds.), *Atlas of Advanced Endoscopy*, https://doi.org/10.1007/978-981-97-2732-2_72

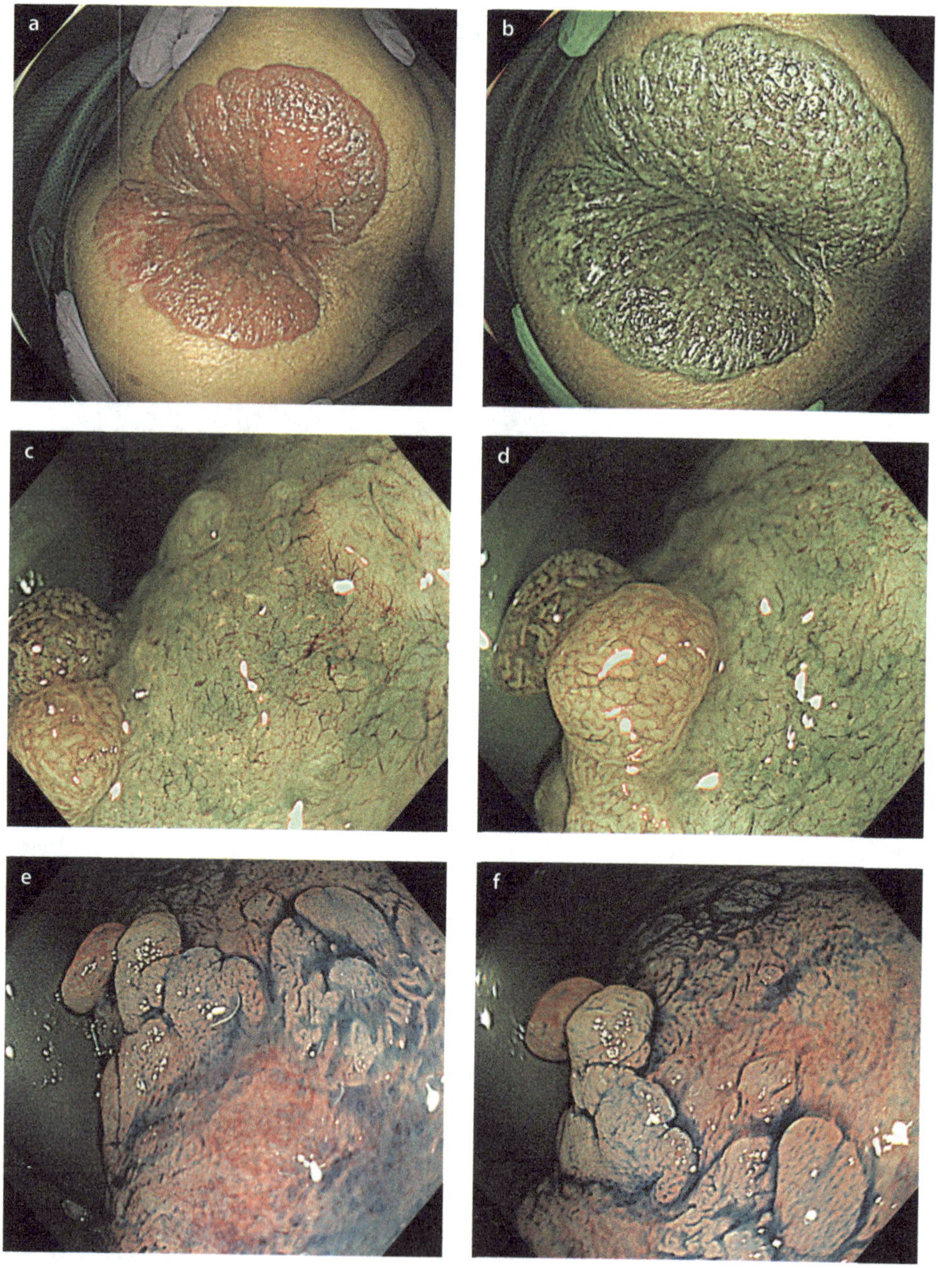
a
b
c
d
e
f

Pathological Diagnosis

Perianal Paget's disease as spread from a rectal adenoma.

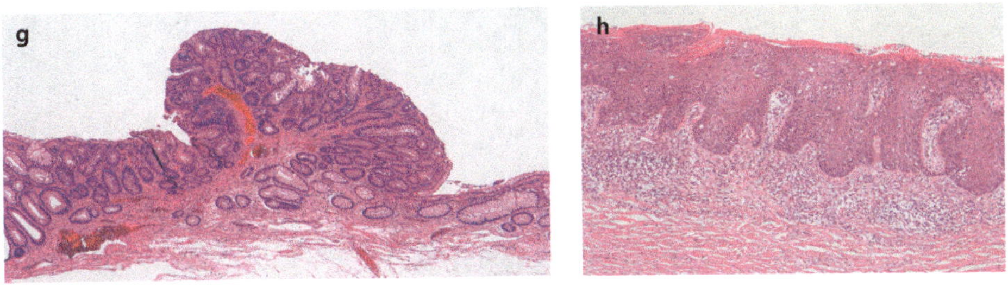

Explanation

The patient had been aware of perianal skin change for 1 year prior to diagnosis. The lesion was seen as a red-colored thickening of the skin in the surrounding area of the anus (Fig. a). Colonoscopy found the whole circumferential irregular mucosa in the lower rectum (Fig. b). The mucosa of the lower rectum and anal canal showed flat or partially protruded. With NBI magnification, vessels showed irregular distribution, though thickened vessels were not seen (Figs. c and d). The surface structure was unclear. JNET classification was difficult because of unusual endoscopic findings. In the nodular part, a regular meshed vessel pattern was observed and evaluated as JNET type 2A. Chromoendoscopy with indigo carmine dye depicted uneven surface structure of the mucosa of the lower rectum and anal canal (Figs. e and f).

Biopsy specimen from the perianal skin revealed cytologically malignant cells infiltrating the squamous epithelium singly or in small clusters. Immunohistochemically, atypical cells were positive for CK20 and CDX2, suggesting secondary pagetoid spread from the colorectum. Radical excision was performed. Histological examination of the resected specimen revealed that the lesion was a secondary Paget's disease of the perianal skin arising in association with a rectal adenoma (Figs. g and h). An adenoma was seen at the dentate line and showed pagetoid spread to anal squamous epithelium widely. The adenoma showed low-grade dysplasia and was confined to the mucosa. The tumor extension was 106 × 67 mm. There were no lymph node metastases.

Tips

Perianal Paget's disease is categorized into two types: primary or secondary. Primary type is a cutaneous malignancy in which the tumor cells show adnexal gland differentiation. Secondary type is intraepithelial spread of an underlying or adjoining visceral neoplasm. Since perianal Paget's disease may arise in association with invasive and/or in situ colorectal neoplasia, it is important to confirm the presence or absence of colorectal lesions by colonoscopy.

Colitis-Associated Dysplasia

Kaoru Takabayashi and Makoto Naganuma

Y. Sano et al. (eds.), *Atlas of Advanced Endoscopy*, https://doi.org/10.1007/978-981-97-2732-2_73

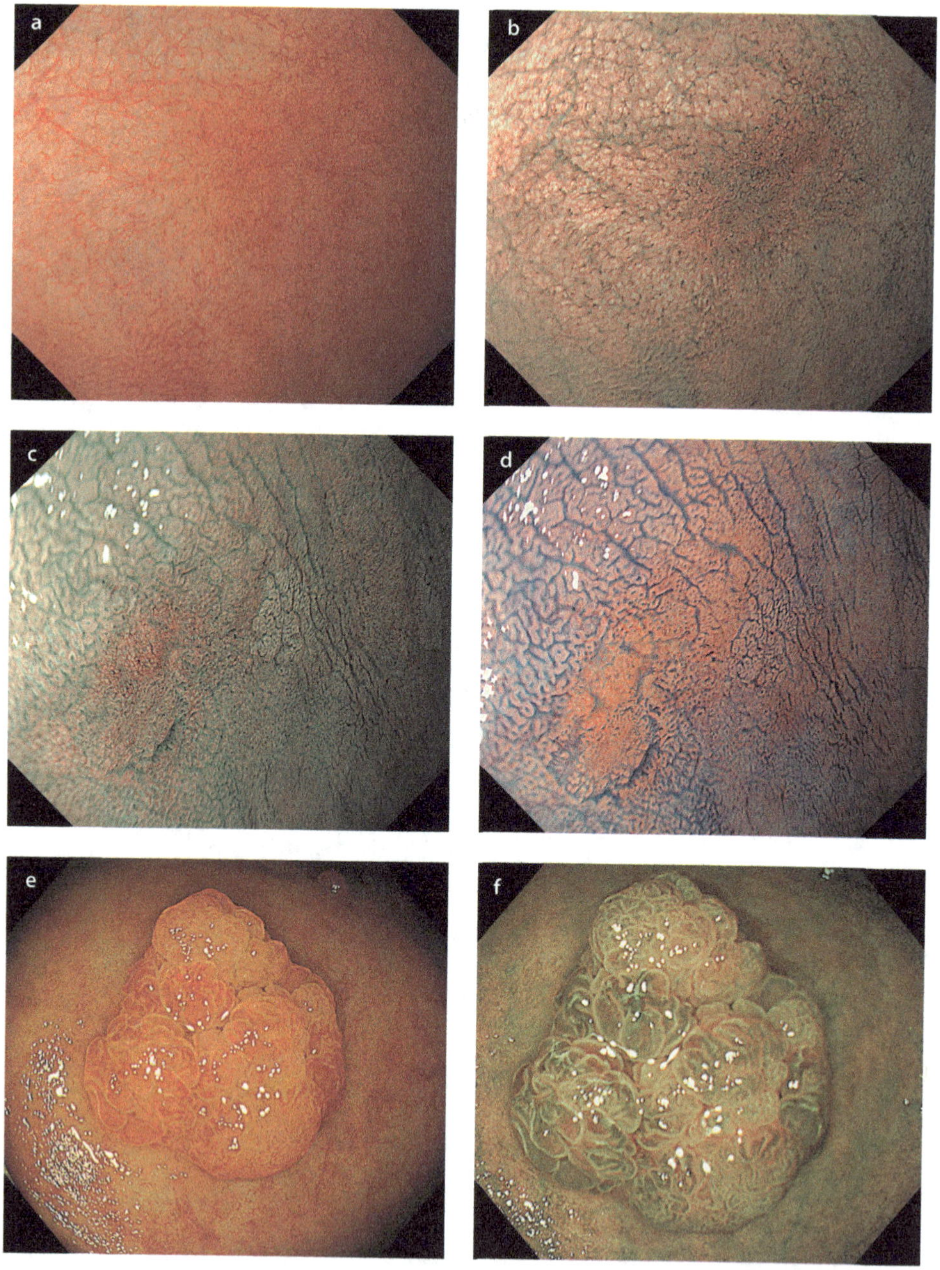

a Finding of a flat lesion of UC-associated dysplasia (white light imaging); **b** Finding of a flat lesion of UC-associated dysplasia (narrowband imaging); **c** Finding of a flat lesion after switching to high-magnification NBI; **d** Finding of a flat lesion using high-magnification chromoendoscopy; **e** Finding of a sessile lesion in patients with remitting UC (white light imaging); **f** Finding of a sessile lesion after switching to high-magnification NBI

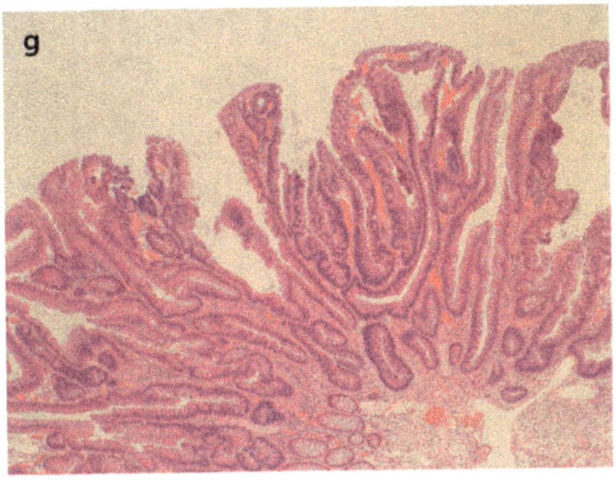

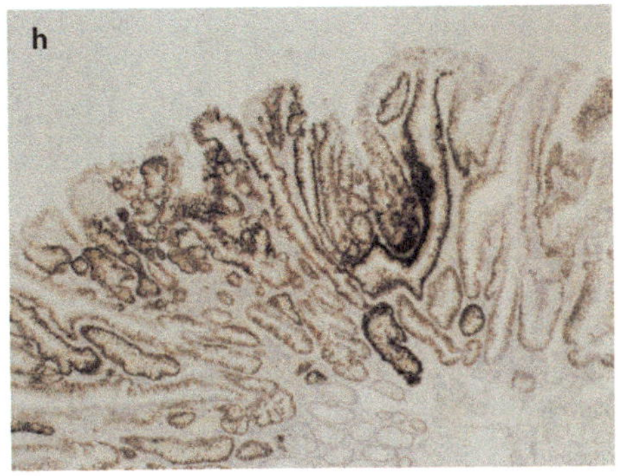

g Histological finding of HGD (hematoxylin and eosin staining); **h** Histological finding of HGD (immunohistochemical staining for p53)

Case

White light imaging revealed endoscopic remission and a flat lesion with slightly reddish mucosa compared to the surrounding area in the rectum (Fig. a). Non-magnifying narrow band imaging (NBI) identified a brownish lesion (Fig. b). These findings create uncertainty with regard to diagnosing between inflammation and dysplasia. Switching to high-magnification NBI enhanced margin visibility (Fig. c), revealing small round surface structures in the brownish area. High-magnification chromoendoscopy using indigo carmine dye enhanced the visualization of the margin and surface of the lesion and identified a small round pattern in the reddish area, as well as mucosa displaying a gyrus pattern (upper left side of the lesion) and a ripple-like appearance (right side of the lesion) (Fig. d). Histological findings indicated high-grade dysplasia (HGD) in the biopsied specimen of the small round pattern; in contrast, no HGD was observed in the biopsied specimen of the ripple-like appearance.

A sessile polypoid lesion measuring approximately 10 mm was detected in the sigmoid colon of a patient with remitting UC (Fig. e). High-magnification NBI revealed a pine cone appearance with Kudo's type IV pit pattern (Fig. f). The diagnosis was a traditional serrated adenoma, and endoscopic submucosal dissection was performed. The histological diagnosis was HGD owing to atypical glands found in the mucosa (Fig. g) and p53 immunohistochemistry indicating p53 overexpression (Fig. h).

Tips

Chronic inflammation in UC increases the risk of colonic dysplasia and cancer. The SCENIC consensus statement classifies UC-associated dysplasia's endoscopic manifestations as pedunculated, sessile, superficial elevated, flat, or depressed lesions. Detecting flat lesions, commonly seen in UC-associated dysplasia, can be challenging. Recognizing slightly reddish mucosa compared to the surrounding tissue is critical for dysplasia detection. NBI can aid in identifying flat and depressed lesions. Distinguishing between HGD and non-dysplastic lesions is facilitated by identifying the small round pattern using magnifying NBI or chromoendoscopy. Although the indication for endoscopic resection of UC-associated dysplasia remains controversial, performing such resections can provide histological clarity on the presence of UC-associated dysplasia or sporadic adenoma.

Natural History of Diminutive 0-IIc

Takahiro Fujii and Takahiro Fujimori

Supplementary Information The online version contains supplementary material available at https://doi.org/10.1007/978-981-97-2732-2_74. The videos can be accessed individually by clicking the DOI link in the accompanying figure caption or by scanning this link with the SN More Media App.

Y. Sano et al. (eds.), *Atlas of Advanced Endoscopy*, https://doi.org/10.1007/978-981-97-2732-2_74

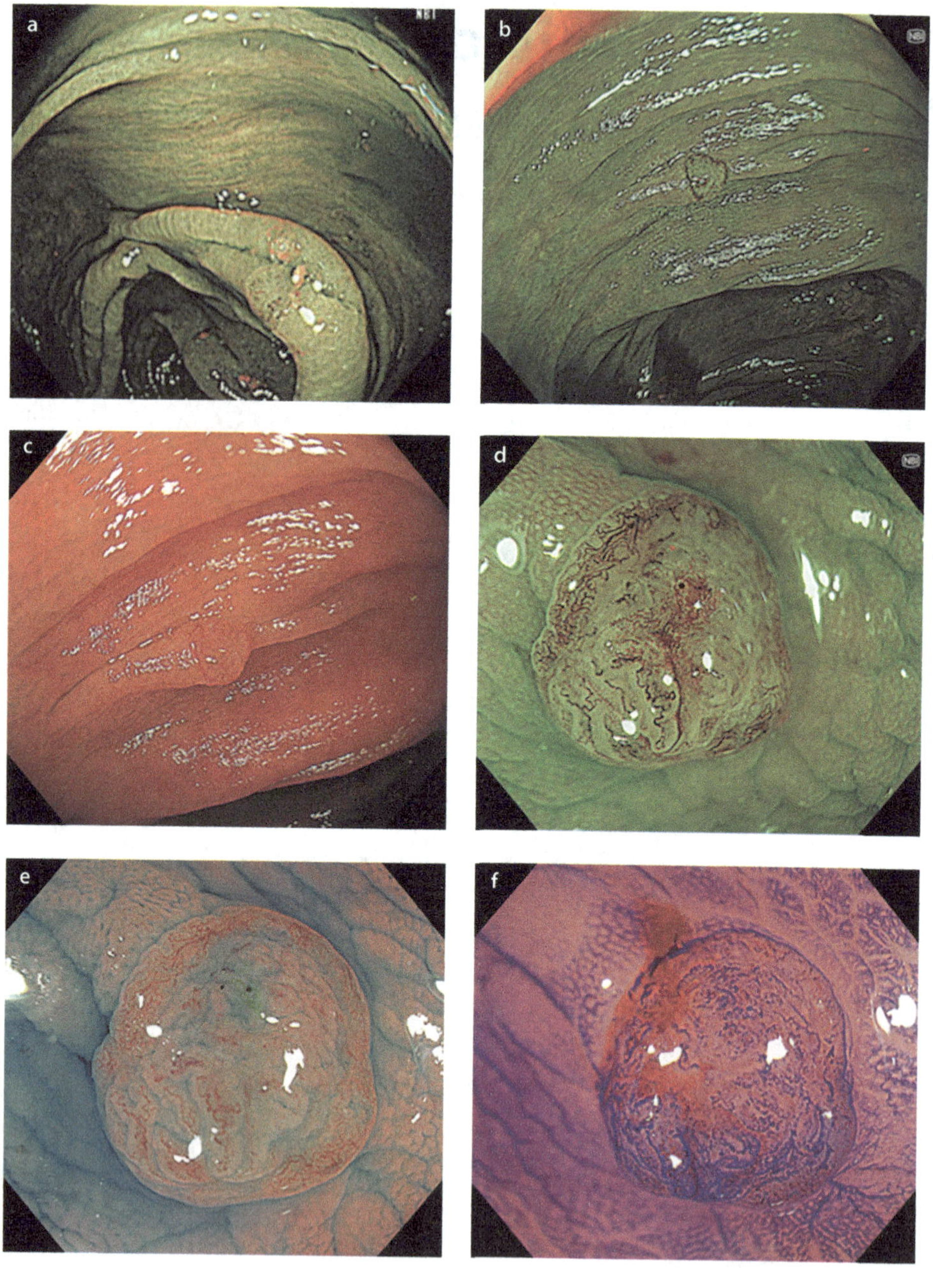
a
NBI
b
c
d
e
f

Pathological Diagnosis

Well-differentiated adenocarcinoma, pTis, M, ly0, v0, pHM0, pVM0

The first histopathological section showed only a negligible carcinoma (Fig. g). After further serial sections, the 250th section revealed the main part of the carcinoma and was confidently diagnosed as a well-differentiated adenocarcinoma (Figs. h, i and j).

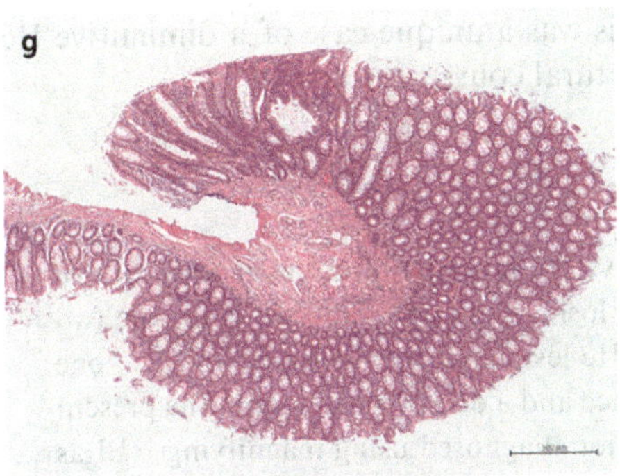
g

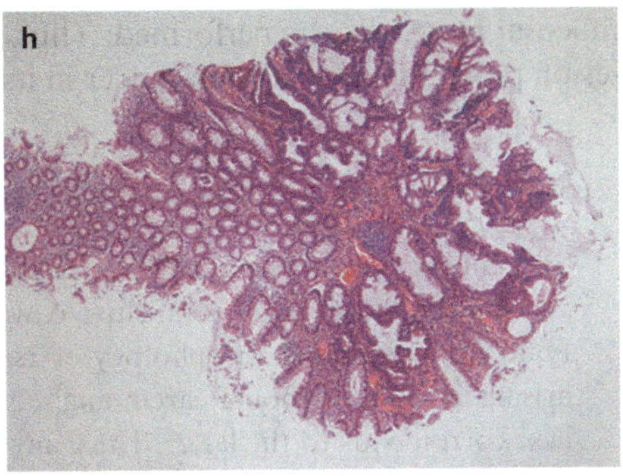
h

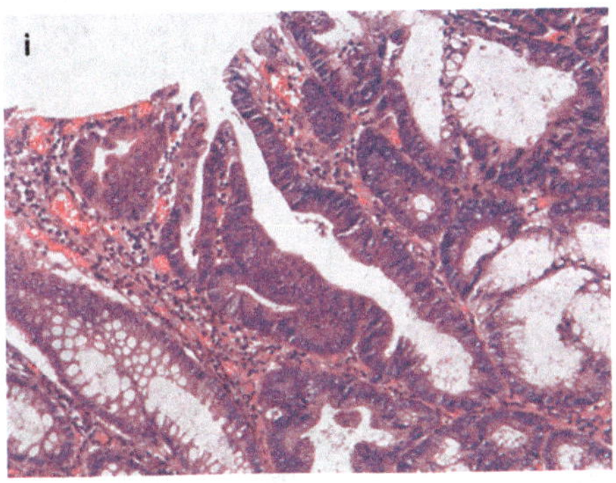
i

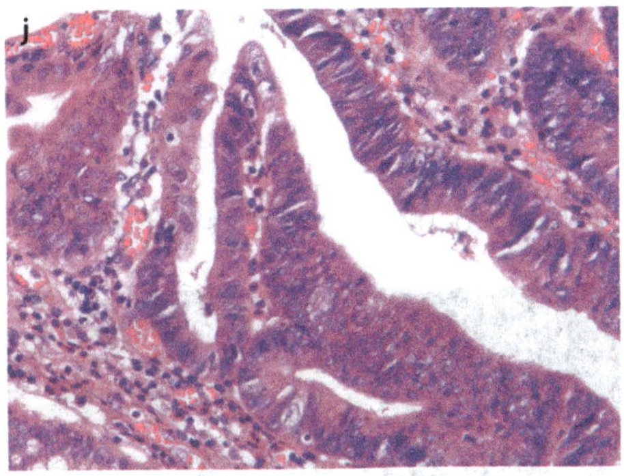
j

Explanation

80 years old, male: A diminutive IIc lesion was found at the hepatic flexure during withdrawal of colonoscopy with narrow band imaging (NBI) observation from the cecum, with its depressed area and marginal elevation depicted as whitish and brownish, respectively, which constituted what we proposed as the "O-ring sign" (Fig. a). While the lesion was lost to sight due to poor observation conditions associated with poor bowel preparation and increased peristalsis, colonoscopy 1 year later revealed a Is + IIc lesion about 5 mm at the same site, which was deemed to be the same lesion that had been lost to sight. The O-ring sign of the lesion was highlighted (Fig. b) with a nodular change in the depressed area, suggesting a morphological change from a IIc to a Is + IIc lesion. While recognized as a small, discolored sessile polyp on white light imaging (Fig. c), the lesion was diagnosed based on vascular findings on magni-

fying NBI as type 2B according to the Japan NBI Expert Team (JNET) classification (Fig. d) and was deemed an indication for crystal violet (CV) staining. Magnifying chromoendoscopy with indigo carmine dye spraying showed no definite findings as to whether the lesion had a type V pit pattern, but CV staining revealed a type V_I high-grade pit across the entire nodular area, suggesting an invasive pattern (Figs. e and f). However, the nodular area occupied by the type V pits was so small (3 mm) that a diagnosis of intramucosal carcinoma could not be ruled out and endoscopic mucosal resection was performed. Thus, this was a unique case of a diminutive IIc lesion progressing to Is + IIc cancer in its natural course.

Tips

The authors recommend marking diminutive IIc lesions found under unfavorable viewing conditions with metal clips to avoid losing sight of these lesions. Of the two types of macroscopic morphology of Is + IIc lesions in their progression, i.e., one arising from the adenoma-carcinoma sequence and a de novo-like lesion, the present case corresponds to the latter. Thus, any lesion diagnosed using magnifying NBI as type 2B in the JNET classification calls for a precise pit pattern diagnosis using non-traumatic catheter with CV staining.

▪ Video

A short video is available as ◘ Fig. 1.

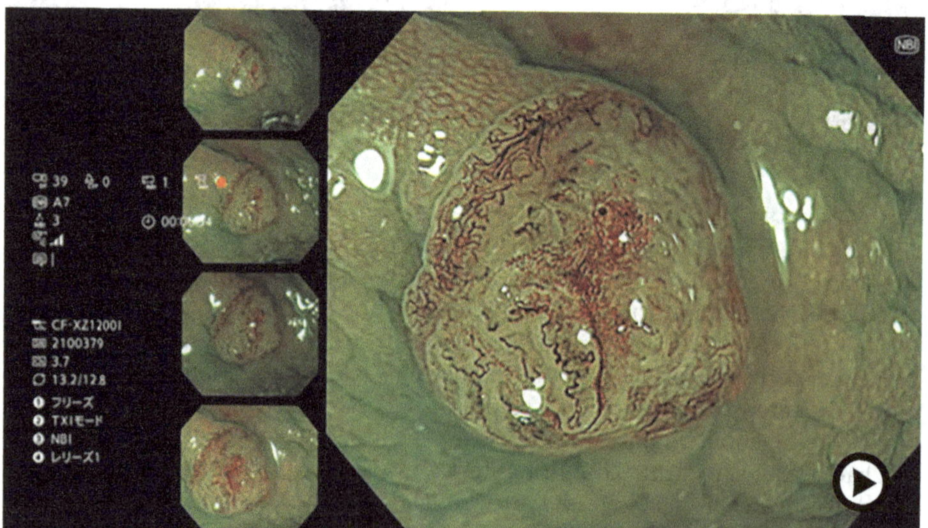

◘ Fig. 1 This lesion was found by NBI observation. In NBI, it looks brownish color around the edge of the lesion, and we call this finding as the O-ring sign. Marking clip is placed to be not lose sight of the lesion. The NT catheter has a spherical tip so as not to damage the mucosa. During magnifying observation, NT catheter is used to hold the surrounding mucosa of the lesion and perform magnifying observation to facilitate focusing. In addition, indigo carmine dye spraying, crystal violet staining, and washing with water are also performed from this catheter (▶ https://doi.org/10.1007/000-c8y)

Supplementary Information

Y. Sano et al. (eds.), *Atlas of Advanced Endoscopy*, https://doi.org/10.1007/978-981-97-2732-2

Appendix: ANBIIG Education Programs

What Is ANBIIG?

We are an academic interest group known as the Asian NBI Group Limited. The group consists of world-class healthcare professionals (HCP) in Asia, with the objectives to education, development, and association. The group was established as a nonprofit-making organization in 2014 and changed its name into The Asian Novel Bio-Imaging and Intervention Group (ANBIIG) with an aim to include more diversity in the content of the training, ranging from endoscopic diagnosis for early GI cancer to therapeutic endoscopy (Fig. A.1).

In 2020, ANBIIG begins e-learning by offering webinars in different formats for various countries as well as the Webinar Series – Journey to Advancement of Early GI Cancer Management that consists of 15 sessions during July to December. This program has been endorsed by various GI societies in the region including the following:

- Asian Pacific Society of Digestive Endoscopy (APSDE)
- Gastroenterological Society of Singapore (GESS)
- Thai Association of Gastrointestinal Endoscopy (TAGE)
- Society of Gastrointestinal Endoscopy of India (SGEI)
- Hong Kong Society of Digestive Endoscopy (HKSDE)

Fig. A.1 The Asian Novel Bio-Imaging and Intervention Group (ANBIIG)

What Can We Offer You?

We offer a two-prong approach in training beginners through basic local workshops and equip experienced endoscopists to train local beginners through the train-the-trainers program. We also provide lecture videos, clinical case atlas, and endoscopic images and quiz to educate more endoscopists. These are available free of charge here: ▶ https://anbig.org/resources-center/clinical-case-atlas/# (registration is required).

Please take a look at an example of the clinical case atlas (◻ Fig. A.2; ▶ https://youtu.be/rhe5w4UHPt8).

As shown in ◻ Table A.1, more than 30 case presentations are available on the ANBIIG website. In addition to these case studies, more than 60 lecture videos and guide posters are also available (◻ Figs. A.3, A.4 and ◻ Table A.2). Their case studies and lectures will inspire you and help you to improve your endoscopic skills.

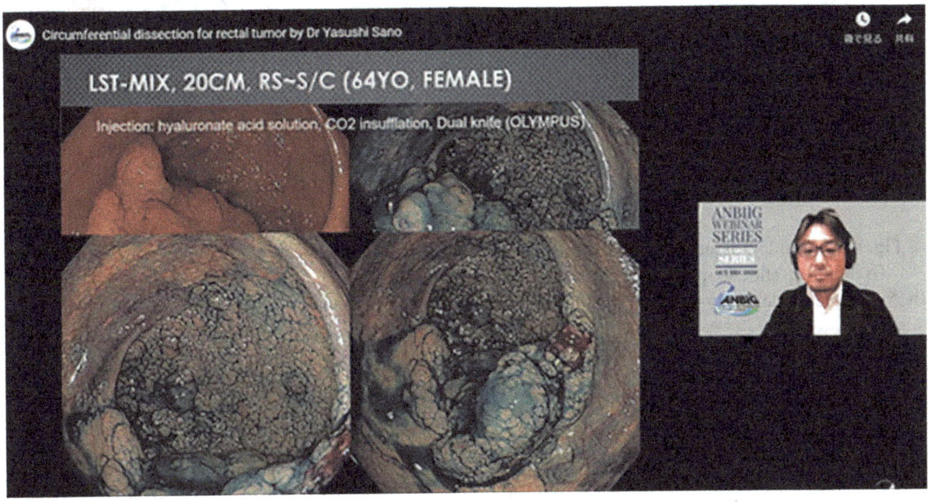

◻ **Fig. A.2** Clinical case atlas (▶ https://anbig.org/resources-center/clinical-case-atlas/#)
This video is also available here (▶ https://youtu.be/rhe5w4UHPt8)

◻ **Fig. A.3** Lecture videos (▶ https://anbig.org/resources-center/lecture-videos/)

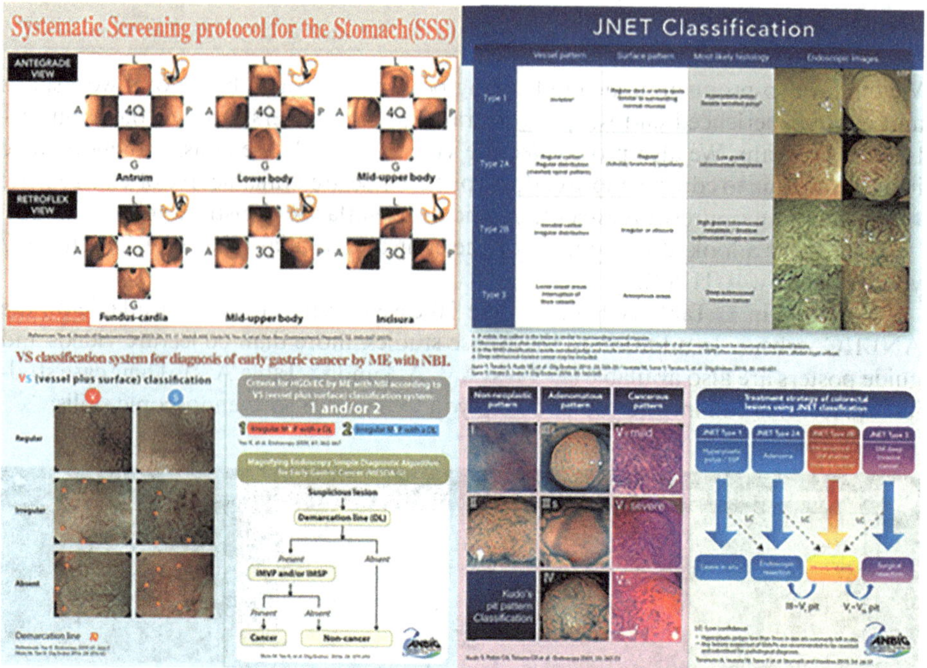

Fig. A.4 Guide posters (▸ https://anbig.org/resources-center/guide-for-nbi-diagnosis/)
To facilitate endoscopists in applying NBI classification for early GI cancer, ANBIIG has developed the reference guide for gastric and colorectal diagnosis and classification

Table A.1 List of clinical case atlas

Video no.	Webinar Theme	Case	Speaker
C001	ESD for Lower GI	JNET 2A, benign adenoma, EPMR	Jeong-Sik Byeon
C002		JNET 2B, Kudo type VI pit pattern, superficial SM cancer	Jeong-Sik Byeon
C003		JNET 2B, superficial SM cancer, planned EPMR	Jeong-Sik Byeon
C004		Adenocarcinoma, pT3 (SS)	Shinji Tanaka
C005		Circumferential dissection for rectal tumor	Yasushi Sano
C006	EMR and ESD for Upper GI Neoplasia	High-grade adenoma	Naomi Kakushima
C007		Low-grade adenoma	Naomi Kakushima
C008		Squamous cell carcinoma	Daisuke Kikuchi
C009	Techniques of Polypectomy and EMR for Lower GI	Rectal carcinoid tumor	Hyun-Gun Kim
C010		Rectal carcinoid	Takuji Kawamura

Table A.1 (continued)

Video no.	Webinar Theme	Case	Speaker
C011	POEM	Achalasia type 3 POEM	Hwoon-Yong Jung
C012		Achalasia type 2 POEM	Mohan Ramchan-dani
C013		GERD	Mohan Ramchan-dani
C014	POET (STER)	Subepithelial Tumor	Hon-Chi Yip
C015	Endoscopic Full Thickness Resection	EFTR on gastric subepithelial tumour	Hon-Chi Yip
C016		EFTR on gastric subepithelial tumour	Hon-Chi Yip
C017	NBI Diagnosis & Classification of Squamous Esophageal Neoplasia	Early esophageal squamouscan-cer, T1a-LPM, s/p ESD	Lee Ching-Tai
		Squamous dysplasia, HG	Hwoon-Yong Jung
		Submucosal cancer	Hwoon-Yong Jung
		Squamous cell carcinoma SCC pT1b SM1	Chen-Ya Kuo
C018	NBI Diagnosis and Classification of Early Gastric Cancer	Gastric Cancer, Differentiated-type	Hisashi Doyama
		Gastric cancer, Undifferentiated-type	Hisashi Doyama
		Early gastric caner, differentiated type, intestinal type	Kenshi Yao
		Early gastric cancer, 0-IIc, Undifferentiated type	Kenshi Yao
C019		Sigmiod colonic lesion	Li Xiao-Bo
		Rectal lesion	LI Xiao-Bo
		Tumor in sigmoid colon	Shinji Tanaka
		Tumor in acending colon, adenocarcinoma	Shinji Tanaka
C020	Endocytoscopy for Upper GI	Esophageal endocytoscopy	Hitomi Minami
		Esophageal endocytoscopy	Hitomi Minami
		Gastric endocytoscopy	Hon-Chi Yip

(continued)

Table A.1 (continued)

Video no.	Webinar Theme	Case	Speaker
C021	NBI Diagnosis of Barrett's Neoplasia	Barrett's mucosa with low and high grade dysplasia	Ang Tiing Leong, Tan Chin Kimg & Wang Lai Mun
		Barrett's mucosa with basal crypt low grade dysplasia	Ang Tiing Leong, Tan Chin Kimg & Wang Lai Mun
		Barrett's esophagus	Rajvinder Singh
		Barrett's esophagus	Rajvinder Singh
C022		Low-grade adenoma	Kohei Takizawa & Tetsuya Suwa
		High-grade adenoma	Kohei Takizawa & Kazuo Shiotsuki
		Duodenal adenoma with LGD	Rapat Pittayanon
C023	Computer-aided Diagnosis/AI-assisted Endoscopy	Diminutive polyp—target mode	David Hewett
		Low grade adenoma	Masashi Misawa
		T1b (SM massive) cancer or beyond	Masashi Misawa
C024	Handling of Endoscopic Resected Specimen and GI pathology in Japan	Adenocarcinoma with tubulovillous adenoma	Yasushi Sano
		Intramucosal adenocarcinoma	Shin Ichihara
		Superficially serrated adenoma (SuSA)	Shin Ichihara
C025	EMR and ESD for Upper GI Neoplasia	Adenocarcinoma in the Barrett esohagus	Mitsuhiro Fujishiro
		Tubular adenocarcinoma, submucosal deep invasion	Mitsuhiro Fujishiro
		Squamous cell carcinoma	Seiichiro Abe
		Moderately to well differentiated tubular adenocarcinoma	Seiichiro Abe
C026	ESD for Lower GI	High grade adenoma-T1a	Kenichiro Imai
		pT1a cancer (360um)	Kenichiro Imai
		Tubulovillous adenoma (LGD)	Han-Mo Chiu
		Tubulovillous adenoma with HGD	Han-Mo Chiu
C027	POEM	Esophageal spasm & G-POEM	Hwoon-Yong Jung
		Supine reflux, G-POEM	Hwoon-Yong Jung
		Sigmoid achalasia & Dysphagia	Chainarong Phalanusitthepha
		Dysphagia	Chainarong Phalanusitthepha

Table A.1 (continued)

Video no.	Webinar Theme	Case	Speaker
C028	Endoscopic Full Thickness Resection	Gastrointestinal stromal tumor (GIST)	Satoki Shichijo & Yoji Takeuchi
		Glomus tumor	Satoki Shichijo & Yoji Takeuchi
		Gastric schwannoma EFTR	Philip Chiu
		Endoscopic resection Gastric GIST	Ang Tiing Leong
C029	Polypectomy and EMR for Lower GI	Well differentiated adenocarcinoma	Takuji Kawamura
		Well differentiated adenocarcinoma	Takuji Kawamura
		Cold snare EMR & FTR	Rajvinder Singh
		FTR	Rajvinder Singh
C030	Laparoscopic—Endoscopic Collaborative Surgery (LECS)	Spindle cell tumor arising from muscularis propria	GV Rao
		GIST arising from submucosa	GV Rao
		GIST/Leiomyoma, EFTR	Amol Bapaye & Gajanan Wagholikar
		Gastric SET, EFTR	Amol Bapaye & Gajanan Wagholikar
C031	ANBIIG Gastro Pathology for Upper GI	Malignant melanoma of esophagus	Shelly Ni & Hon-Chi Yip
C032		Intramucosal adenocarcinoma	Shelly Ni & Hon-Chi Yip
C033		Collagenous gastritis	Wang Lai Mun
C034		Eosinophilic gastroenteritis & collagenous gastritis	Wang Lai Mun
C035		Villous adenoma with high grade dysplasia	Anuradha Sekaran
C036		Tubulovillous adenoma with low grade dysplasia	Anuradha Sekaran
C037		Well differentiated adenocarcinoma	Anuradha Sekaran
C038		Cronkhite-Canada syndrome	Zhou Wei-xun

Table A.2 List of lecture videos

Video no.	Issue	Topic	Speaker
1	Diagnostic	Mucosal Imaging—How I do it	Han-Mo Chiu
2	Diagnostic/ therapeutic	Dysplasia in IBD: Detect, characterize, resect	Gottumukkala S. Raju
3	Diagnostic	Quality indicators for screening colonoscopy	George Shiaw-Hooi Ho
4	Diagnostic	Gastric atrophy, intestinal metaplasia and dysplasia in the stomach	Noriya Uedo
5	Diagnostic	Mucosal Imaging—How I do it	Noriya Uedo
6	Diagnostic	Role of IEE for colorectal polyp	Yutaka Saito
7	Therapeutic	How to choose appropriate polyp removal technique: Polypectomy EMR and ESD?	Yutaka Saito
8	Therapeutic	Endoscopic treatment for upper GI cancer	Daisuke Kikuchi
9	Diagnostic	Role of IEE for premalignant lesions and early gastric cancer detection	Daisuke Kikuchi
10	Diagnostic	Basic Principle of NBI	Daisuke Kikuchi
11	Diagnostic	How to assess chronic gastritis of background gastric mucosa including IM by NBI	Noriya Uedo
12	Diagnostic	How to make a differential diagnosis of cancer and non-cancer according to VS classification system and MESDA-G	Kenshi Yao
13	Diagnostic	Barrett's Oesophagus (Imaging): what we should be doing in 2020	Rupa Banerjee
14	Therapeutic	Barrett's Oesophagus (Treatment): what we know & how to push the boundaries?	Rajvinder Singh
15	Diagnostic	Image Enhanced Endoscopy for detection of superficial esophageal neoplasia	Philip Chiu
16	Diagnostic	Endoscopic characterization & classification of superficial esophageal neoplasia	Kenichi Goda
17	Diagnostic	Basics and NICE/JNET classification	Han-Mo Chiu
18	Diagnostic	How to diagnose serrated lesions?	Yasushi Sano
19	Diagnostic	Endoscopic diagnosis of colorectal adenoma and sessile serrated lesion	Yoji Takeuchi
20	Diagnostic	JNET and pit pattern diagnosis for early colorectal neoplasia/cancer	Yutaka Saito
21	Diagnostic	How to diagnose duodenal neoplasm: Tips and Tricks	Rapat Pittayanon

Table A.2 (continued)

Video no.	Issue	Topic	Speaker
22	Therapeutic	Endoscopic treatment of Non ampullary duodenal neoplasms	Rajvinder Singh
23	Diagnostic	General Detection and Detailed EC Observation	Hitomi Minami
24	Diagnostic	Endocytoscopy in esophageal lesions	Shannon Chan
25	Diagnostic	The use of Endocytoscopy for Colorectal Neoplasms	Masashi Misawa
26	Diagnostic	The Application of AI Technology into Colonoscopy	Yuichi Mori
27	Therapeutic	Indication of colorectal ESD and my technical strategy	Shinji Tanaka
28	Therapeutic	Indications of colorectal ESD and EPMR: case-based discussion	Jeong-Sik Byeon
29	Therapeutic	EMR and ESD for Upper GI Neoplasia—Stomach	Kohei Takizawa
30	Therapeutic	EMR and ESD for Upper GI Neoplasia—Duodenum	Naomi Kakushima
31	Therapeutic	Techniques of polypectomy and EMR for lower GI	Hyun-Gun Kim
32	Therapeutic	Cold resection techniques: sessile serrated lesions and beyond	David Hewett
33	Therapeutic	POEM: Basics and Adverse Events	Hwoon-Yong Jung
34	Therapeutic	POEM: Difficult situations: How to Manage?	Mohan Ramchandani
35	Therapeutic	Indications, principles, planning, technical steps of STER	Amit Maydeo
36	Therapeutic	Complications, precautions, special situations while performing STER	Amol Bapaye
37	Therapeutic	Endoscopic Full Thickness Resection—Indications, Principles and Various Closure Techniques (except suturing)	Amol Bapaye
38	Pathology	Japanese histological principles in the Upper GIT	Shin Ichihara
39	Pathology	Western histological principles in the Upper GIT	Andrew Ruszkiewicz
40	Pathology	Japanese histological principles in the Lower GIT	Hiroshi Kawachi

(continued)

Table A.2 (continued)

Video no.	Issue	Topic	Speaker
41	Pathology	Western histological principles in the Lower GIT	Andrew Ruszkiewicz
42	Diagnostic	NBI diagnosis & classification of early colon cancer	Kenichiro Imai
43	Diagnostic	Update on image-enhanced endoscopy: TXI, RDI and its application	Supakij Khomvilai
44	Diagnostic	Importance of colorectal cancer screening	Han-Mo Chiu
45	Diagnostic	Diagnosis of colorectal cancer and classification + video case	Yoji Takeuchi
46	Therapeutic	Treatment of early colorectal cancer (ESD, EMR, polypectomy) + video case	Han-Mo Chiu
47	Diagnostic	New technology on image enhanced endoscopy (TXI, RDI) and its application + video case	Rajvinder Singh
48	Diagnostic	How does endocytoscopy improve detection of early lesions	Yuichi Mori
49	Diagnostic/ therapeutic	AI-assisted colonoscopy + video case	Yuichi Mori
50	Diagnostic	Application and clinical value of AI in endoscopy and healthcare	Rajvinder Singh
51	Therapeutic	Cold snare polypectomy: tips and tricks	David Hewett
52	Diagnostic	New technology on IEE (TXI, RDI) and its application	Rajvinder Singh
53	Diagnostic	Application and clinical value of AI in endoscopy and healthcare	Rajvinder Singh
54	Diagnostic	Detection of high-risk flat and depressed polyp & M-NBI diagnosis	Mineo Iwatate
55	Diagnostic/ therapeutic	Non ampullary duodenal lesions: classification and treatment	Bronte Holt
56	Diagnostic	Straight up the small bowel with the Spiral Enteroscope	Mohan Ramchandani
57	Diagnostic	Surveillance guidelines in colonoscopy: an Update	Cameron Bell
58	Diagnostic/ therapeutic	Clinical significance of flat lesions: Results from the Japan Polyp Study	Yasushi Sano
59	Diagnostic/ therapeutic	Insertion and withdrawal: my top 10 tips	Cameron Bell
60	Diagnostic	Mucosal imaging for polyps in 10 min—all you need to know	Yasushi Sano

Table A.2 (continued)

Video no.	Issue	Topic	Speaker
61	Therapeutic	Hot tips, Cold snare	Spiro Rafto-poulos
62	Therapeutic	Cool tips, hot snare	Bronte Holt
63	Therapeutic	How I do it—Colonoscopy potpourri (Polyloop, tattoo, chromoendoscopy in IBD)	Spiro Rafto-poulos
64	Diagnostic/therapeutic	Triple zero: Emergencies in colonoscopy (acute lower GI bleed, perforation, pseudo-obstruction	Bronte Holt
65	Therapeutic	EMR vs ESD: Can (should) we meet halfway?	Yasushi Sano
66	Education	Training the future Endoscopist	Noriya Uedo
67	Diagnostic/therapeutic	3rd Space Endoscopy: where are we heading	Mohan Ramchandani
68	Education	AI: will it change our practice?	Rajvinder Singh
69	Diagnostic	Diagnosis and Endoscopic Rx of BE	Rajvinder Singh

9789819727315